Receptors and Recognition

General Editors: P. Cuatrecasas and M.F. Greaves

About the series

Cellular Recognition – the process by which cells interact with, and respond to, molecular signals in their environment – plays a crucial role in virtually all important biological functions. These encompass fertilization, infectious interactions, embryonic development, the activity of the nervous system, the regulation of growth and metabolism by hormones and the immune response to foreign antigens. Although our knowledge of these systems has grown rapidly in recent years, it is clear that a full understanding of cellular recognition phenomena will require an integrated and multidisciplinary approach.

This series aims to expedite such an understanding by bringing together accounts by leading researchers of all biochemical, cellular and evolutionary aspects of recognition systems. This series will contain volumes of two types. First, there will be volumes containing about five reviews from different areas of the general subject written at a level suitable for all biologically oriented scientists (Receptors and Recognition, series A). Secondly, there will be more specialized volumes (Receptors and Recognition, series B), each of which will be devoted to just one particularly important area.

Advisory Editorial Board

Receptors and Recognition

Series B

Published

The Specificity and Action of Animal Bacterial and Plant Toxins (B1)
edited by P. Cuatrecasas (Burroughs Wellcome, North Carolina)
Intercellular Junctions and Synapses (B2)
edited by J. Feldman (London), N.B. Gilula (Rockefeller University, New York) and
   J.D. Pitts (University of Glasgow)
Microbial Interactions (B3)
edited by J.L. Reissig (Long Island University, New York)
Specificity of Embryological Interactions (B4)
edited by D.R. Garrod (University of Southampton)
Taxis and Behavior (B5)
edited by G.L. Hazelbauer (University of Uppsala)
Bacterial Adherence (B6)
edited by E.H. Beachey (Veteran's Administration Hospital and University of
   Tennessee, Memphis, Tennessee)
Virus Receptors Part 1 Bacterial Viruses (B7)
edited by L.L. Randall and L. Philipson (University of Uppsala)
Virus Receptors Part 2 Animal Viruses (B8)
edited by K. Lonberg-Holm (Du Pont, Delaware) and L. Philipson (University of
   Uppsala)
Neurotransmitter Receptors Part 1 Amino Acids, Peptides and Benzodiazepines (B9)
edited by S.J. Enna (University of Texas at Houston) and H.I. Yamamura (University
   of Arizona)
Neurotransmitter Receptors Part 2 Biogenic Amines (B10)
edited by H.I. Yamamura (University of Arizona) and S.J. Enna (University of Texas
   at Houston)
Membrane Receptors: Methods for Purification and Characterization (B11)
edited by S. Jacobs and P. Cuatrecasas (Burroughs Wellcome, North Carolina)
Purinergic Receptors (B12)
edited by G. Burnstock (University College, London)
Receptor Regulation (B13)
edited by R.J. Lefkowitz (Duke University Medical Center, North Carolina)
Histocompatibility Antigens: Structure and Function (B14)
edited by P. Parham (Stanford University School of Medicine, California) and J.
   Strominger (Harvard University, Massachussetts)

Receptors and
Recognition

Series B Volume 15

Receptor-Mediated Endocytosis

Edited by
P. Cuatrecasas

Wellcome Research Laboratories,
Research Triangle Park,
North Carolina, U.S.A.

and

T. Roth

University of Maryland Baltimore County,
Catonsville,
Maryland, U.S.A.

Springer-Science+Business Media, B.V

First published 1983
by Chapman and Hall Ltd
11 New Fetter Lane, London EC4P 4EE

Published in the USA by
Chapman and Hall
733 Third Avenue, New York, NY 10017

© *Springer Science+Business Media Dordrecht 1983*
Originally published by Chapman and Hall 1983
Softcover reprint of the hardcover 1st edition 1983

British Library Cataloguing in Publication Data

Receptor-mediated endocytosis. – (Receptors and
    recognition: series B; v. 15)
    1. Ingestion
    I. Cuatrecasas, P.   II. Roth, T.F.   III. Series
    596'.01'32   QP145
ISBN 978-94-009-5977-4   ISBN 978-94-009-5975-0 (eBook)
DOI 10.1007/978-94-009-5975-0

Library of Congress Cataloging in Publication Data

Receptor-mediated endocytosis.
    (Receptors and recognition. Series B; v. 15)
    Includes index
    1. Endocytosis. 2. Cell receptors. 3. Protein
binding.   I. Cuatrecasas, P.   II. Roth, T. F. (Thomas F.)
III. Series. [DNLM: 1. Endocytosis. 2. Receptors, Endogenous substances. 3. Protein
binding. W1
RE107MA v. 15 / QH 631 R2949]
QH634.R43 1983   574.87'6   83-1944

Contents

Contributors

Charles E. Chandler, Department of Neurobiology, Stanford University School of Medicine, Stanford, California 94305, U.S.A.

Pedro Cuatrecasas, Department of Molecular Biology, Wellcome Research Laboratories, Research Triangle Park, North Carolina 27709, U.S.A.

A. Christie King, Department of Molecular Biology, Wellcome Research Laboratories, Research Triangle Park, North Carolina 27709, U.S.A.

John Lenard, Rutgers Medical School, University of Medicine and Dentistry of New Jersey, Piscataway, New Jersey 08854, U.S.A.

Carol D. Linden, Virology Division, U.S. Army Medical Research Institute of Infectious Diseases, Ft. Detrick, Frederick, Maryland 21701, U.S.A.

Douglas K. Miller, Rutgers Medical School, University of Medicine and Dentistry of New Jersey, Piscataway, New Jersey 08854, U.S.A.

James E. Niedel, Department of Medicine, Duke University Medical Center, Durham, North Carolina 27710, U.S.A.

Sjur Olsnes, Norsk Hydro's Institute for Cancer Research and The Norwegian Cancer Society, Montebello, Oslo 3, Norway.

Ira H. Pastan, Laboratory of Molecular Biology, National Cancer Institute, Bethesda, Maryland 20205, U.S.A.

Thomas F. Roth, Department of Biological Sciences, University of Maryland Baltimore County, Catonsville, Maryland 21228, U.S.A.

Kirsten Sandvig, Norsk Hydro's Institute for Cancer Research and The Norwegian Cancer Society, Montebello, Oslo 3, Norway.

Eric M. Shooter, Department of Neurobiology, Stanford University School of Medicine, Stanford, California 94305, U.S.A.

Philip Stahl, Washington University School of Medicine, 660 South Euclid Avenue, St. Louis, Missouri 63110, U.S.A.

Richard J. Stockert, Department of Medicine, U-517, Albert Einstein College of Medicine, Bronx, New York 10461, U.S.A.

Ronald D. Vale, Department of Neurobiology, Stanford University School of Medicine, Stanford, California 94305, U.S.A.

Mark C. Willingham, Laboratory of Molecular Biology, National Cancer Institute, Bethesda, Maryland 20205, U.S.A.

John W. Woods, Department of Biological Sciences, University of Maryland Baltimore County, Catonsville, Maryland 21228, U.S.A.

Preface

This volume focuses exclusively on those endocytic processes that sequester proteins by a selective, receptor-mediated mechanism. In such an endocytic process, cell surface receptors specifically bind protein ligands and localize them to specialized invaginations of the plasma membrane. These regions are coated pits, so named because they are lined on the cytoplasmic face with an ordered array of the protein, clathrin. It is this 'coat' which provides their characteristic electron microscopic image. Subsequently, these regions pinch off to form coated vesicles which rapidly lose their 'coat' and then fuse with other organelles or the plasma membrane. The hallmarks of this process are the specific receptors, coated pits, coated vesicles and an ordered sequence of transit events leading to delivery to selected locations. Receptor recognition, specific disposition of the endocytosed ligand and the existence of receptor–ligand complexes at highest density in coated pits define the process as selective and concentrative.

This topic has received ever increasing attention during the past few years. The evolving mechanisms are especially exciting because they come at a time when the conventional views based on thermodynamic arguments suggest that proteins should not be able to cross into the cell. Receptor-mediated endocytosis, however, reconciles the view that biological membranes should be impervious to macromolecules with the evidence that certain macromolecules do gain entrance into the cell.

During the last few years this field has been stimulated by studies on the uptake and processing of low density lipoproteins (LDL) by cells. Such LDL uptake serves to deliver cholesterol to the sites of cholesterol biosynthesis, which in turn regulates cholesterol metabolism. These studies have been particularly attractive because of the availability of cell lines from patients with clinical disorders characterized by severe hypercholesterolemia and premature arteriosclerosis. In these cell lines, the genetic defects are associated with specific steps in receptor-mediated endocytosis. Because of the numerous recent reviews on LDL, a specific chapter has not been devoted to it in this volume, instead the authors, in many instances, compare or contrast their system with that which processes LDL.

The rapid progress in this field has in no small part been spurred by the availability of new techniques, many of which are described. In general, the approach is both morphological and biochemical. Because of the central role that coated vesicles serve in the fundamental endocytic processes, special emphasis has been given to their structure and function. Other chapters deal with specific examples of macromolecules that enter the cell by receptor-mediated endocytosis, with special emphasis placed on the specific mechanisms used and their relation to the ultimate functions and fates of the proteins endocytosed. For instance, in the cases of asialoglycoproteins and maternal-to-fetal transport of IgG, the functions appear clearly related to the endocytic process, although the details of the mechanisms following endocytosis differ vastly. For example, proteins such as asialoglycoproteins and LDL are internalized and degraded in the lysosomes. In contrast, IgG or IgA are transported intact across an entire cellular layer, a process that involves traversing two separate membrane barriers and specifically avoiding compartments which might degrade the proteins. In the case of lysosomal enzyme secretion and re-uptake in macrophages and neutrophils, some parts of the process are obvious while other aspects are much more speculative. In like manner, it is not at all clear why peptide hormones need to be internalized, although in the case of the mitogenic (e.g. epidermal growth factor) or differentiation-related factors (e.g. nerve growth factor) some highly provocative reasons are suggested by the available data. The processing of chemotactic signal peptides by neutrophils is yet another exciting instance of a response mechanism whose overall regulation is modulated, at least in part, by endocytosis. Finally, some provocative examples of 'illicit' transport are illustrated by certain enveloped viruses and bacterial toxins which penetrate cells by 'piggy backing' on endocytosed molecules to exert their infective or toxic functions. In these cases physiological mechanisms are cleverly subverted while others are adopted for ulterior ends which do not bode well for the cell. Thus, understanding these and other examples of RME provide insights that are helpful in understanding normal metabolic, pathogenic and developmental processes in a wide variety of systems that share in common at least the early events in receptor-mediated endocytosis. In this volume, many of the specific elements and examples of RME are reviewed that provide the experimental foundation for this rapidly emerging field.

Pedro Cuatrecasas
Thomas F. Roth

1 Receptor-Mediated Endocytosis: General Considerations and Morphological Approaches

MARK C. WILLINGHAM and IRA H. PASTAN

Receptor-Mediated Endocytosis
(*Receptors and Recognition*, Series B, Volume 15)
Edited by P. Cuatrecasas and T. F. Roth
Published in 1983 by Chapman and Hall, 11 New Fetter Lane, London EC4P 4EE
© 1983 Chapman and Hall

Cells carefully regulate their internal environment by controlling which molecules are allowed to enter and which are not. Proteins and other large molecules enter cells mainly via receptor-mediated endocytosis. After these molecules enter a cell, they proceed from one membrane-limited compartment to another. Each of the major intracellular membrane compartments has a specific function. In this chapter we will describe the organelles (or compartments) that the cell uses to carry out receptor-mediated endocytosis and describe some of the new methods that have been used to carry out these studies. In our own laboratory we have focused on the ligand α_2-macroglobulin (α_2M) and the route by which fibroblastic cells take it up, but we have also applied these methods to the study of a few other ligands including epidermal growth factor (EGF), low-density lipoprotein (LDL) and some viruses. Some aspects of this process have been recently reviewed (Pastan and Willingham, 1981).

To carry out morphological studies on ligand internalization, it is useful to have a ligand–cell system with a few special properties. The cell should have many receptors, the ligand should be immunogenic, antibodies to it should be readily available, and the ligand should be able to be chemically modified without significant loss of binding activity. A useful cell type for such studies is the cultured fibroblast and a useful ligand is α_2M. Fibroblastic cells have many types of receptors on their surface, but the α_2M receptor is often present in very large numbers. On many types of fibroblasts there are about 600 000 receptors per cell (Dickson et al., 1981d). α_2M is a glycoprotein made up of four subunits. Using rhodamine or fluorescein isothiocyanate, up to 15 molecules of the fluorescent probe can be covalently attached to a single α_2M molecule without significant loss of its capacity to bind to cells. For electron microscopic studies, α_2M can be readily coupled to antibodies (Dickson et al., 1981a) or horseradish peroxidase (Dickson et al., 1981a), or adsorbed to colloidal gold (Dickson et al., 1981c).

In addition to receptors for α_2M, fibroblastic cells contain specific receptors for low-density lipoprotein (LDL) (Goldstein et al., 1979), epidermal growth factor (EGF) (Carpenter et al., 1975), insulin (Thompoulos et al., 1976), transferrin (Octave et al., 1979), tri-iodothyronine (T_3) (Cheng et al., 1980), platelet-derived growth factor, somatomedin, lysosomal enzymes, various toxins and viruses (reviewed in Pastan and Willingham, 1981). Thus, these cells can be used to compare the binding and internalization of many different ligands.

The existence of specific receptors on the surface of cells was initially demonstrated using hormones and proteins labeled with radioactive iodine. To determine where in the cell these ligands are located requires morphological analysis. This can be done at the light microscope level with fluorescently labeled ligands and at the electron microscopic level using electron dense probes.

3

1.1 FLUORESCENCE

Fluorescence microscopy has been of immense value in the analysis of the location of proteins in fixed (dead) cells using antibodies to specific cellular proteins. In such cases, usually the cell contains a large number of molecules to which the antibodies can bind. In studying the uptake of fluorescently labeled ligands many fewer molecules are involved and the signal is much weaker. In addition, if one wishes to examine living cells, the level of illumination must be diminished to prevent cell damage and bleaching of the fluorescent probe (for discussion see Willingham and Pastan, 1983).

To circumvent these difficulties we have employed a microscope fitted with a silicon intensifier target TV camera. This TV camera amplifies the fluorescent signal about 10 000 times and enables one to detect and record on a video tape fluorescent signals that can barely be seen by the dark-adapted eye. Using this approach it has been possible to follow the uptake of fluorescent derivatives of $\alpha_2 M$, EGF, insulin and tri-iodothyronine by living cells (Cheng *et al.*, 1980; Willingham and Pastan, 1978; Maxfield *et al.*, 1978). These studies showed that many ligands bound to receptors that were rather uniformly distributed all over the cell surface. The receptors were not concentrated in large patches as for example the acetylcholine receptor is on some muscle cells. When cells were allowed to internalize the ligand for a few minutes at 37°C, the ligand was observed to move into intracellular vesicles that moved about in the cytoplasm by saltatory motion (Willingham and Pastan, 1980). It has been known for some time that lysosomes move by saltatory motion (Rebhun, 1972). Most lysosomes are easily visualized by light microscopy because their high content of protein makes them phase dense (Willingham and Yamada, 1978). The organelle in which the recently ingested ligands we studied was located was not phase dense and therefore probably not a lysosome. Its identity was later established by electron microscopy (Willingham and Pastan, 1980).

Using rhodamine-labelled $\alpha_2 M$ which gives a strong signal (Fig. 1.1), it was possible to follow the progress of $\alpha_2 M$ in living cells with video intensification microscopy (VIM). In Swiss 3T3 cells the ligand moved from small vesicles in the peripheral region of the cell to the Golgi region over a period of about 15–30 min. After about 30–45 min ligand began to appear in phase-dense lysosomes. Because fibroblastic cells have many types of receptors on their surface, it is possible to follow the entry of two different ligands in the same cell at the same time by labeling one with rhodamine and the other with fluorescein. In such a manner it was shown that $\alpha_2 M$, EGF, insulin (Maxfield *et al.*, 1978), tri-iodothyronine (Cheng *et al.*, 1980) and LDL (Via *et al.*, 1982) enter the cell in the same vesicles. We suggested that the vesicle participating in receptor-mediated endocytosis be called a receptosome to distinguish this type of vesicle from those participating in non-receptor-mediated endocytosis

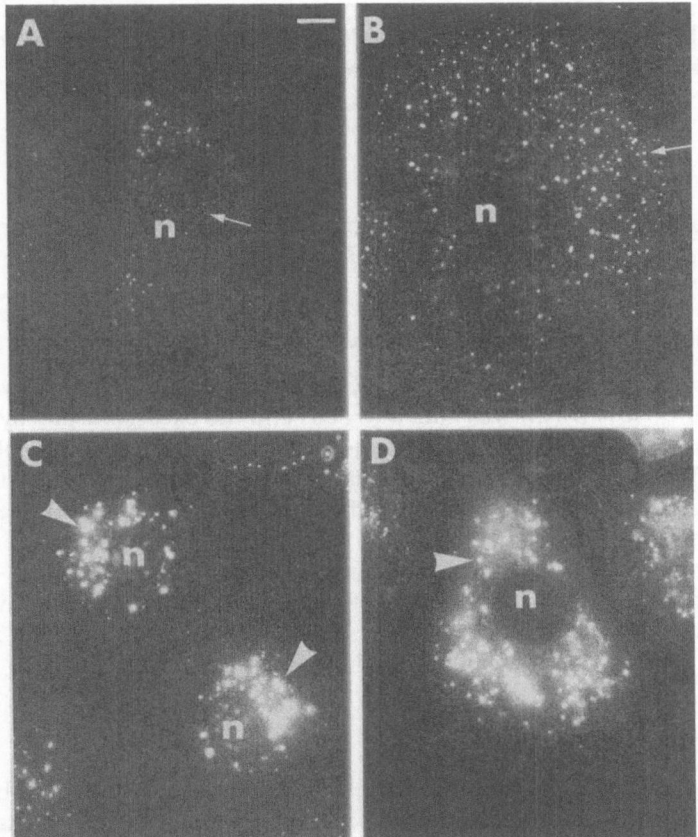

Fig. 1.1 Endocytosis of rhodamine-labeled α_2-macroglobulin by Swiss 3T3 fibroblasts.
Swiss 3T3 cells were incubated at 37°C with 300 μg of rhodamine–α_2M/ml of
serum-free medium for 5 min (A, B), 15 min (C) or 15 min followed by washing in
serum-free medium and incubation for 2 h (D). Rhodamine–α_2M can be found
clustered in very small punctate spots on the cell surface (coated pits), here shown just
barely visible when focused on the surface over the nucleus (A, arrow), in a continuous
5 min incubation at 37°C. At this time, however, most of the cell-associated α_2M has
been internalized into intracellular receptosomes visible over the entire cell outline,
visible as brighter punctate spots (B, arrow). After 15 min at 37°C, much of the
internalized α_2M in these receptosomes has redistributed from the cell margins back to
the perinuclear Golgi region of the cell (C, arrowheads). By 2 h after exposure to α_2M
at 37°C, almost all of the α_2M can be found in larger, discrete lysosomes in the cytoplasm
(C, arrowhead). (Magnification = × 440; bar = 10μm; Kodak Tri-X film developed
in Diafine.)

(pinosomes), and also to emphasize its role in receptor-mediated endocytosis (Willingham and Pastan, 1980). As determined by electron microscopy, receptosomes have a characteristic appearance (see below).

1.2 ELECTRON MICROSCOPY METHODS

To trace the precise pathway by which α_2M enters cells, a variety of conjugates of α_2M have been prepared. Each type of conjugate has its special uses.

1.2.1 Intact antibody conjugate – three-step labeling

Step 1. α_2M was bound to cells at 4°C and the excess washed away. Step 2. Rabbit antibody to α_2M was added to cells and the excess washed away. Step 3. A horseradish peroxidase conjugate of affinity-purified anti-(rabbit IgG (Cappel)) was used to complete the labeling. After being washed, the cells were warmed to 37°C and were fixed at various times to determine the location of the peroxidase conjugates. This is a convenient method for studying internalization and has been applied to other ligands such as β-galactosidase (Willingham *et al.*, 1981a) to which antibodies are available. It is important in such a study that the second-step antibody does not interfere with the binding of the ligand after the ligand is bound to the cell.

1.2.2 Hybrid antibody conjugates

One possible difficulty with the three-step method with intact antibodies is that in the third step, the first- and second-step molecules are cross-linked together on the cell surface. Because of the possibility that cross-linking might alter the pathway by which ligands are internalized, we have prepared hybrid antibodies which consist of one combining site that recognizes α_2M and a second that recognizes peroxidase or, alternatively, ferritin. In these experiments the cells are first exposed to α_2M, next to the hybrid antibody and last to native peroxidase or native ferritin (Dickson *et al.*, 1981a). With such hybrid reagents it was found that the morphological pathway of entry was the same whether or not intact antibodies or hybrid antibodies were used (Dickson *et al.*, 1981a).

1.2.3 Direct conjugate

A simple and readily controlled method of making conjugates linked by a disulfide bond has been described by Terouanne *et al.* (1980) and has been used by us to make α_2M-peroxidase (Dickson *et al.*, 1981a). In this method SH groups are introduced separately into α_2M and peroxidase by reaction with methylmercaptobutyrimidate (MMB). Next, one of the ligands is reacted with (DTNB) to activate its SH group. Finally, the two molecules are mixed together and a disulfide bond is formed between them by an exchange reaction. This method has been used to make α_2M–peroxidase (Dickson *et al.*, 1981a) and also EGF-peroxidase (Willingham and Pastan, 1982).

1.2.4 Colloidal gold

The most readily visualized of the labels used for electron microscopy is colloidal gold. Many proteins will adsorb very tightly to gold under appropriate conditions (reviewed by Horisberger, 1981). We have previously prepared α_2M–gold in which the gold particles are 150 nm in diameter and each particle contains about 400 α_2M molecules (Dickson *et al.*, 1981c).

1.3 ELECTRON MICROSCOPY RESULTS: PREFIXATION EXPERIMENTS

To determine the distribution of the unoccupied α_2M receptor, cells have been prefixed with paraformaldehyde. Then α_2M has been bound to the cell and its location established (Willingham *et al.*, 1979). These experiments showed that the unoccupied α_2M receptor is diffusely distributed over the cell surface. There is no particular concentration in coated pits (Willingham *et al.*, 1979). This result differs from studies reported for LDL which showed that a substantial fraction of the unoccupied LDL receptors were found preclustered in coated pits (Anderson *et al.*, 1976).

1.4 EARLY EVENTS

When α_2M is bound to living cells at 4°C under saturating conditions and its location determined by electron microscopy, some of the ligand is found to be clustered in coated pits on the cell surface, but the vast majority of the molecules are randomly distributed on the cell surface. Because of the prefixation results described above, we believe that at 4°C some α_2M–receptor complexes have diffused from outside the pits into the pits and been trapped and concentrated there. It is known that proteins can diffuse in membranes at 4°C.

We have also studied the binding of EGF (Willingham and Pastan, 1982; Willingham *et al.*, 1983) and adenovirus to KB cells at 4°C (FitzGerald *et al.*, 1983). Neither of these ligands is found concentrated in coated pits when the cells are maintained at low temperature and ligand allowed to bind. However,

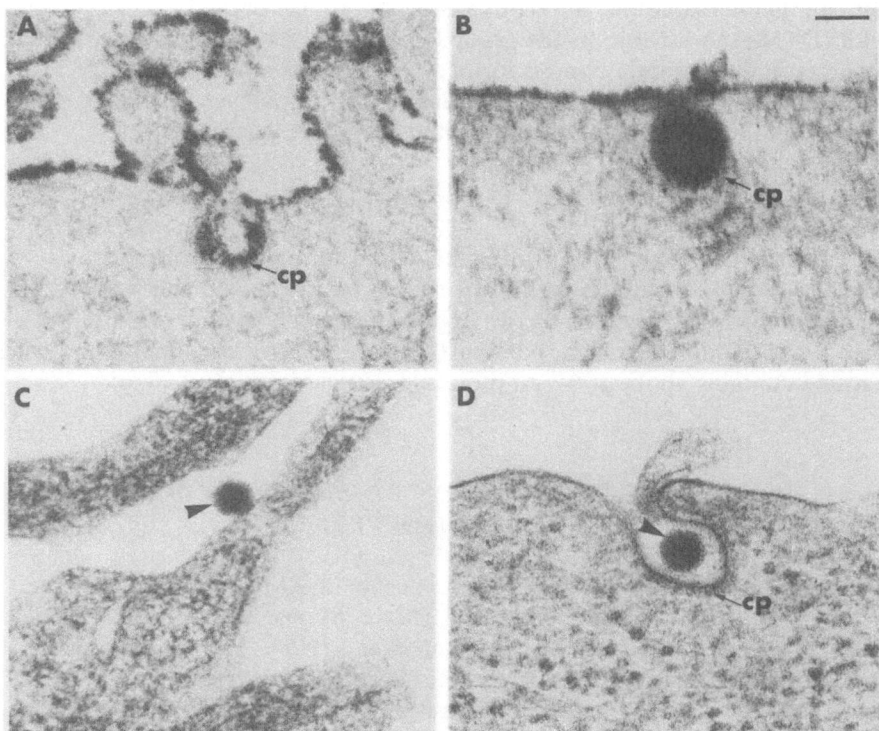

Fig. 1.2 Clustering of receptor–ligand complexes in clathrin-coated pits on the cell surface. KB cells were incubated at 4°C with either EGF–horseradish peroxidase (A) or adenovirus (C) for 60 min, washed at 4°C and fixed in glutaraldehyde (A, C) or warmed to 37°C for 1 min (B, C) prior to fixation. EGF–horseradish peroxidase is totally diffusely distributed on the cell surface on binding at 4°C (A), but on warming to 37°C, rapidly clusters into coated pits (cp) (B). Similarly, adenovirus binds at 4°C to diffusely distributed receptors unrelated to coated pits (C), but with warming to 37°C, viral particles (arrowheads) accumulate in coated pits prior to endocytosis (D), although the clustering of the viral particles is generally slower and more asynchronous than EGF. (Magnification = 70 200; bar = 0.1 μm.)

EGF bound to its receptor rapidly migrates to coated pits when the cells are placed at 37°C (Fig. 1.2). Adenovirus also moves to coated pits when the cells are raised to 37°C, but not nearly as rapidly as EGF (Fig. 1.2).

1.5 CLUSTERING AND INTERNALIZATION

Cells maintained at 0–4°C do not internalize significant amounts of ligands. When the cells are placed at 37°C, the internalization process begins. Within

2 minutes, ligands have moved from the cell surface into a specialized intracellular vesicle that has been termed a 'receptosome' to emphasize its role in receptor-mediated endocytosis. In the case of α_2M, EGF, LDL and β-galactosidase [which enters the cell via the mannose 6-phosphate (M6P)-recognition system] ligand–receptor (L–R) complexes have been found clustered in coated pits prior to their transfer to receptosomes. Viruses such as Semliki Forest virus (Helenius *et al.*, 1980), vesicular stomatitis virus (VSV) and adenovirus (Dales, 1973) have also been found to enter cells via coated pits. On the surface of viruses are molecules that bind to components on the cell surface; the component on the cell surface has been termed a receptor. In only a few cases has the viral receptor been well-characterized.

A virus should be thought of as a multivalent ligand and, although one virus at a time enters a coated pit, each virus probably is bound to a number of cellular receptors and clusters these receptors in pits.

1.6 INTERNALIZATION: RECEPTOSOME, GOLGI, LYSOSOMES

Within 1–2 min after ligands enter cells by receptor-mediated endocytosis, they are found in 'receptosomes.' Typical receptosomes containing a variety of different ligands are shown in Fig. 1.3. When observed by electron microscopy in glutaraldehyde-fixed cells, receptosomes are about 200–400 nm in diameter, have a continuous membrane and are often irregular in shape. Associated with the inner surface of the membrane is a proteinaceous material that appears to be ligand. The center of receptosomes appears to be empty. Receptosomes move to the Golgi region over a period of 10–20 min depending on the cell type studied. Then in a series of steps that are not yet clearly understood, various ligands (α_2M, EGF, β-galactosidase, LDL) have been observed to be transferred to lysosomes. Using conjugates in which peroxidase was attached to either EGF or β-galactosidase it has been possible to follow these ligands from receptosomes into the Golgi and thence on to lysosomes. It is the reticular portion of the Golgi and not the Golgi stocks that is transversed by EGF (Willingham and Pastan, 1982). Present on the reticular portion of the Golgi are small coated pits (~80 nm) about one-half of the diameter of those at the cell surface (Gonatas *et al.*, 1977). In KB cells EGF has been found to concentrate in the Golgi coated pits prior to its transfer to lysosomes. We believe that coated pits of the Golgi may function in a manner analogous to those at the plasma membrane to concentrate ligands prior to transfer to the next compartment. In the case of this part of the Golgi, the next compartment is the lysosome. How this transfer takes place is not known.

Not all ligands that enter cells via coated pits and receptosomes end up in lysosomes. Many viruses are able to move into the cytosol and some eventu-

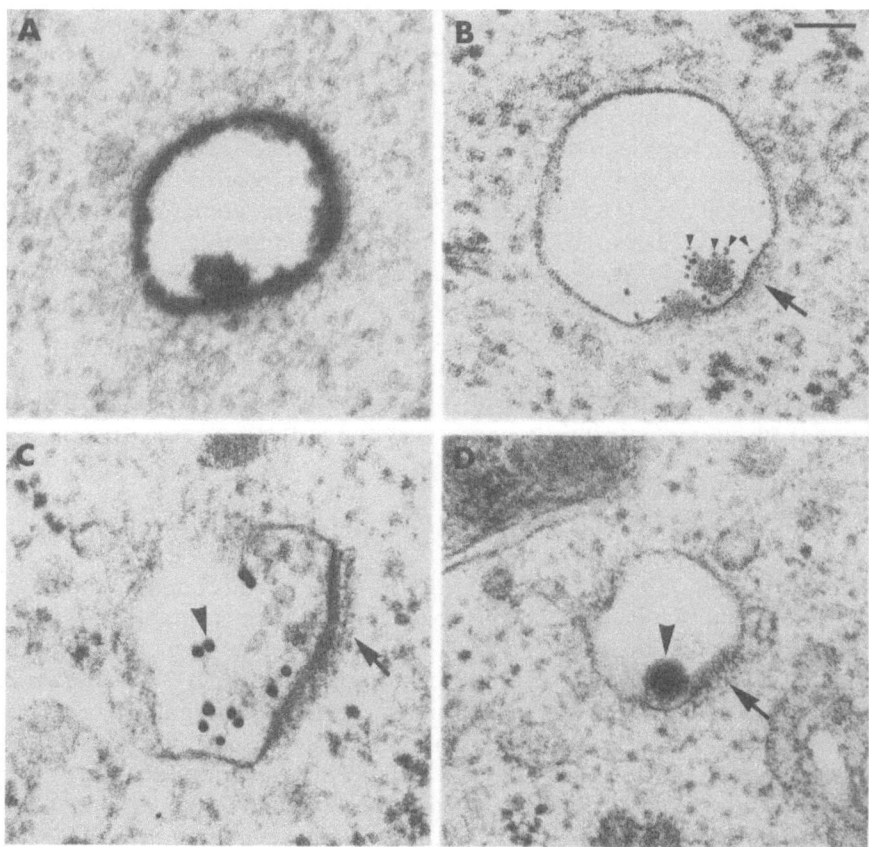

Fig. 1.3 Receptosomes containing internalized ligands derived from coated pits on the cell surface. KB cells (A, B, D) or Swiss 3T3 cells (C) were labeled at 4°C with EGF–horseradish peroxidase (A), ferritin-labeled EGF (B), α_2M–gold (C) or adenovirus (D). After binding at 4°C, the cells were warmed to 37°C for 2–5 min and fixed in glutaraldehyde. Receptosomes containing ligand derived from clustering in surface coated pits can be seen by using peroxidase labeling with EGF–horseradish peroxidase in (A), ferritin labeling for ferritin–EGF in (B) (arrowheads), colloidal gold labeling for α_2M–gold in (C) (arrowhead), or by visualizing individual viral particles in (D) (arrowhead). The characteristic features of receptosomes are their relatively empty appearance with a small intralumenal vesicular structure, concentrated ligand often associated with the membrane, and a fuzzy border on the cytosol face of the membrane (arrow in B, C, D). (Magnification = × 81 000; bar = 0.1 μm.)

ally enter the nucleus. The hormone tri-iodothyronine also eventually may be concentrated in the nucleus. The organelle from which these substances enter the cytosol is not yet clearly defined although it has been suggested that viruses are 'activated' or 'uncoated' in lysosomes prior to emergence in the cytosol. Recently, it has been found that receptosomes have a low pH (Tycho and Maxfield, 1982). This low pH may promote the exit of substances directly from receptosomes into the cytoplasm before they are transported to lysosomes.

1.7 MECHANISM OF RECEPTOSOME FORMATION

Until recently it has been believed that coated pits containing ligands pinch off from the cell surface to form 'coated vesicles' (reviewed by Pastan and Willingham, 1981). These 'coated vesicles' subsequently uncoat to form uncoated vesicles (receptosomes) and the coat (clathrin) recycles to the cell surface to form a new coated pit. Recent experiments using new electron microscopic approaches in combination with the injection of antibodies to clathrin into living cells have provided an alternative explanation for how receptosomes form.

When cells are fixed at 4°C and electron-dense molecules that bind to the plasma membrane are incubated on the outside of the cell, virtually all of the large coated structures near the cell surface are labeled (Willingham *et al.*, 1981b). Therefore, these coated structures are in direct anatomical communication with the cell surface. The communication is through an uncoated neck. It is sometimes difficult to see the communication with the surface because the neck is often tortuous and may not be in the proper plane of section. If the electron microscopic section misses the neck, the pit appears to be an intracellular coated vesicle when in fact it is a pit.

When cells are fixed at 37°C, about one-quarter of the coated pits become inaccessible to electron-dense labels added to the cell exterior (Willingham *et al.*, 1981b). These pits become accessible upon cooling to 0°C. This indicates that at 37°C some coated pits are physiologically sealed from the cell exterior. However, it does not mean they are anatomically separated from the surface. To account for the inability of electron-dense material to label coated pits in cells kept at 37°C, we have suggested that the neck of the coated pit functionally closes and opens as part of the mechanism by which receptosomes form from coated pits (see below).

Another approach in investigating how receptosomes form has utilized the microinjection of antibodies into living cells. If coated pits were to pinch off to form coated vesicles, then the clathrin must recycle to the surface to form new coated pits. Therefore, the injection of anti-clathrin antibodies into living cells should trap and precipitate the clathrin or coated vesicles and arrest endocytosis. Such experiments have been performed and the injection of antibodies to clathrin into living cells has not been found to precipitate intracellular clathrin or arrest endocytosis (Wehland *et al.*, 1981). Instead, the anti-clathrin antibody has been detected bound to the exterior of the coated pits without altering their function or disassembling the clathrin that surrounds the pits. Thus, two lines of evidence have shown that coated pits do not pinch off from the surface to form coated vesicles during endocytosis.

1.8 QUANTITATIVE ASPECTS OF UPTAKE BY COATED PITS

Because electron microscopic studies show that virtually all of the uptake of $\alpha_2 M$, EGF and LDL as well as some other ligands occurs via coated pits, it is possible to calculate how frequently a coated pit transfers its contents into a receptosome. In this way one can obtain a minimal estimate of how often a receptosome forms. To do this one needs to know the maximum rate the ligand is taken up by receptor-mediated endocytosis, the number of coated pits per cell (~ 1000), and the number of ligand molecules that can be clustered in a coated pit. Using $\alpha_2 M$ as a typical ligand we have calculated that coated pits fill and empty 2–3 times per minute (Pastan and Willingham, 1981).

One limiting factor in the entry of $\alpha_2 M$ is the accumulation of ligand–receptor complexes in coated pits. Ligand–receptor (L–R) complexes diffuse about in the plasma membrane at about the same rate as some other membrane proteins (Schlessinger *et al.*, 1978; Maxfield *et al.*, 1981). By a mechanism that is not understood the L–R complexes become trapped and clustered in coated pits. To account for clustering there must be a component in the coated pits that recognizes the ligand–receptor complex and keeps it bound or trapped in the coated pit. A clue to the possible nature of that interaction has come from inhibitor studies. A number of agents have been found that inhibit $\alpha_2 M$ uptake by preventing the accumulation of L–R complexes in coated pits. These include dansylcadaverine (and other primary amines) (Davies *et al.*, 1980), bacitracin (Davies *et al.*, 1980), *N*-benzyloxy-carbonyl-5-diazo-5-oxonorvaline *p*-nitrophenyl ester (Levitzki *et al.*, 1980), and α-bromo-4-hydroxy-3-nitroacetophenone (Dickson *et al.*, 1982). The mechanism by which these agents inhibit clustering in pits is unknown. Because some of the inhibitors are also inhibitors of transglutaminase, an enzyme that cross-links proteins, it has been suggested that they may act by inhibiting the activity of a 'transglutaminase-like' enzyme involved in the clustering mechanism in the coated pits. For example, a receptor could be cross-linked to one of the molecules that make up the coated pit. However, to date no direct evidence of cross-linking has been obtained. Alternative suggestions have also been made to account for the inhibitory action of some of these agents. Because primary amines are known to be able to raise the pH of lysosomes, it has been suggested that the inhibitory action of primary amines on endocytosis might be by their action on pH. To investigate this, Maxfield and co-workers (Anderson *et al.*, 1982) have recently measured the pH within the lysosomes and receptosomes of dansylcadaverine-treated cells; they have not found a significant rise in pH under conditions where dansylcadaverine inhibited α_2M entry. Thus, at least for dansylcadaverine, pH does not seem to be a factor. Another suggestion has been that SH groups are important for ligand internalization and that some of the inhibitors may be modifying SH groups (Levitzki *et al.*, 1980). It seems likely that many enzymes are involved in the clustering process and that different inhibitors may be acting at different biochemical steps.

1.9 EFFECTS OF MONOVALENT IONOPHORES

Another group of substrates has been found that inhibits the entry of α_2M (Dickson *et al.*, 1981b). These agents are monovalent ionophores and include monensin, nigericin, dianemycin, carbonyl cyanide *p*-trifluoromethoxyphenyl hydrazone (FCCP) and 3,3′,4′,5-tetrachlorosalicylanilide (TCSA). These ionophores allow the α_2M–receptor complexes to accumulate in coated pits but somehow inhibit the transfer of the ligand to the receptosome. The simplest explanation for the action of these agents is that they prevent or slow down receptosome formation. However, it is also possible that receptosomes form at a normal rate but the transfer of the ligand–receptor complexes is prevented. Experiments to distinguish between these two mechanisms have not been successful. Because proton-specific ionophores (FCCP and TCSA) are as effective as those affecting protons together with Na$^+$ and K$^+$, it has been suggested that coated pits contain a proton or ion pump which is necessary for receptosome formation. One way this could occur is shown in Fig. 1.4. In this formulation the neck of the coated pit transiently closes and forms a seal that allows an ion pump to increase the concentration of ions within the coated pit, resulting in an inflow of H$_2$O with increasing hydrostatic pressure, causing the pit to balloon out membranes adjacent to the pit and form receptosomes. If proton flow is the first step in this process, the protons would have to be exchanged for an ion such as Na$^+$. If the proton pump were transferred from the coated pit to the receptosome, it could help account for the low pH in receptosomes (Tycko and Maxfield, 1982).

   In addition to blocking α_2M entry, monensin has been found to inhibit the uptake of LDL (Basu *et al.*, 1981), EGF and (VSV) (Schlegel *et al.*, 1981). In the case of LDL, monensin has a large effect on receptor turnover and with

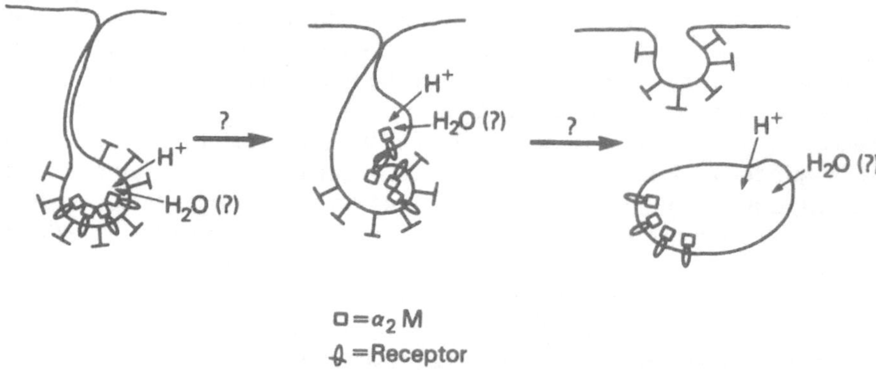

$$\square = \alpha_2 M$$
$$\text{\Lambda} = \text{Receptor}$$

Fig. 1.4 Schematic diagram of a hypothetical model of receptosome formation involving a proton pump.

time decreases the number of LDL receptors on the cell surface (Basu *et al.*, 1981). Monensin also slightly decreases the number of $\alpha_2 M$ and EGF receptors on the cell surface (Dickson *et al.*, 1981b). Thus, as with many drugs, the action of this class of inhibitors is complex. Nevertheless, they do give important clues as to how ligands are internalized.

1.10 GENERAL COMMENTS

The major pathway by which $\alpha_2 M$ and other ligands enter the cell by receptor-mediated endocytosis is summarized in Fig. 1.5. It involves binding of the ligand to cell surface receptors, concentration in coated pits in the plasma membrane and transfer into receptosomes. Receptosomes move by saltatory motion and carry the ligand to the Golgi. In KB cells it has been possible to do a pulse–chase type of experiment with EGF–horseradish peroxidase and follow the ligand through the Golgi. When EGF leaves the receptosome compartment, it appears in the reticular portion of the Golgi and is particularly concentrated in the small coated pits associated with this portion of the Golgi. Then, a few minutes later, the ligand appears in small lysosomes in the Golgi region.

In the studies described here only the fate of a label attached to the ligand can be followed. Therefore, ancillary data are needed to be certain that the ligand itself remains attached to the label. Radioautography using ^{125}I-labeled ligands is very useful in this regard and has been widely used by Orci, Gorden and collaborators (Gorden *et al.*, 1978) as well as Posner, Bergeron and collaborators (Bergeron *et al.*, 1979). Another useful approach is subcellular fractionation of homogenates prepared from cells incubated with ^{125}I-labeled

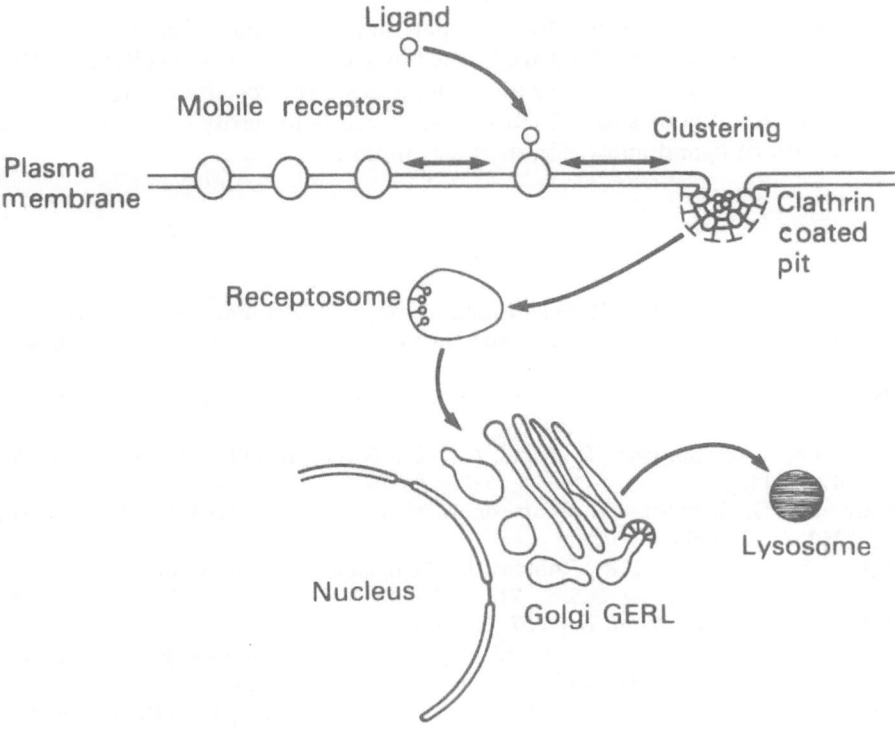

Fig. 1.5 Schematic diagram of the morphological pathway of receptor-mediated endocytosis in cultured cells.

ligands. With this approach it should be possible to isolate the organelle or fraction which contains the ligand of interest.

Not all substances that enter cells via coated pits traverse the entire pathway and end up in lysosomes. Recent studies indicate that viruses such as VSV or adenovirus may escape from the receptosome into the cytoplasm (FitzGerald *et al.*, 1983). One intriguing possibility is that the low pH present in the receptosome has an important role in this process.

It is assumed that the receptor accompanies the ligand into the cell during receptor-mediated endocytosis. However, direct experiments in which the receptor and ligand have been followed at the same time have only recently been initiated. By using either affinity purified or monoclonal antibodies, the receptors for transferrin and lysosomal enzymes (Willingham *et al.*, 1982) on cultured cells have been studied. Both receptors have been found to enter receptosomes along with their ligands (manuscript in preparation). When receptors enter cells with their ligands, it is likely that within the cell the ligand and receptor become separated. In some cases the receptors probably return to the cell membrane to participate in a new round of endocytosis. In other cases they may be destroyed. It is this area of control of cellular traffic that is currently being studied with great interest. The recent development of monoclonal antibodies to cellular receptors will help clarify where in the cell separation of ligand and receptor is accomplished.

REFERENCES

Anderson, P., Tycko, B., Maxfield, F. and Vilcek, J. (1982), *Virology*, **117**, 510–515.

Anderson, R.G.W., Goldstein, J.L. and Brown, M.S. (1976), *Proc. Natl. Acad. Sci. U.S.A.*, **73**, 2434–2439.

Basu, K., Goldstein, J.L., Anderson, R.S.W. and Brown, M.S. (1981), *Cell*, **24**, 493–502.

Bergeron, J.J.M., Sikstrom, R., Hand, A.R. and Posner, B.I. (1979), *J. Cell Biol.*, **80**, 427–443.

Carpenter, G., Sembach, K.J., Morrison, M.M. and Cohen, S. (1975), *J. Biol. Chem.*, **250**, 4297–4304.

Cheng, S.-y., Maxfield, F.R., Robbins, J. Willingham, M.C. and Pastan, I. (1980), *Proc. Natl. Acad. Sci. U.S.A.*, **77**, 3425–3429.

Dales, S. (1973), *Bacteriol. Rev.*, **37**, 103–135.

Davies, P.J.A., Davies, D.R., Levitzki, A., Maxfield, F.R., Milhaud, P., Willingham, M.C. and Pastan, I.H. (1980), *Nature (London)*, **283**, 162–167.

Dickson, R.B., Nicholas, J.-C., Willingham, M.C. and Pastan, I. (1981a), *Exp. Cell Res.*, **132**, 488–493.

Dickson, R.B., Schlegel, R., Willingham, M.C and Pastan, I. (1981b), *J. Cell Biol.*, **91**, 409a.

Dickson, R.B., Willingham, M.C. and Pastan, I. (1981c), *J. Cell Biol.*, **89**, 29–34.

Dickson, R.B., Willingham, M.C. and Pastan, I. (1981d), *J. Biol. Chem.*, **256**, 3454–3459.

Dickson, R.B., Schlegel, R., Willingham, M.C. and Pastan, I.H. (1982), *Exp. Cell Res.*, **140**, 215–225

FitzGerald, D.J.P., Padmanabhan, R., Pastan, I.H. and Willingham, M.C. (1983), *Cell*, **32**, 607–617.

Goldstein, J.L., Anderson, R.G.W. and Brown, M.S. (1979), *Nature (London)*, **279**, 679–685.

Gonatas, N.K., Kim, S.U., Stieber, A. and Avrameas, S. (1977), *J. Cell Biol.*, **73**, 1–13.

Gorden, P., Carpentier, J.L., Cohen, S. and Orci, L. (1978), *Biochem. Biophys. Res. Commun.*, **102**, 992–998.

Helenius, A., Kartenbeck, J., Simons, K. and Fries, E. (1980), *J. Cell Biol.*, **84**, 404–420.

Horisberger, M. (1981), *Scanning Electron Microsc.*, **11**, 9–31.

Levitzki, A., Willingham, M. and Pastan, I. (1980), *Proc. Natl. Acad. Sci. U.S.A.*, **77**, 2706–2710.

Maxfield, F.R., Schlessinger, J., Shechter, Y., Pastan, I. and Willingham, M.C. (1978), *Cell*, **14**, 805–810.

Maxfield, F.R., Willingham, M.C., Pastan, I., Dragsten, P. and Cheng, S.-y. (1981), *Science*, **211**, 63–65.

Octave, J.-N., Schneider, Y.-J., Hoffman, P., Trouet, A. and Crichton, R.R. (1979), *FEBS Lett.*, **108**, 127–130.

Pastan, I.H. and Willingham, M.C. (1981), *Science*, **214**, 504–509.

Rebhun, L.I. (1972), *Int. Rev. Cytol.*, **32**, 93–137.

Schlegel, R., Willingham, M.C. and Pastan, I. (1981), *Biochem. Biophys. Res. Commun.*, **102**, 992–998.

Schlessinger, J., Schechter, Y., Cuatrecasas, P., Willingham, M.C. and Pastan, I. (1978), *Proc. Natl. Acad. Sci. U.S.A.*, **75**, 5353–5357.

Terouanne, B., Nicholas, J.-C. Descomps, B. and de Paulet, C. (1980), *J. Immunol. Methods*, **35**, 267–275.

Thompoulos, P., Roth, J., Lovelace, E. and Pastan, I. (1976), *Cell*, **8**, 417–423.

Tycko, B. and Maxfield, F. (1982), *Cell*, **28**, 643–651.

Via, D.P., Willingham, M.C., Pastan, I., Gotto, A.M., Jr. and Smith, L.C. (1982), *Exp. Cell Res.*, **141**, 15–22.

Wehland, J., Willingham, M.C., Dickson, R. and Pastan, I. (1981), *Cell*, **25**, 105–120.

Willingham, M.C. and Pastan, I. (1978), *Cell*, **13**, 501–507.

Willingham, M.C. and Pastan, I. (1980), *Cell*, **21**, 67–77.

Willingham, M.C. and Pastan, I. (1983), *Methods Enzymol.*, (in press).

Willingham, M.C. and Pastan, I.H. (1982), *J. Cell Biol.*, **94**, 207–212.

Willingham, M.C. and Yamada, S.S. (1978), *J. Cell Biol.*, **78**, 480–487.

Willingham, M.C., Maxfield, F.R. and Pastan, I.H. (1979), *J. Cell Biol.*, **82**, 614–625.

Willingham, M.C., Pastan, I.H., Sahagian, G.G., Jourdian, G.W. and Neufeld, E.F. (1981a), *Proc. Natl. Acad. Sci. U.S.A.*, **78**, 6967–6971.

Willingham, M.C., Rutherford, A.V., Gallo, M.G., Wehland, J., Dickson, R.B., Schlegel, R. and Pastan, I.H. (1981b), *J. Histochem. Cytochem.*, **29**, 1003–1013.

Willingham, M.C., Pastan, I. and Sahagian, G.G. (1982), *J. Histochem. Cytochem.*, **30**, 104–123.

Willingham, M.C., Haigler, H., Fitzgerald, D. and Pastan, I.H. (1983), *Exp. Cell Res.*, (in press).

2 The Structure of Coated Vesicles

CAROL D. LINDEN* and THOMAS F. ROTH

---

* The views of this author do not purport to reflect the positions of the Department of the Army or Department of Defense.

Acknowledgements

The authors thank Drs John Woods and Jack Daiss for their helpful discussions, Ms Beverly Smith for preparation of the manuscript and Mrs Phebe Summers Angel for editorial assistance. This work was supported in part by NIH grants HD 09549 and HD 11519 from the National Institute of Child Health and Human Development and NSF grant PCM 8118717 (T.F.R.).

Receptor-Mediated Endocytosis
(*Receptors and Recognition*, Series B, Volume 15)
Edited by P. Cuatrecasas and T. F. Roth
Published in 1983 by Chapman and Hall, 11 New Fetter Lane, London EC4P 4EE

2.1 INTRODUCTION

In the almost 20 years since Roth and Porter (1964) proposed that coated vesicles mediate the specific transport of proteins, researchers from various fields have joined in investigating both the structure and function of these organelles. Interest in the function of coated vesicles has intensified since it became apparent that coated pits and coated vesicles are the organelles responsible for the initial steps in the receptor-mediated endocytosis of a host of polypeptides, hormones, growth factors and viruses. There is ample evidence also that they function in other intracellular pathways, including transport of proteins from one cell surface to another, from the endoplasmic reticulum to the Golgi and from the Golgi to the cell surface.

Coated vesicles are ubiquitous in cells throughout the plant and animal kingdoms. They are derived by invagination from coated pit regions of membranes, which are distinguished morphologically by an ordered array of protein on the cytoplasmic face. This ordered polygonal array is structurally diagnostic for coated pits and coated vesicles. Clathrin, the principal protein constituent of the coat, has a subunit molecular weight of 180 000 on SDS (sodium dodecyl sulphate)–polyacrylamide gels and is at this time the principal biochemical marker for coated vesicles. Because the structure and composition of coated vesicles appear to be strictly conserved throughout all cell types, efforts to understand their function have focused primarily on understanding their structure. While current molecular and biochemical approaches have yielded much information regarding the architecture of coated vesicles, what is lacking at this time is an understanding of the factors, be they enzymes or other regulators, that control their assembled state and functioning. Equally unclear is the mechanism by which these organelles are targeted to a given site within the cell such that their contents are delivered with appropriate specificity.

Ideally, as we summarize what is known currently about coated vesicle structure, the gaps in our knowledge will serve to stimulate further investigations of these unique and important organelles.

2.2 ULTRASTRUCTURE OF COATED VESICLES

Coated vesicles appear in electron micrographs of thin sections of virtually all eucaryotic cells as discrete spherical structures covered by a lattice-like network (Fig. 2.1). They are observed most commonly in the cortical region of the cell underlying the plasma membrane as well as in the vicinity of the Golgi apparatus. Roth and Porter (1964) orginally observed that coated vesicles form from the progressive invagination and pinching off of coated pits.

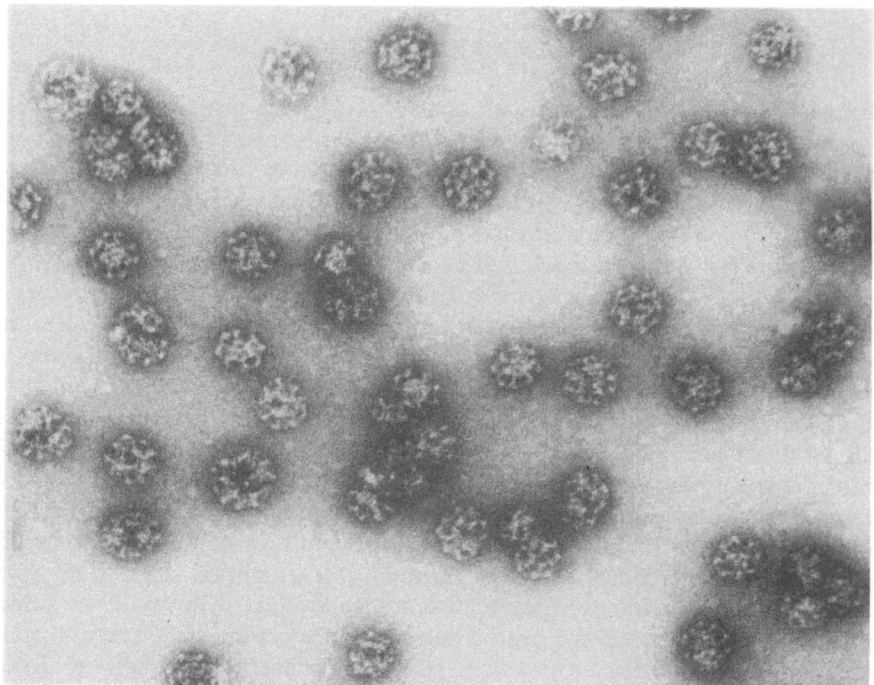

Fig. 2.1 Negatively stained coated vesicles isolated from chicken oocytes. Regular arrays of pentagons and hexagons comprise the coat structure in these 60–80 nm coated vesicles. This size class is most like those found associated with the endoplasmic reticulum–Golgi in all cells.

Recent interest has focused on understanding the dynamics of the transition from the almost flat structural network covering the coated pit to the spherical structure enclosing a vesicle. While it is only recently that coated pits have been visualized in a manner that permits structural analysis (Heuser, 1980), several studies of isolated coated vesicles have elucidated the general organization of the coat structure.

Kanaseki and Kadota (1969) reported the first partial purification of coated vesicles, thus facilitating direct observation of their structural features. Their preparation contained, in addition to membranous contaminants, both coated vesicles and coat structures which appeared devoid of an internal vesicle ('empty coats'). Empty coats in particular have proved useful for studies of the structure of the coat lattice. The coated vesicles had an average diameter of 120 nm with a vesicle diameter of approximately 50 nm. The coat lattice appeared to be organized into arrays of regular hexagons and pentagons with equal side lengths of 14 nm. The surface structures of the empty coats and of coated vesicles were essentially identical. Close examination of coated vesicles revealed no apparent connecting structure between the coat lattice and the vesicle, suggesting that empty coats might result from the loss or destruction of the internal vesicle during the isolation procedures.

Kanaseki and Kadota (1969) proposed a model for the structure of the coat lattice that is consistent with the geometric constraint that a spherical structure composed primarily of hexagonal subunits must contain a minimum of 12 pentagonal units in order to achieve proper curvature. The model suggests that hexagonal and pentagonal units are organized into a pattern like those of a soccer ball. Images of six-sided and eight-sided coat structures observed in the electron microscope are consistent with this model. It was proposed also that stepwise assembly of hexagons and pentagons to a coated pit 'powered' the invagination to form a coated vesicle. This study was a stimulus for subsequent more detailed studies of the coat structure.

Pearse (1975) developed a procedure which yielded a relatively pure population of coated vesicles and empty coats having average diameters of 55–88 nm. In characterizing this preparation, she determined that the principal component of the coat structure was a protein with a subunit molecular weight of 180 000, which was named clathrin. Electron micrographs of purified coated vesicles were subjected to computer analysis to generate models of various coat geometries (Crowther *et al.*, 1976). Three coat structures belonging to a small size class were identified. All were composed of 12 pentagons plus a variable number of hexagons. Larger structures could be constructed using the same packing principles of pentagons and hexagons. Sedimentation analysis of empty coats gave an estimated molecular weight of 22×10^6. A simple model of the coat structure was proposed based on the observed coat geometry and the estimated molecular weights of a coat and of clathrin. The geometry of packing hexagons and pentagons into a spherical structure dictates that three edges of individual polygons form a vertex. It was proposed that each vertex was formed from three clathrin molecules. Since an edge is defined by two vertices, each edge would be composed of two clathrin molecules. It was assumed in this model that edges were formed from clathrin disposed in a rod-like configuration. By counting the number of edges in a small coat, an estimate of the number of constituent clathrin molecules could be obtained. Thus, a small coat would contain approximately 100 clathrin molecules, yielding a minimum molecular weight of 1.8×10^7, which corresponds reasonably well to the estimated molecular weight of a coat. While the question of the precise molecular structure of the coat was unresolved in this model, recent studies (*vide infra*) have further clarified this point.

Ockleford (1976), using a similar technique of image analysis, also concluded that the coat was composed of a polygonal array of hexagons and pentagons. In contrast to others, however, he claimed that the edges of the hexagons and pentagons were thick ridges contiguous with the internal vesicle. Most workers in the field now appear to favor a model where the elements of the coat are in a rod-like configuration. Woods *et al.* (1978) studied the structure of coated vesicles of different diameters and found that they were all composed of hexagons and pentagons. They showed that certain well-oriented structures had rotational symmetry, with the smallest coats having 6-fold symmetry and the largest they observed, a 10-fold symmetry. They also concluded that the individual structural elements of the coat polygons were in the form of rods, rather than ridges contiguous with the vesicle.

What is clear from these and the many other morphological studies of coated vesicles is that they fall into at least two size classes. In sectioned material, coated vesicles in the immediate vicinity of the plasma membrane are most often 120 nm, a diameter equal to many of the deeply invaginated coated pits found along the cell membrane. The number of this size class of coated vesicles is rarely equal to the number of smaller 80 nm-diameter coated vesicles that predominate in the region of the Golgi apparatus (Friend and Farquhar, 1967; Jamieson and Palade, 1971). In most methods used to isolate coated vesicles, the smaller size class usually exceeds 95% of the total. Why this is the case is not clear. However, it does represent the approximate distribution of the coated vesicle size classes in most tissue as seen in thin section images. Other factors such as the relative stability of the larger coated vesicles might also be a factor since the percentage of the large class can be enhanced if a structural stabilizing agent such as D_2O is used in the preparatory steps. This particular method also increases the yield of intact coated vesicles and decreases the number of coats that lack vesicles (Pearse, 1982).

The most visually elegant study of coat structure and the dynamics of coat formation is that of Heuser (1980). Frozen, etched replicas of coated pits on the cytoplasmic surfaces of fibroblasts were examined (Fig. 2.2). This technique has the advantage that only one surface of the forming coated vesicle is visualized, as opposed to the technique of viewing stained images of purified coated vesicles, where a three-dimensional structure is observed in two dimensions. Heuser's results show that coated pits may form by the stepwise addition of individual elements at the edge of a coated patch of membrane. Initially, when the coated pit is shallow and flat, the coat lattice is composed primarily of hexagonal elements. As the membrane invaginates and the coat becomes progressively more curved, increasing numbers of pentagons are observed in the array. Heptagons adjacent to pentagons seem to be intermediate structural elements in the formation of a spherical, discrete coated vesicle. At the edges of coated pits, incomplete polygons are observed. The freeze–etch technique was also applied to purified empty coats. These structures were composed of hexagons and pentagons, the exact number of each depending on the size of the coat. While Heuser's studies give us greater insight into the mechanics of coat assembly, it is still unknown whether the formation of a curved coat structure is the motive force for coated pit invagi-

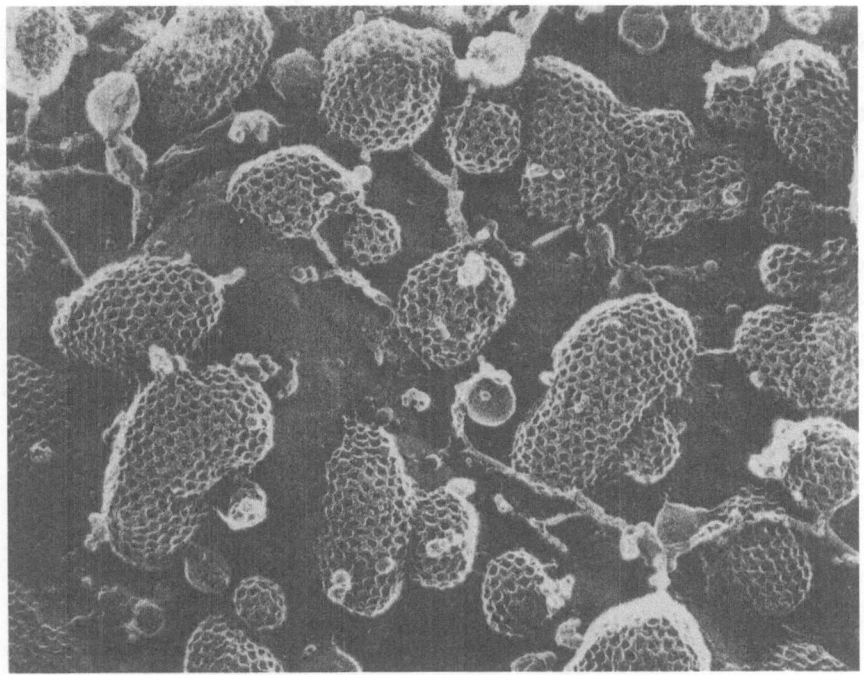

Fig. 2.2 Surface view of the cytoplasmic face of a developing chicken oocyte. Numerous coated pits occur at a high density, reflecting the great amount of protein being internalized by this mechanism in these cells. Note their pleiomorphic shapes which become altered to give rise to a nearly uniform class of 120–140 nm coated vesicles. Picture courtesy of Dr. John Heuser.

nation or whether other forces drive the invagination of the membrane and dictate the formation of the curved coat.

2.3 COMPOSITION OF COATED VESICLES

Rigorous characterization of the components of coated vesicles is ultimately dependent on the purity of the preparation analyzed. To date, coated vesicles have been purified by various differential centrifugation procedures from a variety of cell types and tissues. Pearse (1975) first performed SDS–polyacrylamide gel electrophoresis on coated vesicles purified from pig brain using sucrose gradients. She found that the principal protein component of coated vesicles had a subunit molecular weight of 180 000. Coated vesicles treated with trypsin or pronase lost their coat structure as determined by electron microscopy and lost the 180 000-mol.wt. band on gels. Pearse thus named the 180 000 protein 'clathrin' since it appeared to be the protein that formed the characteristic cage-like coat structure. Subsequently, other investigators have found that several additional proteins are consistently present in coated vesicles prepared from various sources. These include families of proteins in the molecular weight ranges 100–120K, 50–55K and 33–36K (Fig. 2.3). Although clathrin comprises the bulk of the protein in any coated vesicle preparation, the stoichiometry of all of the coated vesicle proteins may vary somewhat depending on the tissue source. Woodward and Roth (1978) and Woods *et al.* (1978) estimated the stoichiometry of the proteins in coated vesicles from brain and chicken oocyte. Of the total protein in brain coated vesicles, clathrin accounted for 43–52%, with protein in the families of molecular weight of *ca.* 125K and 55K each accounting for 10–15%. The molar ratio of clathrin:125K-mol.wt. protein:55K-mol.wt. protein families was calculated

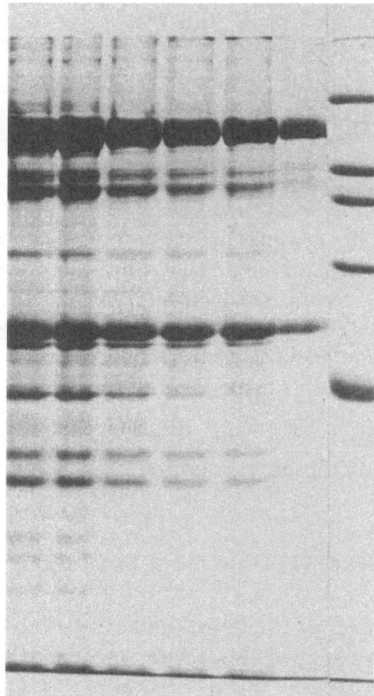

Fig. 2.3 SDS–polyacrylamide gel pattern of coated-vesicle proteins at various concentrations. Note the change in the relative ability to detect each of the various polypeptides and the apparent change in their relative abundance. Molecular weight standards on the right are: myosin, 200K; β-galactosidase, 116K; phosphorylase *b*, 92.5K; bovine serum albumin, 67K; and ovalbumin, 45K.

to be 1.8:1.0:2.2 on a weight basis (Woodward and Roth, 1978). The remaining proteins in the preparation each accounted for less than 5%. In contrast, coated vesicles purified from chicken oocyte contained a protein of 140 000-mol.wt. which amounted to 7–10% of the total, while clathrin accounted for only 20–30% of the total protein. The proteins of *ca.* 110K and 55K mol.wt. comprised 15–20% and 5–10% respectively, with other bands ≤5% (Woods *et al.*, 1978). The contribution of clathrin to the total protein content of coated vesicles is less in these two studies than in those few other studies where quantitation was attempted. For instance, Pearse (1976) stated that clathrin constitutes 70–90% of the total protein; however, no method of quantitation was presented. An inherent problem in quantitating the constituent proteins of coated vesicles is that a variable (16 to 65%) percentage of the structures purified are 'empty baskets' and thus devoid of a membrane vesicle that presumably contains many of the non-clathrin proteins (Woodward and Roth, 1978). The yield of intact coated vesicles may be improved somewhat by use of a recently developed purification technique that utilizes iso-osmotic D_2O/Ficoll gradients (Pearse, 1982).

Recently, coated vesicles have been purified further by techniques unrelated to differential centrifugation. Rubenstein *et al.* (1981) purified coated vesicles by agarose gel electrophoresis. This procedure appears to separate membrane contaminants from a heterogeneous population of empty coats and coated vesicles. SDS–polyacrylamide gel analysis of these preparations revealed the same spectrum of proteins observed in conventional, sucrose-gradient-purified coated vesicles, thus confirming that they are intrinsic constituents of all coated vesicles. Pfeffer and Kelly (1981) purified coated vesicles by chromatography on controlled-pore glass beads and found that minor membranous contaminants could be removed by this method. By comparing the SDS–polyacrylamide gel electrophoretic protein profiles of the coated vesicle fraction and the contaminant fraction, they concluded that the proteins unique to coated vesicles, besides clathrin, included those in the molecular weight range 100–120K, the triplet at 50–55K, and the 33–36K peptides. These proteins were found only in negligible amounts in the membranous material. In common with the coated vesicles, however, the membrane contaminants contained a protein of 33 000 mol.wt and the two larger of the three 50–55K-mol.wt. proteins. By two-dimensional gel analysis, it appeared that two 50–55K-mol.wt. proteins were common to both the contaminant and coated vesicles but that the 33 000-mol.wt. protein found in coated vesicles had a different isoelectric point from that found in the membranes. 'Contaminant' proteins sometimes attributed to coated vesicles in less pure preparations included those of ≃68 000 and ≃40 000 mol.wt.

Clathrin appears to be a highly conserved protein, since the amino acid compositions and peptide maps after cyanogen bromide cleavage or cleavage at cysteine residues are very similar for clathrins derived from several tissue sources (Pearse, 1976). In addition, clathrins from different species are immunologically cross-reactive with polyclonal (i.e. rabbit) antibodies (Fine *et al.*, 1978; Anderson *et al.*, 1978; Keen *et al.*, 1981), but more recent studies using monoclonal anti-clathrin antibodies suggest that there may be subtle differences in clathrins purified from different species (Daiss and Roth, unpublished).

Much less is known about the composition of the membrane vesicles within a coat than about clathrin. Clearly, even a small amount of contaminating membranes in a coated vesicle preparation can influence the analysis and quantitation of the membrane lipids. Pearse (1975) analyzed phospholipid extracts of purified coated vesicles and found 410 nmol of lipid phosphorus/mg of protein (75% protein:25% lipid, w/w). However, coated vesicles purified by agarose gel electrophoresis contained only 150 nmol of lipid phosphorus/mg of protein (Rubenstein *et al.*, 1981). This figure is probably an underestimate because the ratio of empty coats to coats containing membrane vesicles is unknown. Pearse (1976) further showed a phospholipid composition of primarily phosphatidylcholine and phosphatidylethanolamine in brain coated vesicles with a molar ratio of 10:1 phospholipid:cholesterol. She found a lower molar ratio of 3:1 in coated vesicles purified from adrenal medulla. These observations differ from those of Montesano *et al.* (1979) who showed that large cholesterol–filipin complexes appear to be absent from the plasma membrane in the region of a coated pit. Because the membrane component of a coated vesicle would presumably reflect the membrane lipid composition of the source from which it is derived, e.g. plasma membrane or Golgi, etc., there probably is no one phospholipid or cholesterol distribution characteristic of all coated vesicles.

Researchers have long hoped to find a biological or enzymatic activity characteristic of coated vesicles. Such an activity would greatly facilitate the characterization of purified preparations. Tanaka *et al.* (1976) reported a Mg^{2+}- and Ca^{2+}-dependent ATPase activity in coated vesicles purified from rat brain. Subsequently Blitz *et al.* (1977) reported a similar activity and further characterized the putative coated-vesicle ATPase. This latter report was re-evaluated when it was found that coated vesicles purified by agarose gel electrophoresis lacked the Ca^{2+}-dependent ATPase activity (Rubenstein *et al.*, 1981). Rather, the enzyme was associated with the membrane contaminants which were removed by agarose gel electrophoresis. These observations highlight the point that determination of an enzyme or receptor activity associated with coated vesicles is subject to the difficulty of rigorously excluding the possibility of other membrane contaminants. We screened a preparation of coated vesicles purified from brain by sucrose-density-gradient centrifugation for enzyme markers characteristic of plasma membrane, endoplasmic reticulum, Golgi, mitochondria and lysosomes. In the final purified coated vesicle sample there were no organelle marker enzymes uniquely associated with the coated vesicles (Linden and Canonico, unpublished).

The question of what biological substances may be found within coated vesicles or associated with the vesicle membrane is receiving much attention. The current literature contains many reports that various ligands bind to receptors localized in coated pits and are internalized by receptor-mediated endocytosis (RME) within coated vesicles. To date, several hormones, growth factors, proteins [e.g. LDL (low-density lipoprotein) α_2-macroglobulin, IgG, transferrin, vitellogenin], toxins and viruses have been shown to enter cells through a coated pit–coated vesicle-mediated route. Thus one might predict that these substances, and/or their receptors, could be found in coated vesicles. Receptor-mediated endocytosis has been the subject of several recent reviews (see Willingham and Pastan, Chapter 1 and Pastan and Willingham, 1981) and will not be covered in this chapter. There are, however, several instances where specific contents of coated vesicles have been identified and these deserve mention here.

Morphological studies have revealed specific contents associated with coated vesicles participating in secretion and/or organelle biogenesis. Novikoff *et al*. (1980) reported the localization of lysosomal enzymes in Golgi-associated coated vesicles. Franke *et al*. (1976) identified casein within the coated vesicles found in the apical zone of lactating rat mammary epithelial cells. Similarly, Kartenbeck *et al* (1977) observed aggregates of serum lipoproteins in rat liver coated vesicles. In these two cases the material contained within the coated vesicles was presumably destined for secretion. Rothman and Fine (1980) showed biochemically that the spike glycoprotein of vesicular stomatitis virus was transported from the Golgi to the cell surface in coated vesicles. Putative 'contents' have been identified in various purified coated vesicle preparations. These 'contents' include lipovitellin in coated vesicles isolated from oocytes (Woods *et al*., 1978; Pearse, 1978), and transferrin (Booth and Wilson, 1981), ferritin and IgG (Pearse, 1982) in coated vesicles purified from placenta.

Thus far the identification of receptors or characteristic membrane proteins associated with the coated vesicle membrane has lagged behind the identification of contained ligands. However, low-density lipoprotein receptors have been found in coated vesicles purified from bovine adrenal cortex (Mello *et al*., 1980). In order to detect these receptors, the vesicle first had to be disrupted with detergent, an observation consistent with the presumed orientation of the receptor toward the lumen of the vesicle. The search for other receptors associated with isolated coated vesicles continues in several laboratories.

2.4 SUBUNIT STRUCTURE OF THE COAT

2.4.1 Disassembly and reassembly of the coat structure

Blitz *et al*. (1977) observed that 2 M-urea disrupted the coat lattice of coated vesicles and caused the concomitant release in soluble form of several major coated vesicles proteins. These included clathrin, and the 100–200K- and 50–55K-mol.wt. proteins. The urea-insoluble material remaining after the extraction appeared to be an aggregate of smooth membrane vesicles which contained the bulk of the lower molecular weight proteins and little clathrin. These results suggested that other proteins besides clathrin might play a role in the structure of the coat. Kartenbeck (1978) also dissociated coated vesicles in 2 M-urea and noted that after dialysis to remove the urea, the protein reassembled to form empty coat structures. When these coats were isolated by centrifugation and subjected to SDS–polyacrylamide gel electrophoresis, only clathrin was detected. However, it is likely that other proteins were present but not detected owing to their relatively low concentration.

Pearse (1978) solubilized bovine brain coated vesicles in cholate and fractionated the soluble material by gel filtration on BioGel A-15m. The principal peak, which was eluted shortly after the void volume, contained clathrin plus two small proteins with molecular weights of 36 000 and 33 000. As this and later studies show, the two '30K' proteins are intimately associated with clathrin and require harsh conditions (i.e. 6 M-guanidine–HCl) to effect their separation (Keen *et al.*, 1981). When cholate was removed from the clathrin-containing peak, large complexes, showing a surface network similar to that of coated vesicles, were formed. Other investigators have found that completely spherical coat structures can be formed from a mixture of clathrin plus the two 33–36K-mol.wt. proteins (Ungewickell and Branton, 1981).

Woodward and Roth (1978) screened a variety of agents for their ability to disrupt coat structure. Of the agents tested, several caused loss of the characteristic coat structure as determined by electron microscopy. These included 2 M-urea, pH greater than 7.5, $CaCl_2$ or $MgCl_2$ at 0.25 M or greater, and KI, KBr and CsCl at ≥0.6 M. When insoluble material was removed from the urea- or 'pH'-solubilized coated vesicles by centrifugation, the remaining supernatants contained clathrin and the 30K-mol.wt. proteins plus small quantities of proteins in the 100–120K and 50–55K molecular weight ranges. Empty coats could be reassembled from the urea- or 'pH'-soluble supernatants and reisolated on sucrose gradients. When these reassembled coats were examined by SDS–polyacrylamide gel electrophoresis and by electron microscopy, they were found to contain predominantly clathrin and appeared very similar morphologically to untreated coated vesicles. Keen *et al.* (1979) studied solubilization of the coat by various agents. They found that high concentrations of protonated amines at neutral pH solubilized clathrin and the lower molecular weight proteins in the 100–120K, 50–55K and 33–36K range. When these soluble extracts were dialyzed against the isolation buffer to remove the amines, coat structures re-formed. Tris extracts of coated vesicles were fractionated on Sepharose CL-4B. This procedure yielded two pools of material, one with an estimated molecular weight of 1 200 000 containing clathrin (and the 30K-mol.wt. polypeptides) plus trace amounts of a 90 000

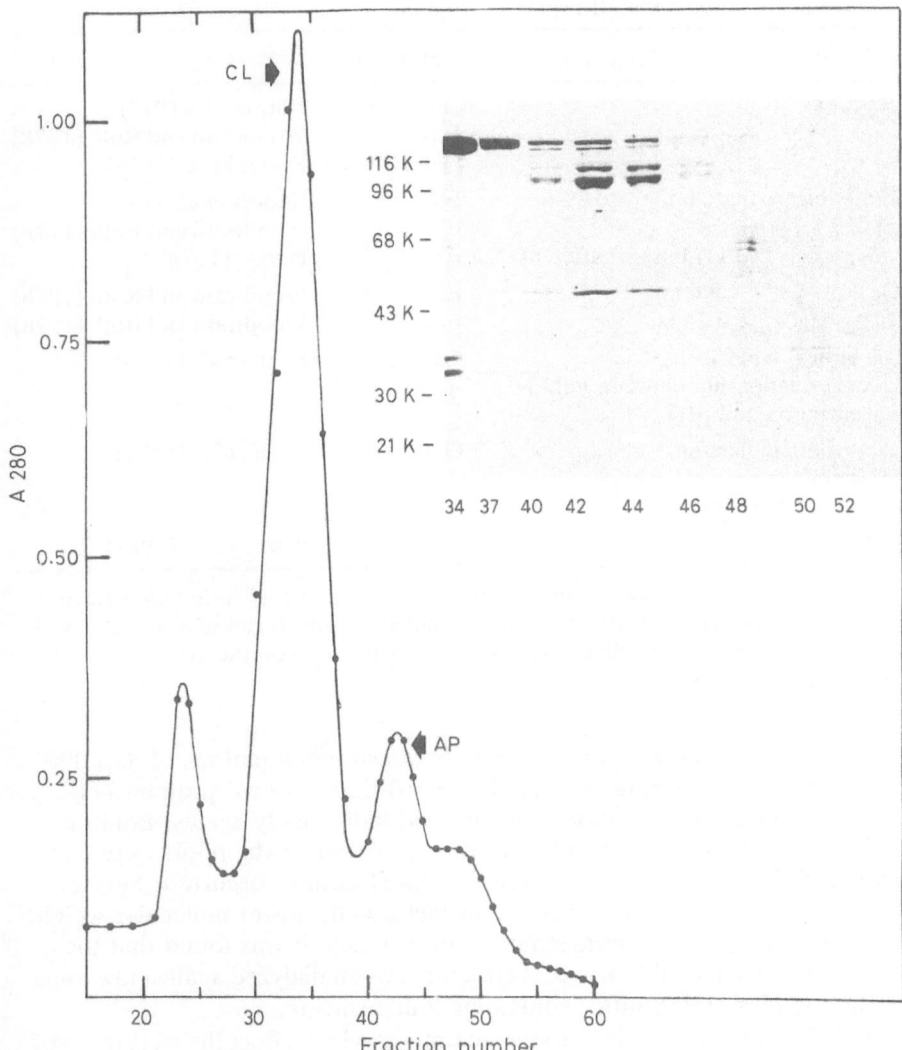

Fig. 2.4 CL-4B gel exclusion profile and associated SDS–polyacrylamide gel electrophoretic pattern of Tris (0.5 M)-dissociated CVs. Triskelions (CL) consisting of the 180K (clathrin) and 34K and 36K (light chains) dalton polypeptides are seen in the included peak at fraction 34. The proteins found in the peak at fraction 42 (AP) consist predominantly of the 100–120K and 50–55K (tubulins and tau) dalton polypeptides. In addition to these associated polypeptides, some clathrin appears to cochromatograph in this peak. The polypeptides in fractions 48–60 are poorly resolved but appear to be distinct from those in peak AP. (Courtesy of Drs S. Zaremba and J. Keen.)

Table 2.1 Treatments that disrupt coat structure

Condition	Assay*	Reference
2 M-urea	Gel	Blitz *et al.* (1977)
	EM	Woodward and Roth (1978)
	Gel, EM	Kartenbeck (1978)
0.25 M-bicarbonate buffer, pH 9.6	EM	Kadota *et al.* (1976)
pH ≥ 7.5 (Tris)	EM	Woodward and Roth (1978)
2.5% cholate in 0.1 M-borate buffer, pH 8.5	EM	Pearse (1978)
KI, KBr, CsCl, ≥0.6 M	EM	Woodward and Roth (1978)
CaCl$_2$, MgCl$_2$, ≥0.25 M	EM	Woodward and Roth (1978)
Tris buffer, ≥pH 7 (inverse relationship between buffer concentration and pH)	Gel	Keen *et al.* (1979)
0.5 M-triethanolamine 0.5 M-ammonium chloride 0.5 M-imidazole chloride	Gel	Keen *et al.* (1979)
ATP plus cytosol	Gel	Patzer *et al.* (1982)

* Assays used to determine the disruption of the coat structure were either electron microscopy (EM) of negatively stained samples or centrifugation of samples with subsequent analysis of pellets and supernatants by SDS–polyacrylamide gel electrophoresis (Gel).

mol. wt. protein and a second peak with an estimated mol.wt. of 420 000 predominantly containing 100–120K and 50–55K mol.wt. proteins (Fig. 2.4). If either pool of material was dialyzed individually against isolation buffer, no coat structures were observed. However, if the pools were combined and then dialyzed, reassembly of coat structures occurred. Further experiments suggested that a heat-labile factor in the lower molecular weight pool was required for coat assembly. Interestingly, it was found that the clathrin pool itself could form coat structures when dialyzed against low ionic strength ($\Gamma/2 = 0.01$) buffer containing 2 mM-calcium.

Table 2.1 summarizes the various treatments which effect the disruption of coat structure and the release of soluble coated vesicle proteins. While these studies suggest that clathrin is the primary structural protein of the coat, they yield little information on the regulation of coat assembly. Greater insight into this question has been derived from studies of coat reassembly *in vitro*.

2.4.2 The triskelion as a functional subunit of the coat

It is now apparent that the standard methods of dissociating the coat lattice yield a structure composed of three clathrin monomers in association with three of the so-called light chains, the 33 000 and 36 000 mol.wt. polypeptides. This macromolecular complex assumes a characteristic pinwheel morphology when examined by electron microscopy and has been termed a triskelion (Ungewickell and Branton, 1981) (Fig. 2.5). Ungewickell and Branton (1981) dissociated coats using 30 mM-Tris plus 2 M-urea and isolated this clathrin–light chain complex by chromatography on Sepharose CL-4B. Cross-linking studies using 3,3-dithiobispropionimidate confirmed previous suggestions that the light chains are intimately associated with the clathrin. The stoichiometry, estimated from relative staining on SDS–polyacrylamide gels, appeared to be one light chain per clathrin molecule. The structural implications of this are unclear, given that the light chains are non-identical, yet together yield a 1:1 ratio with clathrin.

Kirchhausen and Harrison (1981) performed a systematic cross-linking study using three bifunctional reagents. Treatment of triskelions with cross-

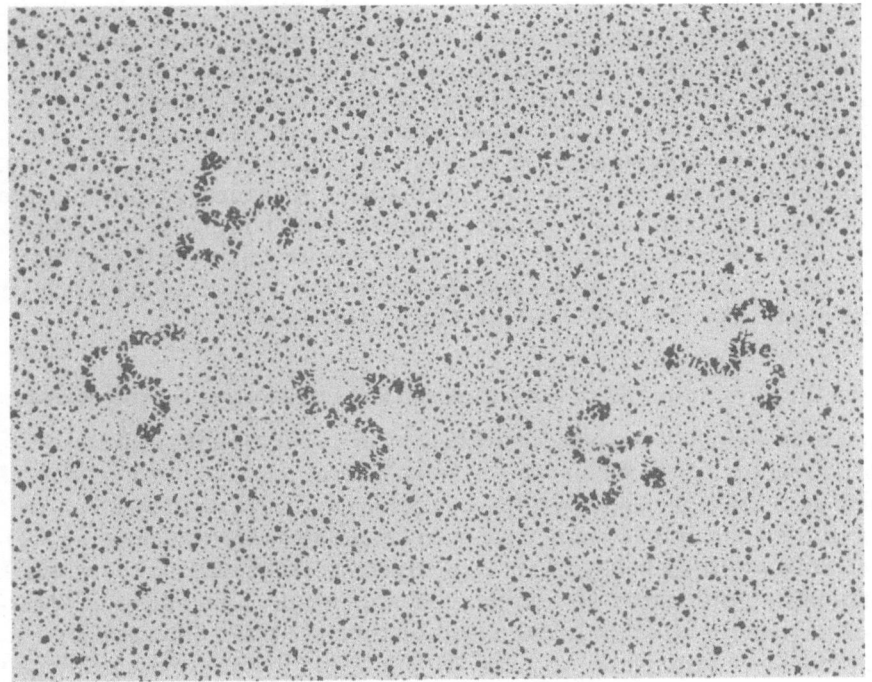

Fig. 2.5 Triskelions from dissociated coated vesicles are readily visualized after low-angle rotary shadowing. Such structures each consist of three clathrins and three light chains. These structures are believed to interlock to form the regular arrays of hexagons and pentagons of the coated vesicles. Micrograph courtesy of Dr. D. Branton.

linkers yielded six different molecular weight species upon electrophoresis of the cross-linked products on SDS–agarose/acrylamide gels. The observed molecular weights were 180K, 215K, 360K, 395K, 430K and 640K. It was thus suggested that the 215K-mol.wt. species represented a cross-linked complex of clathrin plus one light chain, and that the 360K-, 395K- and 430K-mol.wt. species were formed from two molecules of clathrin plus zero, one and two light chains respectively. The 640K-mol.wt. species was thought to be a trimer of the 215K-mol.wt. species, i.e., a cross-linked triskelion. These results suggested that there is little or no contact between light chains themselves, since no multimers of them were observed.

Characterization of the triskelion by analytical ultracentrifugation and electron microscopy has yielded sedimentation coefficients of 8.2 S (Nandi *et al.*, 1980) and 8.4 S (Ungewickell and Branton, 1981) and a molecular weight of approximately 630 000 (Ungewickell and Branton, 1081). The 8.4 S material had a characteristic triskelion morphology when viewed either by rotary shadowing or negative staining. Virtually all triskelions show the same handedness of rotation, with the three bent legs each having a mean length of 44.5 nm. The three legs meet at a vertex with the kink in each leg occurring 19 nm from this point. Other investigators (Crowther and Pearse, 1981; Kirchhausen and Harrison, 1981) have obtained similar measurements of triskelions. The angle of the kink in each arm is approximately 120°. The thickness of the legs appeared to be relatively uniform, between 2 and 4 nm. Triskelions can reassemble to form coat structures virtually identical in size and appearance to the coats of intact coated vesicles (Ungewickell and Branton, 1981). As will be discussed in the following sections, the biophysical properties, dependence on ions and pH, etc., and the role of the light chains in triskelion reassembly have been the subjects of intensive study.

2.4.3 Biophysical properties of triskelions and reassembled coat structures

Extensive biophysical studies of triskelions and coat assembly have been carried out by Edelhoch and co-workers. Various analyses of triskelions all yield an S value of $\simeq 8$; an absorbance coefficient of $A_{280}^{1\%} = 10.9$ and partial specific volume of 0.744 ml/g have been determined (Pretorius *et al.*, 1981). The ability of triskelions to form coat structures was monitored by equilibrium ultracentrifugation and turbidity measurements. Using these measurements, Nandi *et al.* (1980) determined that the 8.2 S triskelion species can reassemble to form empty coats of 150 S and 300 S with molecular weights and mean diameters of 24×10^6 and 64 nm and 100×10^6 and 112 nm respectively. Formation of intermediate sized species was not observed. Both the 150 S and 300 S particles appeared to be similar structurally to the coat lattice of intact coated vesicles when examined by electron microscopy. The dimensions of these particles also correspond fairly well to the two size classes of intact coated vesicles. The reassembly of 8.2 S clathrin to form the 150 S and 300 S species was dependent on pH, ionic strength and calcium concentration, supporting the notion that coat assembly is dependent on ionic interactions. Since the relative amounts of triskelions and 150 S and 300 S coats can be calculated using analytical ultracentrifugation, this technique provides a more accurate estimate than does electron microscopy of the ability of a given clathrin preparation to reassemble under various conditions.

Circular dichroism (CD) was used to determine the secondary structure of the 8.1 S clathrin in various solvents (Pretorius *et al*, 1981). The secondary structure did not change over the pH range of 7.5 to 9.0. The S value showed no change over even a broader range of pH and salt concentration. Alpha-helical structure accounted for 49% of the peptide residues; of the remainder, 17% was beta-structure and the remaining 34% was in a random structure. A comparison of the fluorescence emission spectra of 8.1 S clathrin measured in water and in 6 M-guanidine hydrochloride showed peak shifts which suggested that tryptophan residues are normally shielded by tertiary structure. The CD spectrum of 8.1 S clathrin in 6M-guanidine showed no α-helical structure. Thus both the fluorescence and CD spectra together show that little or no secondary or tertiary structure remained in 6 M-guanidine. Keen *et al*. (1981) showed also that clathrin can be separated from the two light chains by gel filtration chromatography in 6 M-guanidine. Clathrin purified using this chromatographic procedure and subjected to sedimentation equilibrium analysis (Pretorius *et al*., 1981) yielded a monomeric molecular weight of 170 000 ± 26 000. This value is in good agreement with the molecular weight of monomeric clathrin estimated by SDS–polyacrylamide gel electrophoresis (175 000–180 000).

2.4.4 Biochemical factors governing triskelion reassembly

Van Jaarsveld *et al*. (1981) studied the effects of pH, ionic strength, temperature and protein concentration on the rate of 300 S coat formation from triskelions. The reassembly process was monitored by turbidity and light scattering. These measurements were verified by quantitating 8 S, 150 S and 300 S clathrin-containing species using analytical ultracentrifugation. The effect of protein concentration on the rate of reassembly was examined between 0.2 and 1.7 mg/ml at constant pH (6.33). The initial rate of reassembly appeared to be a first-order reaction, and thus, relatively independent of concentration (above 0.2 mg/ml). However, the final ratio of 8 S to 300 S clathrin-containing species after 16 hours did reflect a concentration-dependence, in that at lower protein concentrations, proportionately less assembly occurred. The final extent of reassembly was affected not only by protein concentration but also by pH and temperature. Crowther and Pearse (1981) determined a critical minimum protein concentration for reassembly of 0.05 mg/ml at pH 6.2.

The rate and extent of coat reassembly increased with decreasing pH from

6.8 to 6.0 (Van Jaarsveld *et al*, 1981). Below pH 6.0, clathrin formed large aggregates which precipitated from solution. Analytical ultracentrifugation of clathrin solutions reassembled at various pHs showed once again that intermediate species of coat sizes are not formed. The only observed species were 8 S, 150 S and 300 S. Kinetic analysis of the pH-dependence of reassembly was performed under conditions where reassembly went essentially to completion (pH 6.16, 6.33 and 6.5). The results of this analysis suggested that reassembly was extremely complex and displayed an approximately sixth-order dependence on concentration. Coat reassembly was reversible and could be manipulated by pH adjustments. Although it would seem that such a strong pH-dependence might reflect a strong dependence on proton uptake, measurements of proton consumption as a function of reassembly showed only a small participation of protons in the reassembly process. It was calculated that complete conversion of 8 S to 150 S plus 300 S clathrin species would consume only one proton per mole of 8 S species.

The rate of coat reassembly showed an inverse dependence on salt concentration, that is, with increasing salt concentrations the rate of reassembly progressively decreased. When this salt-dependence was examined in more detail, it was found that the sodium salts of anions (fluoride, acetate, chloride, perchlorate, sulfate) followed the Hofmeister ranking, with the exception of sulfate (which should fall between chloride and acetate). Fluoride had virtually no effect on reassembly, while perchlorate was completely inhibitory. The other anions inhibited reassembly in the order SO_4^{2-}, Cl^-, acetate, F^-. Alkali cations showed a much smaller range of inhibitory effects than anions. Of the alkali chlorides, potassium was most inhibitory, followed by Na^+, Cs^+ and Li^+. The salt effects on reassembly support the proposition that the rate of coat formation is strongly dependent on electrostatic factors. When reassembly was monitored as a function of temperature, it displayed apparent negative enthalpy, with the final extent of coat formation decreasing with increasing temperature. Thus there is evidence that both electrostatic and hydrophobic interactions are critical to coat assembly. It is clear from these results that reassembly of triskelions into coats is complex in both its kinetic and equilibrium parameters. Perhaps the most dependent variable is pH, followed by salt, temperature and protein concentration.

While it is evident that coat assembly can be regulated by a variety of parameters *in vitro*, a study of the effects of divalent cations and various basic biological molecules on clathrin reassembly (Nandi *et al*., 1981) has yielded more insight into potential *in vivo* regulatory agents. In contrast to the inhibitory effects of monovalent alkali cations on coat reassembly, divalent cations were found to increase the initial rate of coat reassembly. Magnesium, manganese and calcium were all effective in promoting reassembly, although magnesium was the least effective cation and was inhibitory at high concentrations (≥ 45 mM). Manganese was approximately 4-fold more effective than calcium in promoting reassembly, with comparable effects observed at 2–4 mM-MnCl$_2$ and 9.5 mM-CaCl$_2$. Since these experiments were performed in the absence of chelating agents, these concentrations presumably reflected the free concentrations of the divalent cations. While it is striking that Ca^{2+} and Mn^{2+} promote coat reassembly under acidic conditions (0.1 M-ammonium acetate, pH 6.3–6.5) where such reassembly occurs anyway, it is of greater interest that both cations promote coat reassembly at physiological pH values and higher (up to pH 7.95) where *in vitro* reassembly does not occur in the absence of the cations. Again, Mn^{2+} is much more effective than Ca^{2+} under these conditions. These results contrast somewhat with those reported by Woodward and Roth (1979), who observed that 10 mM-MgCl$_2$ promoted coat reassembly and inhibited disassembly in Tris buffers at pH 8.0–8.3. Lower concentrations of MgCl$_2$ were ineffective at stabilizing coat structure; 10 mM-CaCl$_2$ or MnCl$_2$ promoted the formation of filaments rather than coats. The reason for the discrepancies in these data regarding divalent cations is not clear but they might be accounted for by the differences in the conditions of pH and ionic strength used in the two studies.

The effects of dansylcadaverine, cadaverine, spermine, spermidine and putrescine on coat reassembly were also investigated by Nandi *et al*. (1981). Of these basic bioamines, only dansylcadaverine and spermine promoted coat reassembly; the other compounds had no effect. Again, reassembly was enhanced by dansylcadaverine or spermine under conditions where it normally would not occur. Basic proteins, including lysozyme, ribonuclease and cytochrome *c*, were tested for their effects on coat reassembly in hopes of discerning the common critical feature among all 'promoters.' Of these basic proteins, only lysozyme enhanced reassembly but with kinetics more complex than those observed with the divalent cations or bioamines. Acidic proteins had no effect on coat reassembly. The authors suggest, on the basis of all these data, that the two critical features for promoters of clathrin reassembly are the availability of at least two monobasic groups plus a hydrophobic moiety. These criteria are fulfilled by dansylcadaverine, spermine and quinacrine, chloroquine, chlorpromazine and trifluoperazine, which also promote reassembly. The ability of lysozyme but not other basic proteins to promote reassembly may reflect a fortuitous fulfilment of the specific conformation requirements of basicity and hydrophobicity by lysozyme.

2.4.5 The role of the light chains in triskelion reassembly

Enzymatic digestion of triskelions has been used as a tool to probe the role of the 33 000- and 36 000-mol. wt. light chains in reassembly. Kirchhausen and Harrison (1981) found that elastase selectively digests the light chains rather than clathrin. When elastase-treated triskelions were exposed to conditions that promote reassembly, aggregates rather than regular individual coat structures formed. However, elastase treatment of previously reassembled coats did not appear to cause any change in their morphology. Similar results were observed by Lisanti *et al.* (1981), who found that chymotrypsin digested the light chains associated with clathrin to yield small polypeptides of ≤15 000 mol. wt. This chymotrypsin treatment of intact coats did not affect their morphology, but such treated preparations would not reassemble following dissociation. Treatment of triskelions with the enzyme yielded material that could not reassemble to form coats but instead appeared to form partially assembled lattices. The results of these studies suggested that the light chains are required for the assembly of triskelions into coats but not for the maintenance of the fully assembled coat. A more recent study (Schmid *et al.*, 1982), however, suggests that this interpretation of the role of the light chains may be oversimplified.

Schmid *et al.* (1982) studied the properties of empty coats, assembled from triskelions and subsequently digested with trypsin or elastase. Trypsin treatment of the coats resulted in the digestion of clathrin with the concomitant appearance of polypeptides with molecular weights of 110 000, 53 000 and 41 000. The light chains appeared to be digested completely. Centrifugation of the digested coats yielded a pellet which contained primarily the 110 000-mol. wt. protein and only a trace of intact clathrin. When examined by electron microscopy, coat structures were observed in this pellet. These 'digested coats' migrated more slowly than untreated coats on agarose gel electrophoresis. Upon exposure to 0.75 M-Tris buffer ('buffer I'), trypsin-digested cages dissociated to yield trimeric structures resembling triskelions. However, these triskelions had shorter arms than normal triskelions and the arms lacked the characteristic kink. The 'truncated' triskelions could also re-form coat structures after dialysis against low ionic strength–low pH buffer. This finding is intriguing because the material used contained no intact clathrin and no intact light chains. The coats formed contained only the 110 000 protein, derived from clathrin, and other tryptic fragments, some of which could presumably arise from the light chains. Even though trypsin treatment of triskelions resulted in the apparent loss of the light chains and removal of the outer arm of the triskelions, this material could reform coats. In contrast, elastase-treated triskelions showed no apparent size change but lost their striking geometry. The light chains were digested but clathrin apparently remained intact. As observed previously (Kirchhausen and Harrison, 1981), the elastase-treated material failed to form 'correct' coats.

The results of Schmid *et al.* (1982) just discussed are intriguing, for they suggest that specific domains of clathrin and/or the light chains are critical for the structural integrity of the coat lattice. The results of these proteolysis studies suggest that further chemical or enzymatic modification of clathrin and light chains may yield vital information on the molecular interactions of coat formation. Some insight is afforded by the data already available. Trypsin apparently cleaves triskelion arms in the region of the kink but does not prevent reassembly. Elastase, on the other hand, destroys light chains and the ability of triskelions to reassemble while leaving clathrin apparently intact. Since trypsin also digests the light chains, we may infer that the trypsin-sensitive region(s) is less important than the elastase-sensitive regions to triskelion reassembly into coats. The available evidence suggests that a domain of the light chains interacts with a clathrin domain in such a way as to fix the geometry of a triskelion. Once this geometry has been fixed, as in the case of coats assembled before treatment with enzymes, the role of the light chain domain in maintenance of the structure is diminished. An unresolved question which might be amenable to study using the approach of domain mapping is the nature of the two different light chains. It is impossible to predict at this time whether they share a common clathrin-recognition domain or whether they each have a unique, specialized function.

2.4.6 Triskelion assembly around membranes

Although the studies of coat reassembly have yielded some information on possible mechanisms of coated vesicle formation, the studies of Unanue *et al.* (1981) on the binding of clathrin triskelions to membranes are of even greater biological relevance. They isolated triskelions from coated vesicles by treatment with Tris buffer, pH 7.0, and radiolabeled them either with tritium, by reductive methylation, or with ^{125}I using the Bolton–Hunter reagent. Uncoated vesicles were generated by treating purified coated vesicles with low ionic strength buffer at pH 8.5. Binding of radiolabeled triskelions to vesicles was then examined using a centrifugation assay. Triskelions bound to 'stripped' vesicles as a function of the amount of residual clathrin remaining associated with the vesicle, but did not bind to intact coated vesicles. Inside-out erythrocyte membranes, known not to contain any clathrin, did not bind triskelions. The optimal pH for triskelion binding to vesicles was between pH 6.0 and 6.5; precipitates formed below pH 6.0. In contrast to the variables strongly affecting triskelion reassembly, neither Ca^{2+}, Mg^{2+} nor temperature affected binding of triskelions to vesicles. The apparent affinity of triskelion binding to vesicles was 5×10^8 M^{-1}, and radiolabeled triskelions appeared to bind as well as unmodified triskelions.

Protease treatments of vesicles were used to examine the requirement of vesicle-associated proteins for triskelion binding. Treatment with increasing amounts of trypsin or elastase proportionally reduced triskelion binding. SDS–polyacrylamide gel analysis showed that trypsin caused a reduction in the amounts of 110 000 and 50 000-mol. wt. peptides associated with the vesicles, whereas elastase was more selective and digested primarily the 110 000-mol. wt. polypeptide. Thus the data suggest that the 110 000-mol.wt. component of vesicles may be required for triskelion binding. Electron microscopy of triskelions bound to vesicles showed that the composite structure closely resembled native coated vesicles; however, no more than 60% of the stripped vesicles were able to re-form coated vesicles regardless of the amount of triskelions added. Excess triskelions formed empty coats under the conditions used. It is unclear why 40% of the vesicles were unable to bind triskelions. The intriguing questions left unanswered by these studies are: (1) which component of the triskelion (clathrin or the light chains) interacts with the protease-sensitive vesicle protein; (2) what is the nature of this interaction; (3) why are there discrepancies in the effects of divalent cations and temperature on the reassembly of triskelions into coats versus the binding of triskelions to vesicles. Clearly, since triskelions can readily self-assemble to form empty coats, the 110 000-mol.wt. protein of vesicles must present a more favorable set of potential interactions to promote triskelion binding to the vesicle.

2.4.7 Arrangement of triskelions in the coat lattice

Crowther and Pearse (1981) obtained well-resolved electron micrographs of negatively stained coat lattices and, with the aid of computer enhancement of the images, were able to propose a model for the packing configuration of triskelions in the coat. The critical feature of their experiments was that they caused triskelions to assemble directly on the surface of the electron microscope grid and thus obtained essentially two-dimensional coat fragments rather than three dimensional coats. Rather clear images of the edges and vertices of individual polyhedral coat elements were obtained from these negatively stained fragments. The model of triskelion packing that best fits the images obtained by electron microscopy is one in which each vertex of a hexagon or pentagon is occupied by the vertex of an individual triskelion (Fig. 2.6). The legs of adjacent triskelions cross over each other. This disposition was favored over one in which the legs were parallel, on the basis of the similarity in appearance of this model to negatively stained material. In addition, the stain-penetration pattern on various polyhedral edges was consistent with this model. Woodward and Roth (1979) had previously suggested cross-over packing based on observations of negatively stained coated vesicles. An individual edge of a hexagon or pentagon in a coat could be formed from overlapping portions of the proximal parts of two triskelion legs and the distal parts of two other legs. Measurements of individual triskelions and of individual polyhedral edges, however, suggest that 8 nm of the leg length of a triskelion is not accommodated in this packing model. The conformation of this distal tip of a triskelion leg is thus a matter of conjecture.

Given that the assembly of triskelions to form coats appears to be a rapid and highly ordered process, packing models such as this one give rise to the prediction that there must be specific recognition sites within each clathrin

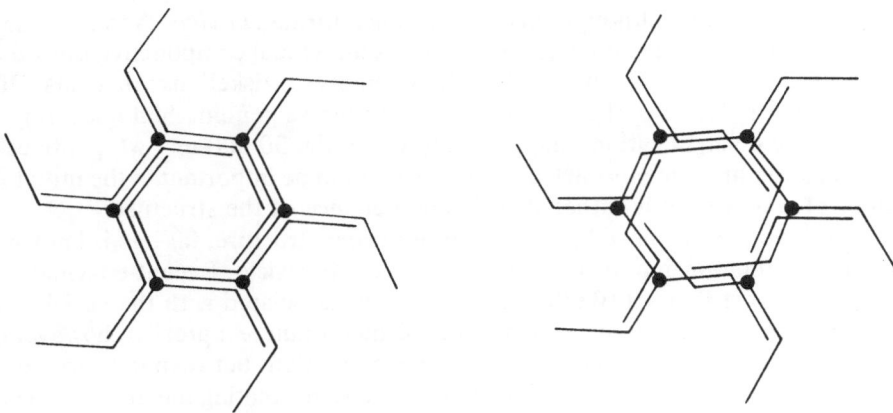

Fig. 2.6 Models illustrating two potential ways in which triskelions can be ordered to yield a hexagonal packing. The left model illustrates simple side-by-side packing, whereas the right one shows a cross-over packing pattern. A small change in packing angle permits the triskelions to be ordered into a pentagon (not shown). Adapted from a drawing by Crowther and Pearse (1981).

molecule. These sites must include, at a minimum, one residing in the proximal portion of each triskelion leg such that two proximal legs will array themselves anti-parallel to each other. *In vivo*, the formation of individual hexagons or pentagons may also be dictated by the interaction of assembling triskelions with underlying membrane.

2.5 DISCUSSION

A principle goal of *in vitro* studies of coated vesicles has been to understand better their regulation and function within a cell. While the studies conducted to date have fallen somewhat short of this goal, we have achieved a better understanding of coated vesicle structure and the factors which may regulate coated vesicle assembly. It is clear that, in a general sense, coated vesicles fall into the broad catagory of biological structures capable of undergoing self-assembly, much like virus capsids and actin. While coated vesicles display greater heterogeneity in their structure than these other self-assembling complexes, their assembly appears to be governed by relatively simple biophysical and chemical interactions. We are still in the process of understanding these interactions.

The cage-like coat of coated vesicles is composed of clathrin and two small polypeptides. The triskelion, composed of three clathrin molecules and three of the *ca*. 30 000-mol.wt. peptides, appears to be the biologically functional subunit of the coat, although this remains unconfirmed *in vivo*. As yet we do not know how a triskelion is formed from its individual components, but we do know that several factors regulate the assembly of triskelions into coats. Of these factors, the ones that seem most promising as genuine biological regulators are divalent cations and pH. The role of the 30 000-mol.wt. peptides in coat structure remains unclear, but they seem to be important in the initial steps of coat assembly rather than in maintenance of the structure.

While we have achieved some insight into coat structure, far less is known about the relationship of the coat to the internal vesicle. Recent enzymatic studies suggest that a 110 000-mol.wt. protein associated with the vesicle mediates the formation of a coat around the membrane. At present, however, there is no information on the nature of the interactions between this protein and clathrin triskelions. The difficulties inherent in isolating membrane vesicles from purified coated vesicles have complicated the characterization of the vesicles themselves.

In the laboratory, coated vesicles can be fractionated into their various components and both empty coats and coated vesicles reconstituted from them. Are the mechanisms by which we accomplish these structural transformations at all related to the mechanisms which operate in a cell? Because divalent cations, in particular calcium, appear to play a role in coat assembly (especially at low clathrin concentrations and physiological pH), it is possible that this ion may be a cellular regulator of coated vesicle structure and perhaps function. Although we are not yet able to propose a complete model of coated vesicles, it is becoming increasingly clear that they might be subject to regulation by some basic cellular mechanisms. Calmodulin, the ubiquitous calcium-dependent regulatory protein, binds to coated vesicles (Linden *et al.*, 1981; Puszkin *et al.*, 1981) and has been found to interact with a restricted set of vesicle proteins as well as with the 33 000-mol.wt. polypeptide of triskelions (Linden, 1982). It is tempting to speculate that calcium and calmodulin participate in the structural assembly or function of coated vesicles. There is increasing evidence that this hypothesis is valid (Salisbury *et al.*, 1980, 1981) and it provides a new avenue of approach for further investigations of coated vesicle structure and function.

REFERENCES

Anderson, R.G.W., Vasile, E., Mello, R.J., Brown, M.S. and Goldstein, J.L. (1978), *Cell*, **15**, 919–933.
Blitz, A.L., Fine, R.E. and Toselli, P.A. (1977), *J. Cell. Biol.*, **75**, 135–147.
Booth, A.G. and Wilson, M.J. (1981), *Biochem. J.*, **196**, 355–362.
Crowther, R.A., Finch, J.T. and Pearse, B.M.F. (1976), *J. Mol. Biol.*, **103**, 785–798.
Crowther, R.A. and Pearse, B.M.F. (1981), *J. Cell Biol.*, **91**, 790–797.
Fine, R.E., Blitz, A.L. and Sack, D.H. (1978), *FEBS Lett.*, **94**, 59–62.

Franke, W.W., Lüder, M.R., Kartenbeck, J., Zerban, H. and Keenan, T.W. (1976), *J. Cell Biol.*, **69**, 173–195.

Friend, D.S. and Farquhar, M.G. (1967), *J. Cell Biol.*, **35**, 357–376.

Heuser, J. (1980), *J. Cell Biol.*, **84**, 560–583.

Jamieson, J.D. and Palade, G.E. (1971), *J. Cell Biol.*, **50**, 135–158.

Kadota, T., Kadota, K. and Gray, E.G. (1976), *J. Cell Biol.*, **69**, 608–621.

Kanaseki, T. and Kadota, K. (1969), *J. Cell Biol.*, **42**, 202–220.

Kartenbeck, J. (1978), *Cell Biol. Int. Rep.*, **2**, 457–464.

Kartenbeck, J., Franke, W.W. and Morré, D.J. (1977), *Cytobiologie*, **14**, 284–291.

Keen, J.H., Willingham, M.C. and Pastan, I. H. (1979), *Cell*, **16**, 303–312.

Keen, J.H., Willingham, M.C. and Pastan, I. (1981), *J. Biol. Chem.*, **256**, 2538–2544.

Kirchhausen, T. and Harrison, S.C. (1981), *Cell*, **23**, 755–761.

Linden, C.D. (1982), *Biochem. Biophys. Res. Commun.*, **109**, 186–193.

Linden, C.D., Dedman, J.R., Chafouleas, J.G., Means, A.R. and Roth, T.F. (1981), *Proc. Natl. Acad. Sci. U.S.A.*, **78**, 308–312.

Lisanti, M.P., Schook, W., Moskowitz, N., Ores, C. and Puszkin, S. (1981), *Biochem. J.*, **201**, 297–304.

Mello, R.J., Brown, M.S., Goldstein, J.L. and Anderson, R.G.W. (1980), *Cell*, **20**, 829–837.

Montesano, R., Perrelet, A., Vassalli, P. and Orci, L. (1979), *Proc. Natl. Acad. Sci. U.S.A.*, **76**, 6391–6395.

Nandi, P.K., Pretorius, H.T., Lippoldt, R.E., Johnson, M.L. and Edelhoch, H. (1980), *Biochemistry*, **19**, 5917–5921.

Nandi, P.K., Van Jaarsveld, P.P., Lippoldt, R.E. and Edelhoch, M. (1981), *Biochemistry*, **20**, 6706–6710.

Novikoff, A.B., Novikoff, P.M., Rosen, O.M. and Rubin, C.S. (1980), *J. Cell Biol.*, **87**, 180–196.

Ockleford, C.D. (1976), *J. Cell Sci.*, **21**, 83–91.

Pastan, I.H. and Willingham, M.C. (1981), *Annu. Rev. Phsiol.*, **43**, 239–250.

Patzer, E.J., Schlossman, D.M. and Rothman, J.E. (1982), *J. Cell Biol.*, **93**, 230–236.

Pearse, B.M.F. (1975), *J. Mol. Biol.*, **97**, 93–98.

Pearse, B.M.F. (1976), *Proc. Natl. Acad. Sci. U.S.A.*, **73**, 1255–1259.

Pearse, B.M.F. (1978), *J. Mol. Biol.*, **126**, 803–812.

Pearse, B.M.F. (1982), *Proc. Natl. Acad. Sci. U.S.A.*, **79**, 451–455.

Pfeffer, S.R. and Kelly, R.B. (1981), *J. Cell Biol.*, **91**, 385–391.

Pretorius, H.T., Nandi, P.K., Lippoldt, R.E., Johnson, M.L., Keen, J.H., Pastan, I. and Edelhoch, H. (1981), *Biochemistry*, **20**, 2777–2782.

Puszkin, S., Moskowitz, N. and Schook, W. (1981), *J. Cell Biol.*, **91**, 92a.

Roth, T.F. and Porter, K.R. (1964), *J. Cell Biol.*, **20**, 313–332.

Rothman, J.E. and Fine, R.E. (1980), *Proc. Natl. Acad. Sci. U.S.A.*, **77**, 780–784.

Rubenstein, J.L.R., Fine, R.E., Luskey, B.D. and Rothman, J.E. (1981), *J. Cell Biol.*, **89**, 357–361.

Salisbury, J.L., Condeelis, J.S. and Satir, P. (1980), *J. Cell Biol.*, **87**, 132–141.

Salisbury, J.L., Condeelis, J.S., Maihle, N.J. and Satir, P. (1981), *Nature (London)*, **294**, 163–166.

Schmid, S.L., Matsumoto, A.K. and Rothman, J.E. (1982), *Proc. Natl. Acad. Sci. U.S.A.*, **79**, 91–95.

Tanaka, R., Takeda, M. and Jaimovich, M. (1976), *J. Biochem.* (*Tokyo*), **80**, 831–837.

Unanue, E.R., Ungewickell, E. and Branton, D. (1981), *Cell*, **26**, 439–446.

Ungewickell, E. and Branton, D. (1981), *Nature (London)*, **289**, 420–422.

Van Jaarsveld, P.P., Nandi, P.K., Lippoldt, R.E., Saroff, H. and Edelhoch, H. (1981), *Biochemistry*, **20**, 4129–4135.

Woods, J.W., Woodward, M.P. and Roth, T.F. (1978), *J. Cell. Sci.*, **30**, 87–97.

Woodward, M.P. and Roth, T.F. (1978), *Proc. Natl. Acad. Sci. U.S.A.*, **75**, 4394–4398.

Woodward, M.P. and Roth, T.F. (1979), *J. Supramol. Struct.*, **11**, 237–250.

3 Adsorptive Pinocytosis of Epidermal Growth Factor: Studies of its Relevance to Mitogenesis

A. CHRISTIE KING and PEDRO CUATRECASAS

Receptor-Mediated Endocytosis
(*Receptors and Recognition*, Series B, Volume 15)
Edited by P. Cuatrecasas and T. F. Roth
Published in 1983 by Chapman and Hall, 11 New Fetter Lane, London EC4P 4EE
© 1983 Chapman and Hall

3.1 GROWTH REGULATION BY MITOGENIC PEPTIDES

Cell growth *in vitro* depends on an adequate supply of minerals, nutrients, growth factors and growth modulators. Peptide hormones and growth factors are chemical mediators of the growth response (Bradshaw, 1978; Smith and Temin, 1974; Gospodarowicz and Moran, 1976; Chen and Buchanan, 1975), but in spite of the concerted efforts of many, the molecular details of their modes of action remain substantially undefined. Little evidence has accumulated to implicate receptor-coupled membrane transduction mechanisms like the stimulation of adenylate cyclase or redistribution of divalent cations as primary intracellular effectors. Indeed, the very nature of delayed responses like cell growth may indicate that alternative forms of regulation are applicable.

Recently, it has been widely accepted that peptide hormones and growth factors are internalized into target cells through endocytic processes mediated by specific receptors (for review, see King and Cuatrecacas, 1981a). The internalized hormones are sequestered into a variety of cytoplasmic organelles (lysosome, Golgi, endoplasmic reticulum and nucleus) (Josefsberg *et al.*, 1979; Carpenter *et al.*, 1979; Marchisio *et al.*, 1980; Gorden *et al.*, 1978; McKanna *et al.*, 1979); however, the exact role, if any, that hormone internalization plays in mediation of physiological responses remains obscure. Although the time course for the internalization process is too slow to explain immediate metabolic changes evoked by peptide hormones, its importance cannot be excluded from mediation of responses that exhibit a considerable delay, such as enzyme activation, cell growth and differentiation. This review deals with one of the best studied hormone systems whose regulation is intimately associated with the processes of internalization and lysosomal degradation. In addition, we believe that epidermal growth factor (EGF) warrants careful consideration in this volume because of increasing evidence that the internalization process is linked to mediation of its biological effects on cell growth.

3.2 RESPONSES OF CULTURED CELLS TO EGF

3.2.1 EGF evokes diverse responses

Studies of the mode of action of EGF have focused on the characterization of a diverse set of temporally regulated responses that culminate in cell growth in sensitive fibroblastic and epidermal cell lines (Carpenter and Cohen, 1979). Some of the more immediate effects include the tyrosine-specific phosphorylation of membrane (Carpenter *et al.*, 1979; Ushiro and Cohen, 1980; Fernadez-Pol, 1981) and cytoplasmic proteins (Cooper and Hunter,

1981a,b), an enhanced influx of Ca^{2+}, which in turn activates phospholipase C and increases phosphatidylinositol turnover (Sawyer and Cohen, 1981), stimulation of K^+ influx (Rozengurt and Heppel, 1975) and an enhanced transport of cellular nutrients (Hollenberg and Cuatrecasas, 1975; Barnes and Colowick, 1976). Concomitant with these biochemical changes, several morphological responses take place, including pseudopod formation and cell rounding (Chinkers *et al.*, 1979, 1981; Brunk *et al.*, 1976). Each of these changes occurs rapidly and is dependent on plasma-membrane-associated enzymes; therefore, it is probably safe to assume that these are mediated through membrane transduction mechanisms. However, a component of the stimulation of aminoisobutyric acid transport by EGF is inhibited by cycloheximide (Hollenberg and Cuatrecasas, 1975). This suggests that activation of nutrient transport may involve both longer-term and shorter-term events. After a delay of 2 to 4 hours, EGF exposure results in increased protein and RNA synthesis (Hollenberg and Cuatrecasas, 1975; Cohen and Stastry, 1968; Hoober and Cohen, 1967). Several apparently unrelated enzyme systems are activated, including phosphofructokinase and ornithine decarboxylase (Dipasquale *et al.*, 1978; Moriorite *et al.*, 1981; Diamond *et al.*, 1978). An enhanced rate of DNA synthesis takes place after 16 to 20 hours of exposure. In most cell types, these events culminate in cell division (Rose *et al.*, 1975; Carpenter and Cohen, 1976a). It is not known how EGF mediates any of these metabolic changes, or how their co-ordinate control is regulated. Furthermore, the relevance of the expression of certain early membrane events, such as tyrosine phosphorylation, to the mitogenic response is also unknown.

3.2.2 Second messengers

Early studies of the interaction of radioiodinated EGF with cultured fibroblasts demonstrated a single receptor population of extraordinarily high affinity ($\sim 10^{-10}$M) (Hollenberg and Cuatrecasas, 1975; Aharonov *et al.*, 1978a,b; Vlodavsky *et al.*, 1978; Carpenter and Cohen, 1976b). This suggested that the diverse responses produced by EGF were caused through interaction with a single type of cell surface receptor and were not due to interactions with multiple receptor types. In spite of the implication of this finding, no unique second messenger of EGF action has been found. Cyclic nucleotides may modulate EGF action, but do not appear to be directly responsible for any of the metabolic changes (Hollenberg and Cuatrecasas, 1973). Cyclic AMP (cAMP) levels may decrease transiently when fibroblasts are stimulated to grow (Seifert and Rudland, 1974), and cAMP levels are decreased for exponentially growing cells and for transformed cells, while resting cells have higher concentrations (Otten *et al.*, 1971, 1972; Kram *et al.*, 1973). Dibutyryl cAMP, phosphodiesterase inhibitors and cholera toxin reversibly inhibit DNA synthesis and nutrient uptake (Seifert and Rudland, 1974; Otten *et al.*,

1971, 1972; Kram *et al.*, 1973). Thus, while cAMP plays an important regulatory part in the growth response, it does not meet the requirements of a second messenger of EGF action.

3.2.3 Continued EGF exposure is a requirement for mitogenesis

Receptor occupation by EGF is not sufficient to induce cell growth (Carpenter and Cohen, 1979; Aharonev *et al.*, 1978a,b; Vlodavsky *et al.*, 1978). Binding of ^{125}I-EGF to specific, high-affinity receptors (Hollenberg and Cuat-

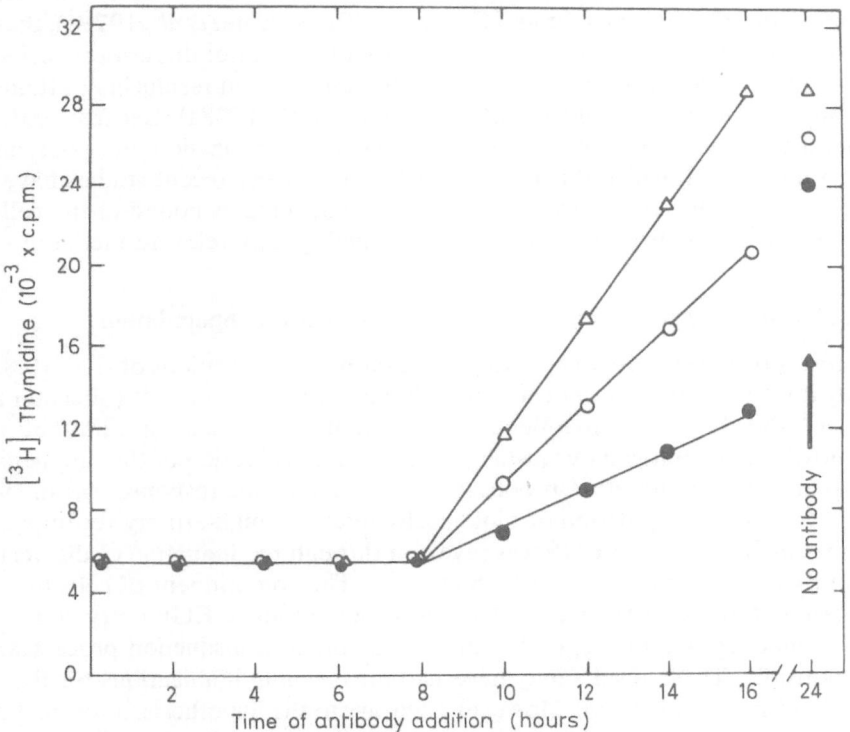

Fig. 3.1 Effect of addition of anti-EGF antibody at increasing time intervals on the DNA synthesis stimulated by mouse EGF in human fibroblasts. Human fibroblasts (HF) were incubated with three different concentrations of EGF for 24 hours. The incorporation of [^{3}H]thymidine (1 μCi/ml) was then measured during a 1-hour pulse. The values are averages of three replicate determinations. The antibody was added at the indicated times during the incubation with EGF. EGF concentrations were 0.166 nM (●), 0.833 nM (○) and 1.66 nM (△). The amount of antibody added was 5, 10 and 40 μl respectively. The extent of stimulation at the various EGF concentrations in the absence of antibody is shown on the right. Data from Shechter *et al.* (1978a).

recasas, 1975) does not obey simple rules of receptor dynamics (Carpenter and Cohen, 1976b). The affinity constant correlates well with the mitogenic response ($Kd \sim 10^{-10}$ M), but the response exhibits a delay of 16 to 24 hours (Carpenter and Cohen, 1976b; Aharonov *et al.*, 1978a), even though binding is maximal after a 1 hour period of incubation (Carpenter and Cohen, 1976b). Thus, there is no temporal correlation between the initial process of EGF binding to cell surface receptors and the generation of a mitogenic response. Furthermore, the continuous presence of EGF in the cell medium for at least 6 to 8 hours is required before cells are committed to respond. If cells are washed free of EGF (Carpenter and Cohen, 1976a; Savion *et al.*, 1980; King *et al.*, 1981), or if EGF is immunoinactivated by specific antisera prior to this time (Carpenter and Cohen, 1976a; Shechter *et al.*, 1978a), there is no mitogenic response (Fig. 3.1). Removal of EGF after this critical period of incubation does not completely block the response but results in its attenuation (Carpenter and Cohen, 1976a; Shechter *et al.*, 1978a). Because of the complexities associated with receptor-mediated internalization processes, and the rapidity with which this process takes place, many recent studies have attempted to distinguish whether it is EGF that remains bound at the cell surface or internalized mitogen that is the biologically relevant mediator.

3.2.4 Mitogenic response and high-affinity receptor subpopulation

In one report, exposure of cells to subsaturating concentrations of EGF (0.83 nM) for a short incubation (30 min) stimulated DNA synthesis (Shechter *et al.*, 1978a). In these experiments, the addition of EGF antisera, but not removal of the mitogen by washing, resulted in inactivation of the mitogenic response. It was therefore proposed that the mitogenic response was mediated by a small population of biologically relevant, high-affinity receptors whose interaction with EGF was reversed through the induction of dissociation by a competitive process with antisera. The commitment of cells to respond after incubations considered too short to allow EGF entry and lysosomal degradation suggested that a membrane transduction process was responsible. These results may have had far reaching implications for the mechanism of EGF action. However, contrary to the hypothesis, antisera did not promote a significant alteration in the dissociation of prebound ^{125}I-EGF (Shechter *et al*, 1978a). Indeed, the 'dissociation' reported was consistent with the time course of release of ^{125}I-EGF-degradation products after the internalization and degradation processes were largely complete ($t_{1/2}$ 30 min) and did not represent a reversal of surface binding ($t_{1/2}$ 2–4 hours) (Carpenter and Cohen, 1976b) (Fig. 3.2). These results indicate that even in the presence of the antisera, prebound ^{125}I-EGF is probably internalized, delivered to the lysosomal compartment, and degraded. In addition, ^{125}I-EGF antisera is unable to recognize receptor-complexed EGF after 5–10 minutes of exposure

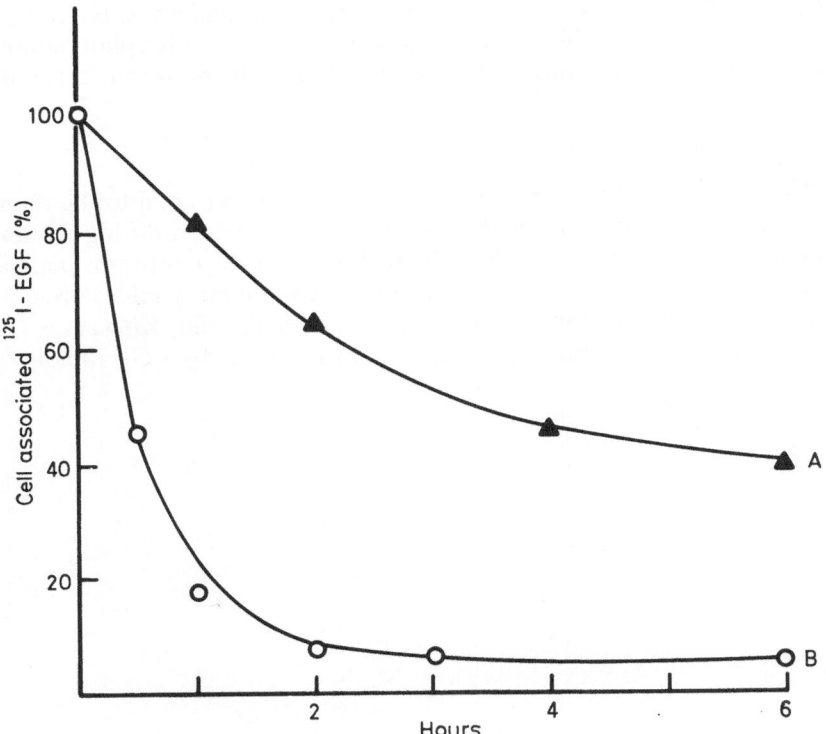

Fig. 3.2 Dissociation of cell-bound ^{125}I-EGF at different temperatures. ^{125}I-EGF (2 ng/ml) was preincubated with confluent fibroblast monolayers for 40 min, at 37°C or at 4°C. The cells were then washed and incubated in 1.0 ml of Krebs + 0.1% BSA. At the indicated times, cell-bound radioactivity was determined. In A, the temperature of the association and dissociation phases was 0°C and in B both incubations were at 37°C. Data from Shechter *et al.* (1978a).

to cells (Carpenter and Cohen, 1976b). Thus, EGF becomes rapidly masked from detection by antisera either by inducing a conformational change in the receptor or by internalization.

No other study has reported a mitogenic response after short incubation periods, and it is now generally agreed that under usual culture conditions, this response requires a longer-term (6–8 hours) exposure to the mitogen. Thus, the primary and dominant effect of anti-EGF is to deplete the medium of the mitogen (King *et al.*, 1981). This indicates that growth potentiation by EGF requires exposure of plasma membrane receptors to exogenous EGF for prolonged periods, a situation different from that normally considered to operate in most hormonally sensitive systems. In particular, the prolonged temporal requirement for EGF suggests that the immediate and delayed actions of EGF are regulated by different molecular pathways. Even more importantly, the early effects mediated by EGF, such as phosphorylation or PI turnover, cannot be sufficient for production of the mitogenic response.

3.2.5 Receptor clearance process

Clearly, the events of the first 6 to 8 hours of EGF exposure must be rigorously defined. One of the most obvious processes that occurs during this time frame is the stimulation by EGF of the removal of its own receptor population (Fig. 3.3a). Although EGF receptors are continuously internalized as a result of basal turnover processes (Aharanov *et al.*, 1978b; King *et al*; 1980a,b; Fox *et al.*, 1980), the occupation of receptors by EGF rapidly

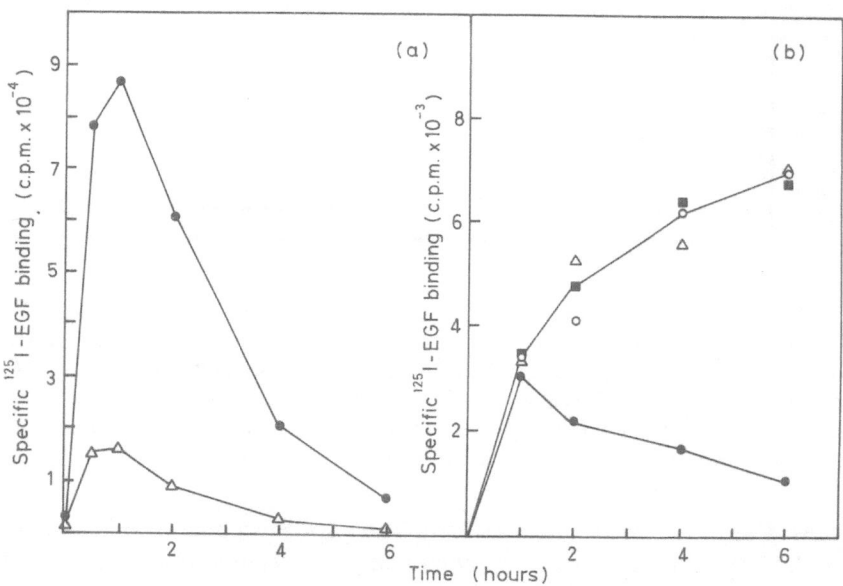

Fig. 3.3 (a) Time course of binding of ^{125}I-EGF. Confluent HF cells (△) were incubated with ^{125}I-EGF at 2 ng/ml in Krebs–Ringer solution + 0.1% BSA, pH 7.2, at 37°C in monolayer cultures. At different times, triplicate wells were washed three times with ice-cold buffer to remove unbound ligand and solubilized in 1 M-NaOH for measurement of radioactivity. KB cells (●) were assayed in triplicate with ^{125}I-EGF at 40 ng/ml. At the indicated times, cells were filtered on 1.0-μm-pore Millipore filters and washed twice with cold buffer. (b) Effect of alkylamines on time course of binding of ^{125}I-EGF. HF cells were preincubated with buffer (●), 10 mM-MeNH$_2$ (△), 50 μM-chloroquine (○) or both amines (■) for 10 min, then ^{125}I-EGF binding was determined at different times as described for (a). Data from King *et al.* (1980a).

(within 10 min) stimulates their rate of removal (King *et al.*, 1980b; Fox *et al.*, 1980). By the process of receptor-mediated internalization (Haigler *et al.*, 1979), both EGF (Carpenter and Cohen, 1976b) and its receptors (King *et al.*, 1980a; Fox and Das, 1979) are delivered to lysosomes (Carpenter and Cohen, 1976b) where they are subject to proteolysis.

A great deal of controversy has surrounded the meaning of this 'down regulation' process. Two proposals have been made to explain its physiological significance. First, that internalization and lysosomal degradation serve to inactivate high-affinity surface receptor complexes more rapidly ($t_{1/2}$ 30 min) than could be accounted for by ligand dissociation ($t_{1/2}$ 4 hours). The reduction in receptor number would ultimately lead to the desensitization of cells to subsequent hormone exposure (Aharanov *et al.*, 1978a,b; Vladarsky *et al.*, 1978; Gavin *et al.*, 1974). However, we have found that prior exposure of cells to EGF for periods sufficient to induce an 80 to 90% reduction in the number of receptors (2 to 4 hours) does not lead to desensitization (King and Cuatrecasas, 1981b). Cells become refractory to EGF but only after very long exposures (20 to 24 hours) (Heldin *et al.*, 1979), suggesting that desensitization may depend on post-receptor mechanisms. The second hypothesis (endocytic activation) proposes that the mitogenic stimulation of cells by EGF depends on the internalization and proteolytic processing of EGF–receptor complexes to generate a mitogenic fragment capable of direct interaction with nuclear components to initiate transcription (King *et al.*, 1981; Fox and Das, 1979; Das and Fox, 1978). As yet, the mechanism for stimulation of quiescent cells by EGF is not known. However, mounting circumstantial evidence has made it clear that binding of EGF to cell surface receptors is not an adequate signal. This implicates the receptor-mediated internalization of EGF as a possible pathway for mediation of its delayed biological effects.

3.3 INTERACTIONS OF ^{125}I-EGF WITH CULTURED CELLS

3.3.1 Time course of ^{125}I-EGF interaction

Incubation of fibroblasts with ^{125}I-EGF reveals a kinetically complex sequence of events. In what is now a classic study, Graham Carpenter and Stanley Cohen showed that subsequent to binding of ^{125}I-EGF to cell surface receptors, the radioactivity associated with fibroblast cultures diminished with time so that after 2 to 4 hours of incubation, only about 15 to 25% remained (Carpenter and Cohen, 1976b). The decline in cell-associated radioactivity is prevented by proteolytic inhibitors, including inhibitors of trypsin-like catalysis or inhibitors of lysosomal degradation, such as chloroquine (Fig. 3.3b). These authors were the first to suggest that EGF complexed to surface receptors is pinocytosed into cells, delivered to lysosomes and degraded.

3.3.2 'Down regulation'

As a consqeuence of exposing cells to EGF, and the delivery of receptors to lysosomes, the number of surface receptors is dramatically reduced. This process, called 'down regulation', represents the removal (and possible degradation) of EGF receptors concomitant with internalization of the mitogen (Aharonov *et al.*, 1978a,b; Vlodavsky *et al.*, 1978; Carpenter and Cohen, 1976b). Cells exposed to EGF for even brief periods of time (1 to 4 hours), clear receptors from their surface. A new, low level of ^{125}I-EGF binding persists throughout the time that EGF is present in the medium (Aharonov *et al.*, 1978a). If EGF is removed from the medium, then the receptors are gradually replaced ($t_{1/2}$ 6 hours) by a process requiring serum and protein synthesis *de novo* (Carpenter and Cohen, 1976a; Fox *et al.*, 1980). The receptor clearance process has a high degree of specificity, since no other peptide hormone can induce the long-term removal of EGF receptors. However, platelet-derived growth factor (PDGF) and fibroblast growth factor (FGF) may cause transient (1 to 4 hour) reductions in EGF receptor concentrations (Fox *et al.*, 1980). This could represent a modulatory role between different growth factors. Transformed, growing, confluent and serum-deprived cells lose EGF receptors after exposure to the hormone (Aharonov *et al.*, 1978; Vlodavsky *et al.*, 1978; Fox *et al.*, 1980). In contrast to confluent cells, however, growing cells rapidly replace EGF receptors even in the presence of EGF (Aharonov *et al.*, 1978a; Vlodavsky *et al.*, 1978).

3.3.3 Internalization of unoccupied EGF receptors

'Down regulation' is dependent on the temperature, EGF concentration, and the duration of the incubation (Aharonov *et al.*, 1978a,b; Vlodavsky *et al.*, 1978). Receptor loss is half maximal at ~1–2 nM-EGF, a concentration that produces a half-maximal mitogenic response (Aharonov *et al.*, 1978a,b; Das and Fox, 1978; Vlodavsky *et al.*, 1978). However, at this concentration of EGF, only about 20% of the total receptor population is occupied (Aharonov *et al.*, 1978a,b; Vlodavsky *et al.*, 1978). Thus, with low concentrations of EGF, unoccupied EGF receptors are presumed to be cleared from the surface of cells along with the occupied forms. Since the complexed receptors are rapidly internalized, this phenomenon may only represent a shift in the over-all binding equilibrium.

3.3.4 Biological importance of down regulation process

Even though the internalization and degradation of ligand–receptor complexes is a candidate for mediation of mitogenic stimulation, the early rapid loss of receptors and massive degradation of EGF does not constitute sufficient criteria to evoke the response (Aharonov *et al.*, 1978a,b; Vlodavsky *et al.*, 1978; Savion *et al.*, 1980; Shechter *et al.*, 1978a). Many non-responsive cells exhibit the phenomenon of 'down regulation'. Further, in fibroblasts, the apparent temporal relationship of the initial receptor clearance does not correlate well with the required exposure period of 6 to 8 hours. Addition of antisera to cells or washing cells free of EGF after the major phases of down regulation and internalization are complete (>4 hours) totally reverses the mitogenic response (Carpenter and Cohen, 1976a; Shechter *et al.*, 1978a). Receptor clearance reaches a maximum after 2 to 4 hours. The residual receptors seem highly resistant to regulation by internalization, and exposure of cells to oversaturating concentrations of EGF (100 ng/ml) for extended periods of time does not result in a further reduction in surface binding activity (Aharonov *et al.*, 1978a). In addition, occupation of only 10 to 25% of receptors is sufficient to induce a maximal mitogenic response. This percentage correlates well with the population of residual receptors remaining on the cell surface after the majority of the initial complement has been cleared. These initial results suggest that at least the early phase of the internalization process is not germane to mitogenesis. Thus, residual receptors present on cells after completion of the down regulation process are candidates for the biologically relevant mediators.

3.3.5 Residual EGF receptors are also internalized

In sensitive cell strains, the resulting low level of ^{125}I-EGF and receptors that appear to resist internalization (Aharonov *et al.*, 1978a,b; Vlodavsky *et al.*, 1978; Shechter *et al.*, 1978a) is not due to the establishment of a new receptor binding equilibrium, but is caused by a steady state condition defined by the continuous binding, internalization and degradation of the complex (King and Cuatrecasas, 1981b; Wiley and Cunningham, 1981). By adding lysosomal inhibitors to cells during various stages of the down regulation process to prevent degradation of ^{125}I-EGF, we observe a net ^{125}I-EGF accumulation rather than the usual loss associated with the down regulation process (Fig. 3.4). Evidence presented in a later section shows that ^{125}I-EGF binds and is internalized normally in the presence of these lysosomal inhibitors. Thus, contrary to previous assumptions (Aharonov *et al.*, 1978a,b; Vlodavsky *et al.*, 1978; Shechter *et al.*, 1978a), the internalization and intracellular processing of EGF–receptor complexes is not limited to the early and very rapid receptor clearance phase (King and Cuatrecasas, 1981b). EGF is continuously endocytosed and degraded throughout the extended incubations required for the potentiation of DNA synthesis (King and Cuatrecasas, 1981b). These results suggest that a continued process of EGF–receptor internalization may account for the prolonged exposure periods required to evoke a mitogenic response. Could cellular activation require a threshold concentration of an activated mitogenic mediator derived from lysosomal processing? Are there alternative pathways for internalization, for example, one that mediates hormone degradation and another that mediates the response? There are still no

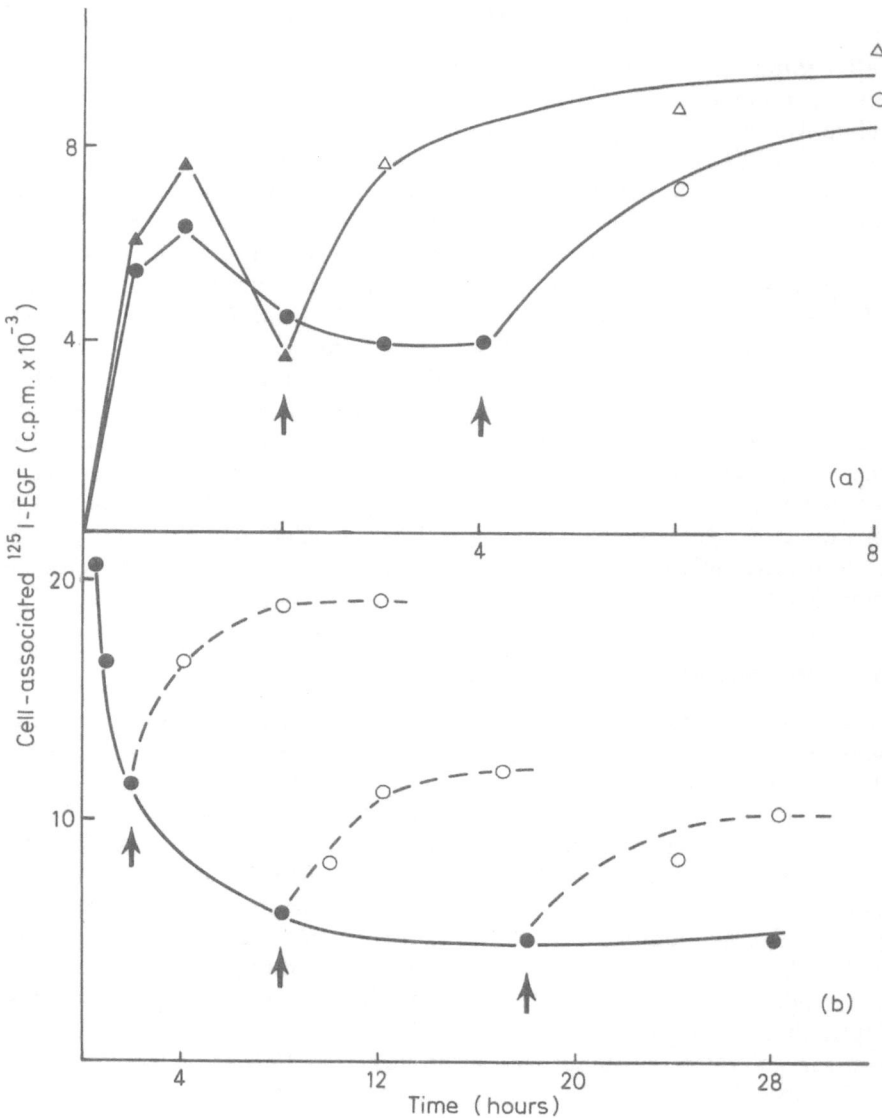

Fig. 3.4 Effect of $MeNH_2$ on association of ^{125}I-EGF with cells. Time courses of the association of ^{125}I-EGF were performed as described in Fig. 3.3 with the addition of 10 ng of ^{125}I-EGF/ml initiating the experiments. The cultures were untreated in the beginning of the experiment except for the radiolabeled EGF, and the association of ^{125}I-EGF was monitored in these untreated cultures (●, ▲). At the times indicated by the arrows, 10 mM-$MeNH_2$ was added to the remaining cultures and the association of ^{125}I-EGF in the presence of $MeNH_2$ was now determined (△, ○). (a) $MeNH_2$ added at 2 and 4 hours. (b) $MeNH_2$ added at 2, 8 and 18 hours. Specificity was determined with 5 μg of unlabelled EGF/ml. Data from King and Cuatrecasas, 1981b.

answers to these questions, but three general experimental approaches (morphological, radiotracer and pharmacological) are enriching our understanding about the possible relevance of the internalization process.

3.4 MORPHOLOGICAL STUDIES

3.4.1 Receptor aggregation

It is now well established that the pathway responsible for degradation of EGF by cultured fibroblasts resides exclusively with internalization of the hormone and delivery to lysosomes (Carpenter and Cohen, 1976a; Aharonov *et al.*, 1978a,b; Vlodavsky *et al.*, 1978; Savion *et al.*, 1980; King *et al.*, 1980a,b, 1981; Shechter *et al.*, 1978a; Fox *et al*, 1980; Haigler *et al.*, 1979; Fox and Das, 1979). The biochemical mechanisms responsible for the translocation process are poorly understood and a matter of general controversy. The low-density lipoprotein (LDL)–cholesterol receptor has served as a prototype for studying receptor-mediated internalization processes (Anderson *et al.*, 1977). In human fibroblasts, LDL receptors are localized over clathrin-coated regions of the plasma membrane. For EGF, surface receptors are diffusely distributed, and, upon binding of EGF, coalesce into coated invaginations of the plasma membrane, or coated pits (Haigler *et al.*, 1979; Maxfield *et al.*, 1979 and see Chapter 1). These are subsequently internalized to form endocytic vesicles ('receptosomes' or 'endosomes') which are translocated and delivered to lysosomes (Anderson *et al.*, 1979; Maxfield *et al.*, 1979). There the ligand is degraded (Carpenter and Cohen, 1976a; King *et al.*, 1980a; Haigler *et al.*, 1979). It is now believed that receptor aggregation into coated membrane regions is a prerequisite for adsorptive pinocytosis (Anderson *et al.*, 1977; Maxfield *et al.*, 1979), but internalization of EGF receptors from non-coated regions of the membrane has been observed with some cell types (Haigler *et al.*, 1979).

Early morphological studies using fluorescence microscopy and rhodamine–lactalbumin–EGF (Rh–L–EGF) showed that EGF receptors are initially mobile in the plane of the plasma membrane, but a massive immobilization of these receptors follows (Schlesinger *et al.*, 1978). By video intensification, the fluorophone aggregated after about 10 to 20 min in 3T3 cells. It was not clear what cellular processes were responsible for 'cluster' formation, but it was believed that EGF–receptor internalization (Maxfield *et al.*, 1979; Schlessinger *et al.*, 1978; Shechter *et al.*, 1978b) would be mediated by clathrin-coated regions of the plasma membrane like those observed for LDL (Anderson *et al.*, 1977). Upon further incubation (20 to 30 min) the aggregates of Rh–L–EGF were inferred to be internalized because of the appearance of fluorescent patches with the saltatory motion that is characteristic of lysosomes (Shechter *et al*, 1978b; Davies *et al.*, 1980).

3.4.2 Transglutaminase mediation of receptor aggregation

It was claimed that the receptor aggregation phase required calcium (the patches did not form and the fluorescence remained diffusely distributed in Ca^{2+}-free buffers) and that inhibitors of the calcium-dependent enzyme, transglutaminase, prevented 'patch' formation (Schlessinger *et al.*, 1978; Maxfield *et al.*, 1979; Davies *et al.*, 1980). Thus, it was suggested that the aggregation of plasma membrane EGF receptors into coated pits required the activity of a calcium-dependent transglutaminase to covalently cross-link receptors. Effective transglutaminase inhibitors (lipophilic primary alkylamines, such as methylamine or dansylcadaverine) inhibit EGF degradation (Carpenter and Cohen, 1976b; King *et al.*, 1980a; Haigler *et al.*, 1979). These agents were claimed to interfere with transglutaminase-mediated cross-linking of EGF receptors and thereby block internalization and targeting of EGF to lysosomes (Maxfield *et al.*, 1979; Davies *et al*, 1980). Tertiary alkylamines are not transglutaminase inhibitors, but also inhibit EGF degradation (Carpenter and Cohen, 1976b; King *et al.*, 1980a; Haigler *et al.*, 1979; Maxfield *et al.*, 1979; Davies *et al.*, 1980). From the fluorescent microscopic studies, tertiary alkylamines were suggested to block degradation, not by interfering with the aggregation process, but by preventing the internalization of the surface path (Maxfield *et al.*, 1979; Davies *et al.*, 1980). Thus, the aggregation and internalization processes were proposed to be dissociated pharmacologically by the use of different alkylamine compounds (Maxfield *et al.*, 1979; Davies *et al.*, 1980).

Careful re-evaluation of the fluorescent microscopic studies in our laboratory (King *et al.*, 1980a, 1981) and others (Yarden *et al.*, 1981) suggests that the formation of visible fluorescent patches is not prevented by transglutaminase inhibitors. The formation of large receptor aggregates after about 30 minutes of incubation are now thought to be caused by the coalescence of minute endocytic vesicles within the cell, and not to be caused by surface receptor aggregation (King *et al.*, 1980a, 1981; Yarden *et al.*, 1981). Electron microscopic studies with ferritin–EGF in A_{431} cells confirmed the aggregation of surface receptors, but indicated that only limited receptor aggregates of 8 to 15 monomers were the rule (Haigler *et al.*, 1979). These small aggregates would not be detected by current fluorescent microscopic techniques. Thus, early studies using fluorescent ligands were probably monitoring processes that take place after the ligand has already entered the cell.

3.4.3 Rate of receptor-mediated internalization

Studies with ferritin–EGF (Fe–EGF) (Haigler *et al.*, 1979) showed that the internalization of EGF occurs more rapidly than initially estimated in the fluorescent studies (Maxfield *et al.*, 1979; Schlessinger *et al.*, 1978; Shechter *et al.*, 1978b; Davies *et al.*, 1980). With A_{431} cells, which have an extraordinarily large number of EGF receptors (2×10^6–3×10^6), internalization of subsaturating concentrations of Fe–EGF is complete well within a 5 to 10 minute period (Haigler *et al.*, 1979). These results are consistent with the time course of loss of anti-^{125}I-EGF binding sites from human fibroblasts pretreated with unlabeled EGF (Carpenter and Cohen, 1976b). As the receptor occupation is increased in A_{431} cells, the rate of internalization is diminished (Wiley and Cunningham, 1981, 1982). Indeed, it has been noted that A_{431} cells down regulate their receptors poorly in response to exposure to supersaturating concentrations of EGF (Wrann and Fox, 1979). Interestingly, A_{431} cells also have unusually high concentrations of LDL receptors which are also internalized poorly (Anderson *et al.*, 1981). These results may indicate that internalization of EGF-receptor complexes takes place at specialized membrane sites like those observed for LDL (Anderson *et al.*, 1977) but that the accumulation of receptor complexes into these sites is a rate-limiting step in the internalization process (Wiley and Cunningham, 1981, 1982). A_{431} cells seem to be unique, and most cell types have fewer receptors and apparent overloads of the internalization process do not occur.

3.5 RADIOTRACER STUDIES OF EGF INTERNALIZATION

3.5.1 Calcium

By using ^{125}I-EGF, it can be shown that the internalization of EGF–receptor complexes is not calcium dependent. The total cellular association of ^{125}I-EGF is slightly increased in Ca^{2+}-free or EDTA-containing buffers. Further, the clearance of surface EGF receptors induced by pretreatment of cells with unlabeled EGF is unaltered in Ca^{2+}-free buffers (Sawyer and Cohen, 1981). This result (and others to follow) places grave doubts on the relevance of calcium-dependent transglutaminases for mediation of EGF internalization.

3.5.2 Effects of lysosomotropic alkylamines on EGF internalization

It is now well documented that internalization of EGF–receptor complexes is unaltered by most transglutaminase inhibitors or tertiary alkylamines. Both primary and tertiary lipophilic alkylamines prevent EGF degradation (Carpenter and Cohen, 1976b; King *et al.*, 1980a, 1981; King and Cuatrecasas, 1981b) by raising the pH in lysosomes (Ohkuma and Poole, 1978; Poole and Ohkuma, 1981). Fe–EGF is readily localized within cytoplasmic vesicles even when cells are incubated with methylamine (Haigler *et al.*, 1979). By using ^{125}I-EGF, we showed that there is a 2–4-fold enhancement in accumulation when a variety of primary or tertiary amines are present (King *et al.*, 1980a, 1981; King and Cuatrecasas, 1981b). Since these agents do not alter the receptor clearance process induced by exposure of cells to EGF, there must result a net intracellular accumulation of ^{125}I-EGF in cells treated with alkylamines. In support of this, ^{125}I-EGF incubated with cells in the presence of alkylamines becomes increasingly insensitive to trypsin and resists dissociation (King *et al*., 1981). Thus, alkylamines enhance the apparent uptake of ^{125}I-EGF by preventing ^{125}I-EGF degradation. Because a similar number of surface receptors are lost from the cell surface, previous ^{125}I-EGF 'binding' assays may have represented a gross underestimate of the overall capacity of the EGF internalization process.

3.5.3 EGF receptors do not recycle

How does one account for the potentiation in ^{125}I-EGF accumulation with lysosomal inhibitors? One possibility is that there exists an intracellular pool of EGF receptors. Alternatively, alkylamines might prevent degradation of the receptor, and therefore more receptors may be available for recycling to the plasma membrane, once they are internalized. It has been known for some time that the association of ^{125}I-EGF with cells is temperature-dependent. Binding of EGF measured at 4°C is less than that measured at 37°C (Carpenter and Cohen, 1976b). One explanation may be that EGF receptors recycle. To investigate this, we measured the number of EGF receptors on KB cells at 4°C in order to prevent internalization of the ligand and compared this to the total uptake capacity of KB cells for ^{125}I-EGF at 37°C when all surface receptors are occupied. We found that the maximum uptake of EGF measured after a 1 hour incubation exactly correlates with the number of receptors (King *et al*., 1980b). Thus, each receptor internalized under saturating conditions carries only a single EGF molecule into the cell. After this, the receptors are presumably proteolytically inactivated, resulting in a reduction in the number of EGF receptors. The maximal uptake capacity is only slightly inhibited by cycloheximide and indicates that most of the early internalization of EGF does not depend on newly synthesized receptors (King *et al*., 1980b). The EGF receptor seems to be unique, since most other receptor-mediated internalization systems are dependent on recycling of the receptor for entry of the biological macromolecule.

3.5.4 Hypo-osmotic lysis of methylamine-treated cells exposes internalized EGF-binding sites

When cells are treated with unlabeled EGF and methylamine, the internalized receptors are masked from detection by ^{125}I-EGF in intact cells but are exposed and detected by the ligand after hypo-osmotic lysis (King *et al*., 1981). This is not true for untreated cells where receptor internalization results in inactivation of all binding activity (intracellular and cell-surface) as well as the hormone (King *et al.*, 1981). The receptor binding activity recovered from hypo-osmotic lysates of cells exposed to alkylamines migrates into denser membrane fractions after equilibrium sucrose density centrifugation than those prepared from untreated cells (King *et al.*, 1981). Morphological studies using Fe–EGF suggested that these membrane vesicles might be large vesicular structures composed of multiple EGF-containing endocytic vesicles (multivesicular bodies). Thus, alkylamines cause an increased intracellular accumulation of not only EGF (King *et al.*, 1980a,b, 1981) but also its receptor (King *et al.*, 1981) probably by preventing the normal modes of lysosomal degradation. This is clear evidence that alkylamines do not prevent the receptor-binding, aggregation or internalization processes (King *et al.*, 1980a,b, 1981; Haigler *et al.*, 1979; King and Cuatrecasas, 1981b).

3.6 TURNOVER OF MEMBRANE EGF RECEPTORS

3.6.1 Turnover rate of surface receptors

An approximation of the turnover rate of surface EGF receptors has been obtained in a variety of cell types by pretreating cultures with concentrations of cycloheximide that block *de novo* protein synthesis (2–10 μg/ml). When cells are treated in this manner, there is a time- and temperature-dependent depletion in the cell surface binding capacity which is consistent with a surface receptor half-life of about 6 hours (Aharonov *et al.*, 1978a; King *et al.*, 1980a,b; Fox *et al.*, 1980). Thus, surface EGF receptors appear to be continuously replaced by a newly synthesized population. If this is true, the corresponding rates of removal of old receptors must be equivalent to that of insertion of newly synthesized receptors, since under normal conditions, the receptor number is constant (King *et al.*, 1980a,b). This information has led to the obvious conclusion that the EGF receptor is probably internalized and degraded in lysosomes even in the absence of binding of EGF to surface receptors.

3.6.2 EGF binding enhances rate of receptor internalization

When EGF and cycloheximide are added to cultures, the rate of loss of surface receptors is accelerated about 10-fold over that observed with cycloheximide alone (Fig. 3.5a). This enhancement is only transient and returns to the basal rate after about 90 minutes (King *et al.*, 1980a,b). Cycloheximide has little or no effect on the initial rate of ^{125}I-EGF internalization (King *et al.*, 1980b), indicating that there is a requirement for substantial *de novo* synthesis of receptors for maximal uptake. Thus, the addition of EGF to cultures potentiates the net internalization of receptors and is principally responsible

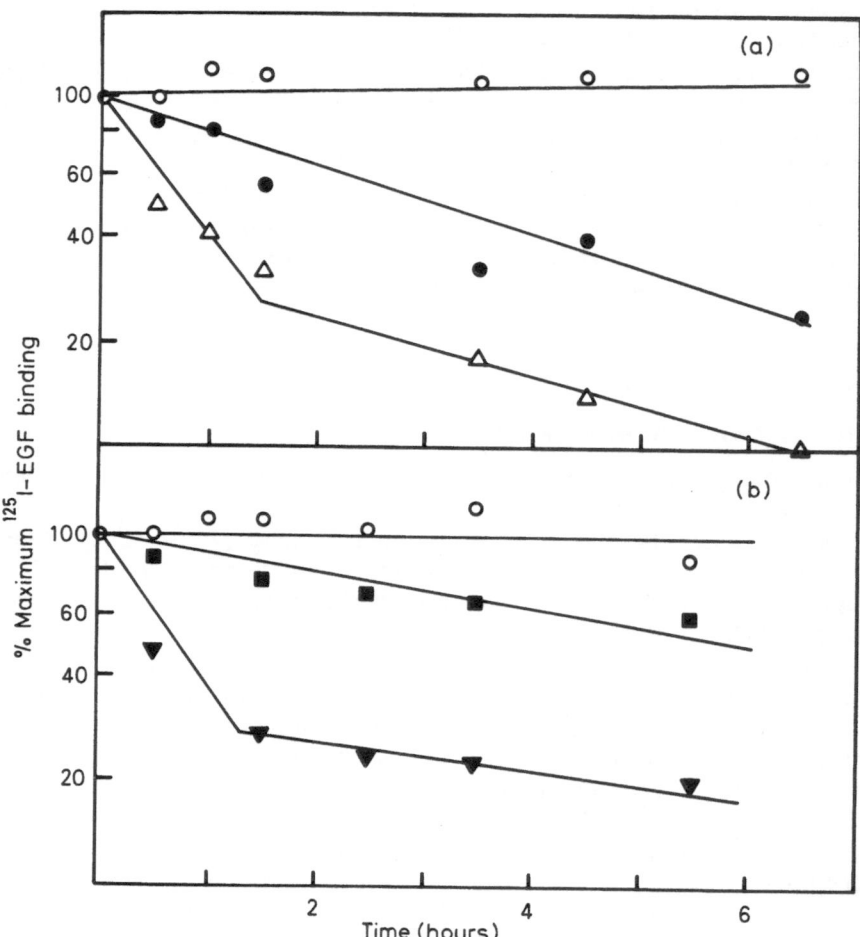

Fig. 3.5 (a) Effect of cycloheximide pretreatment on ^{125}I-EGF binding to intact cells. KB cells were incubated with (●) and without (○) cycloheximide at 10 μg/ml or with both EGF at 10 ng/ml and cycloheximide at 10 μg/ml (△). At the indicated times, ^{125}I-EGF-binding capacity was determined with ^{125}I-EGF at 50 ng/ml for 30 min at 37°C. Unlabeled EGF at 1 μg/ml was included to determine non-specific binding. (b) Effect of MeNH$_2$ pretreatment on ^{125}I-EGF binding to intact cells. Preincubations and binding assays were performed as described for (a), except that the preincubation was with (■) or without (○) 10 mM-MeNH$_2$ or with both EGF at 10 ng/ml and 10 mM-MeNH$_2$ (▼). The preincubation times are shown. Data from King *et al.* (1980a).

for the decreased number of receptors on the cell surface after treatment with the mitogen.

3.6.3 Methylamine also blocks surface receptor expression

Interestingly, compounds, such as methylamine, that prevent degradation of the receptor cause a time-dependent loss of surface receptors identical with that caused by cycloheximide (Fig. 3.5b). This is true whether EGF is present or not. At this time, we cannot account for the mechanism responsible for the loss of EGF receptors in cells treated with alkylamines.

3.6.4 Effects of cycloheximide and methylamine on total receptor pool

The previous studies only document the effects of protein synthesis and degradation inhibitors on the surface receptor pool. They cannot give absolute values of receptor turnover, since treatment of cells with these agents may interfere with the production of an intracellular effector required for maintenance of cell surface receptors. It remains uncertain how methylamine can cause a decrease in the surface receptor pool but enhance the total EGF-uptake capacity of the cell. To address this question in more detail, we pre-

Table 3.1 Total binding of ^{125}I-EGF to lysates prepared from cells pretreated with cycloheximide and methylamine

Preincubation condition	Relative ^{125}I-EGF binding
Untreated	1.0
EGF	0.28
Methylamine (MeNH$_2$)	2.4
EGF and MeNH$_2$	2.4
Cycloheximide	0.35
EGF and cycloheximide and MeNH$_2$	0.96
MeNH$_2$ and cycloheximide	1.1

KB cells were untreated or were incubated for 3 hours at 37°C in Dulbecco's modified Eagle's medium (DMEM), 5% fetal bovine serum (FBS) with combinations of the following as indicated: 20 ng of EGF/ml, 15 mM-MeNH$_2$ or 10 μg of cycloheximide/ml. After this, cells were scraped from the culture flask and washed free of the drug [Krebs buffer + 0.1% bovine serum albumin (BSA)]. The cells were then lysed by hypo-osmotic lysis and membranes prepared as described (King *et al.*, 1980a). The membranes were resuspended in Krebs buffer + 0.1% BSA and incubated with 50 ng of ^{125}I-EGF/ml for 60 minutes at 37°C. Free ^{125}I-EGF was removed by vacuum filtration, and non-specific binding was determined by including 5 μg of unlabeled EGF/ml in parallel incubations.

treated cells with cycloheximide or methylamine with and without EGF and determined the effect on the total receptor pool (i.e. that of binding activity of hypo-osmotic lysates of KB cells). Each of these pretreatment conditions results in a reduction of the surface receptor binding activity assayed in intact cells. However, only pretreatment with EGF or cycloheximide or both results in a loss of total binding activity (Table 3.1). This loss parallels that observed in intact cells. Pretreatment of cells with MeNH$_2$ with or without EGF results in an enhancement of the total binding activity measured in hypo-osmotic lysates. These results are consistent with the interpretation that at least a portion of the EGF receptor pool may turn over at a rapid rate. The increased binding activity measured in hypo-osmotic lysates prepared from cells treated with methylamine exactly correlates with the enhancement in ^{125}I-EGF uptake observed in intact cells. Clearly, all the receptors we assay in hypo-osmotic lysates must be inserted into the plasma membrane and direct the entry of EGF into cells. Thus, cells treated with alkylamines increase their total receptor pool, but all of these receptors are destined to accumulate intracellularly. These receptors may be synthesized *de novo*, since their expression is inhibited by cycloheximide. This could suggest a feedback regulation on receptor synthesis by the process of receptor internalization and lysosomal degradation.

3.7 PHARMACOLOGICAL STUDIES

3.7.1 Lysosomal degradation

Pharmacological studies designed to determine the importance of lysosomal processing to the mitogenic activity of EGF have yielded confusing results at best. Lipophilic lysosomotropic compounds prevent the mitogenic response of EGF, insulin and serum when they are present throughout the incubation (King *et al.*, 1981, and Fig. 3.6). The mechanism of inhibition of lysosomal function appears to be the same for both primary and tertiary alkylamines. Because they are weak bases, they accumulate into intracellular acid compartments such as lysosomes where they are protonated and raise the pH (Ohkuma and Poole, 1978; Poole and Ohkuma, 1981). The efficacy of the alkylamines as lysosomal inhibitors depends on their lipophilic character and ease of penetration through the plasma membrane. Cadaverine, a diamine, does not penetrate cells and is devoid of lysosomal inhibitory activity. A derivative of this compound, dansylcadaverine, a monoamine, readily penetrates cells and is a potent inhibitor. The dose response suggests that mitogenic inhibition occurs at the concentrations of amine that cause significant pH shifts in lysosomes (Ohkuma and Poole, 1978; Poole and Ohkuma, 1981), and both primary and tertiary alkylamines are potent inhibitors. In human fibroblasts, the transglutaminase inhibitor, bacitracin (Maxfield *et al.*,

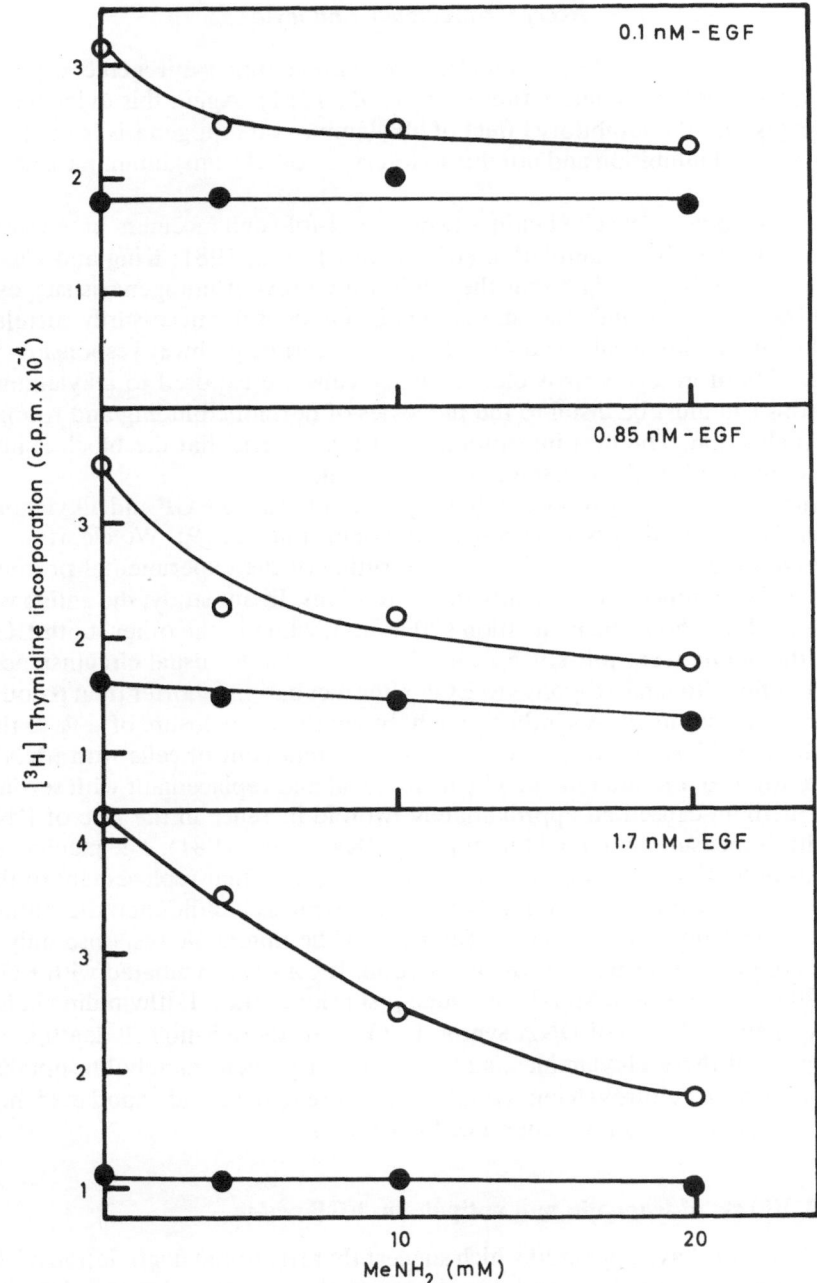

Fig. 3.6 Effect of MeNH$_2$ on basal and EGF-stimulated [^{3}H] thymidine incorporation. Confluent monolayers of HF cells were incubated with (○) and without (●) EGF for 23 hours in the presence of various concentrations of MeNH$_2$ and then pulsed with 1 µCi of [^{3}H] thymidine/ml for 1 hour and the extent of incorporation was determined. Data from King *et al.* (1981).

1979; Davies *et al.*, 1980), is not effective in preventing a mitogenic response nor does it prevent degradation (King *et al.*, 1981). Again this evidence indicates that the inhibitory effect of alkylamines on mitogenesis is the result of lysosomal inhibition and not due to intervention of transglutaminase activity.

All lysosomotropic alkylamines cause a 3–4-fold enhancement in the overall extent of EGF accumulation (King *et al.*, 1980a, 1981; King and Cuatrecasas, 1981b). The fact that they uniformly prevent mitogenesis suggests that the extent of cellular accumulation of EGF does not necessarily correlate well with the biological response. Clearly, the normal pathway responsible for mediation of mitogenesis is blocked when cells are exposed to alkylamines, and this site must be distal to the processes of hormone binding and receptor internalization. The best interpretation at this time is that the block is most likely localized in the lysosomal compartment.

Another laboratory reported that exposure of cells to EGF and alkylamines potentiates the mitogenic response (Maxfield *et al.*, 1979). We clearly observed the opposite effect. Careful scrutiny of the experimental protocol revealed differences in the incubation conditions. In our study, the amine was present throughout the incubation (20 hours), while in the other, both EGF and the amine were present for only 2.5 hours. Under usual circumstances, there is no mitogenic response to EGF after incubations shorter than 6 hours. Thus, increases in DNA synthesis might be caused by exposure of cells to the alkylamine. This turned out to be the case. Treatment of cells with $MeNH_2$ alone for 2 to 4 hours followed by its removal and replacement with serum-free medium caused an approximately twofold increase in the rate of DNA synthesis measured after a 24 hour period (King *et al.*, 1981). The mechanism of this potentiation is unknown, but it was reasoned that replacement of the medium might remove 'chalones' that had previously conditioned the culture. Thus, alkylamines are effective inhibitors of the mitogenic response only if they are present throughout the incubation. If cells are incubated with EGF for 20 to 24 hours and $MeNH_2$ is added just prior to the [³H]thymidine pulse, there is no inhibition of DNA synthesis. Further, there is no cell death in the presence of these alkylamines, and they do not prevent metabolite uptake into fibroblast cultures (King *et al.*, 1981). These data provide good evidence that alkylamines are not general cellular toxins.

3.7.2 Effects of leupeptin and antipain on EGF action

Other reports have appeared which suggest that lysosomal degradation of the EGF–receptor complex may be irrelevant to the mitogenic process. Specific peptidic inhibitors of lysosomal cathepsins (antipain and leupeptin) were found to inhibit the degradation of ¹²⁵I-EGF in granulosa cultures with only minimal inhibitory effects (20 to 30%) on [³H]thymidine incorporation after a 24 hour incubation (Savion *et al.*, 1981). These agents may not disrupt the transmembrane proton gradient that is required for lysosomal function as do alkylamines, thus possibly implicating the low intralysosomal pH as critical to mitogen action rather than proteolytic processing. In support of this, the inhibitory effect of lipophilic alkylamines can be readily reversed either by washing cells free of the amine (which rapidly restores the low intralysosomal pH) or by exposing cells pretreated with EGF and the amine to a reduced pH environment that mimicks that of the lysosome. Thus, compartmentalization of the hormone or receptor in an intracellular acid compartment like that found in lysosomes might be critical for cellular activation.

3.7.3 Inhibition by alkylamines of mitogenic response is reversed by washing cells free of the amine

A curious feature of alkylamine treatment was revealed by further experiments. Exposure of cells to EGF with $MeNH_2$ decreases the exposure period required for EGF to induce a mitogenic response (Table 3.2). For example, while exposure to EGF for 4 hours produces no change in the rate of DNA synthesis assayed after 24 hours, exposure of human fibroblasts to EGF and $MeNH_2$ for 4 hours results in a degree of responsiveness that generally requires 16 to 20 hours of exposure. These results may indicate that a critical level of accumulation of EGF–receptor complexes is required for mitogenic activation. Of course, the mitogenic response occurs only after removal of the alkylamine and the return of normal lysosomal function. Chloroquine, a tertiary amine, causes a similar inhibitory effect to that of primary amines when it

Table 3.2 Reversal of methylamine inhibition of EGF-stimulated mitogenesis results in potentiated response

	[^{3}H]Thymidine incorporation (c.p.m.)	
	Untreated	Washing step
Basal	4 609	5 678
10 mM-$MeNH_2$	4 216	4 473
1 ng of EGF/ml	17 075	5 414
1 ng of EGF/ml + 10 mM-$MeNH_2$	5 668	21 789

Human fibroblast monolayers were treated with 1 ng of EGF/ml with or without 10 mM-$MeNH_2$, or with 10 mM-$MeNH_2$ alone. Untreated cells were incubated under these conditions for 20 hours prior to a 4 hour pulse with [^{3}H]thymidine. Duplicate samples were washed free of the drug and EGF after 4 hours of incubation. The medium was replaced with DMEM + 0.5% serum, and the cells were pulsed with [^{3}H]thymidine for 4 hours after 16 additional hours of incubation.

is present for the entire incubation period (King *et al.*, 1981). Unlike the primary alkylamines, however, reversal of inhibition does not occur (not shown). Chloroquine and other tertiary alkylamines cannot readily be washed out of fibroblast cultures, and their inhibitory effects on lysosomal function are persistent (Ohkuma and Poole, 1978; Poole and Ohkuma, 1981).

3.7.4 Exposure of cells to low extracellular pH reverses methylamine block on mitogenic response

By artificially creating a transmembrane proton gradient across the plasma membrane (the initial site of interaction of EGF with its receptor), we have seemingly tricked cells to respond to EGF even though alkylamines are present (Fig. 3.7). Cells are incubated with EGF and methylamine for periods too short to cause a mitogenic response with EGF alone, washed free of the mitogen but not of methylamine, and subjected to a brief (5 min) exposure to medium of reduced pH (pH 5.8 to 6.5). Cells are then incubated with basal medium containing $MeNH_2$ to ensure that lysosomal degradation of EGF does not occur, and the rate of DNA synthesis is determined 20 hours after the initiation of the experiment. Under these conditions, cells respond to EGF in a manner that is blocked by the addition of EGF antisera, but only if anti-EGF is added prior to the low-pH treatment. After this, cells are committed to respond and the EGF is masked from inactivation by antisera. We believe that we have completely bypassed the requirement for the mitogen and its receptor to be delivered to the lysosome or some other undefined intracellular acid compartment. Low-pH treatment reduces the period that cells need to be exposed to EGF (6 to 2 hours) and results in potentiation of the mitogenic response (King and Cuatrecasas, 1982). These results suggest that mitogenic activation is accomplished at the cell surface by a pH-sensitive modification of plasma-membrane-associated EGF-receptor complexes that under normal physiological conditions requires delivery to an intracellular compartment that actively accumulates protons.

3.7.5 Cytoskeleton

Drugs that disrupt the cytoskeleton of cells have a profound potentiating effect on the DNA-synthetic response evoked by EGF and other mitogens (Savion *et al.*, 1981; Otto *et al.*, 1979). These compounds (colchicine, vinblastine, colcemid and phodophlotoxin) have been suggested to inhibit EGF degradation without interfering with receptor-mediated internalization (Brown *et al.*, 1980). If this is so, their effect on EGF degradation is only minimal (10 to 20%) and might not account for the 5–7-fold enhancement in DNA synthesis. Further, although it was claimed that these drugs exert a synergistic effect on cell growth, under these conditions they do not potenti-

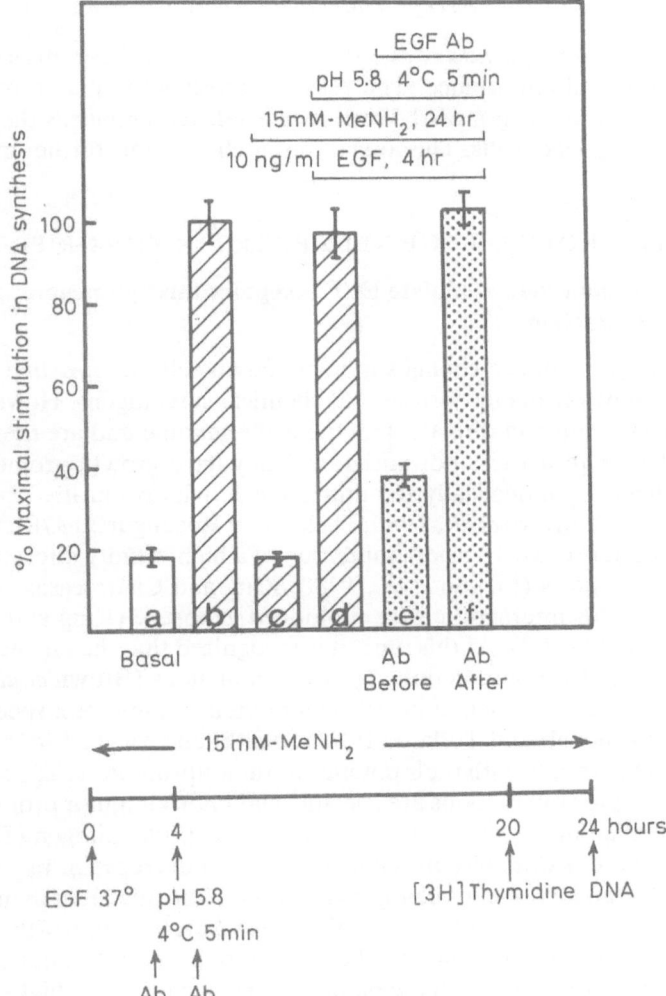

Fig. 3.7 Effect of MeNH₂ and low pH on mitogenic activity of EGF. Confluent and serum-starved cells were untreated (a) or were exposed to 10 ng of EGF/ml for 24 hours (b,c) or for 4 hours (d–f). Some cells were incubated with 15 mM-MeNH₂ throughout the experiment (c–f), and some were exposed to pH 5.8 medium for 5 minutes at 4°C (d–f). The experimental protocol for columns d–f is shown at the bottom of the figure. When the EGF antiserum was added before the pH change (e), it was only present during the 5 minute, pH 5.8 incubation. When it was added after the pH change (f), it remained in the incubation medium throughout the duration of the experiment. Briefly, cells were exposed to 10 ng of EGF/ml for 4 hours, 37°C with 15 mM-MeNH₂. Afterwards, the cells were washed twice with ice-cold DMEM, 0.5% FBS, 15 mM-MeNH₂, 20 mM-Hepes, pH 7.2, and then incubated for 5 minutes in DMEM, 0.1% BSA, 15 mM-MeNH₂, 20 mM-MES, pH 5.8. The cells were washed twice again in pH 7.2 medium and then incubated for a further 16 hours in the same medium prior to the addition of ³[H]thymidine. Data from King and Cuatrecasas (1982a).

ate growth alone (Savion *et al.*, 1981). Potentially, these observations could be of profound importance. The cellular cytoskeleton may provide a constraint to normal cell growth which when relaxed enhances the normal mitogenic signal of cells. This area of research warrants further investigations.

3.8 THE ROLE OF EGF RECEPTORS IN TUMOR PROMOTION

3.8.1 Phorbol esters modulate EGF receptors through membrane transduction

Phorbol esters induce changes in mammalian cells *in vitro* that mimic transformation by oncogenic viruses and chemical carcinogens. However, phorbol esters to not interact directly with the cell's genome and are therefore considered tumor promoters not initiators. They have growth-promoting activity and enhance synergistically the mitogenic activity of purified growth factors such as EGF (Brown *et al.*, 1980; Dicker and Rozengurt, 1978). Interestingly, phorbol esters cause transient inhibition of a high-affinity interaction of EGF with its receptors (Magun *et al.*, 1980; King and Cuatrecasas, 1982b) and potentiate this interaction after prolonged exposure (King and Cuatrecasas, 1982b, and Fig. 3.8). At this time, it is recognized that the efficacy of the early inhibition of EGF interactions by phorbol analogs (Brown *et al.*, 1979; Shoyab *et al.*, 1979; Salomon, 1981) and their affinity for a specific phorbol receptor (Shoyab and Todaro, 1980; Solanki and Slaga, 1981; Horowitz *et al.*, 1981) correlate with their potency as tumor promoters. The alterations in EGF–receptor interactions are specific, and phorbol tumor promoters do not alter the binding characteristics of any other peptide mitogen (Shoyab *et al.*, 1979). There is difficulty in reconciling how a decrease in receptor binding can lead to an increase in biological response. One explanation may be that a high-affinity receptor subtype mediates exclusively the growth response. In support of this, the half-maximal concentration of EGF required to evoke mitogenesis correlates with the dissociation constant of the high-affinity component (Fox and Das, 1979; Das and Fox, 1978). Interestingly, treatment of cells with EGF (Aharonov *et al.*, 1978a; Vlodavsky *et al.*, 1978; Carpenter and Cohen, 1976b) and transformation of cells by both chemical (Fisher *et al.*, 1980; Hollenberg and Cuatrecasas, 1975) and viral agents (Blomberg *et al.*, 1980; Todaro *et al.*, 1976) often lead to a specific decrease in EGF receptor binding. Thus, in each instance, a potentiated growth response correlates with a decreased binding of EGF to surface receptors. These results suggest that tumor promotion and transformation may lead to altered growth properties through modification of the interaction of EGF with cells.

Currently, the detailed mechanisms of regulation of EGF receptors are unknown. It is therefore impossible to evaluate critically drug-induced alterations in EGF binding capacity. A possible explanation for the effect of tumor

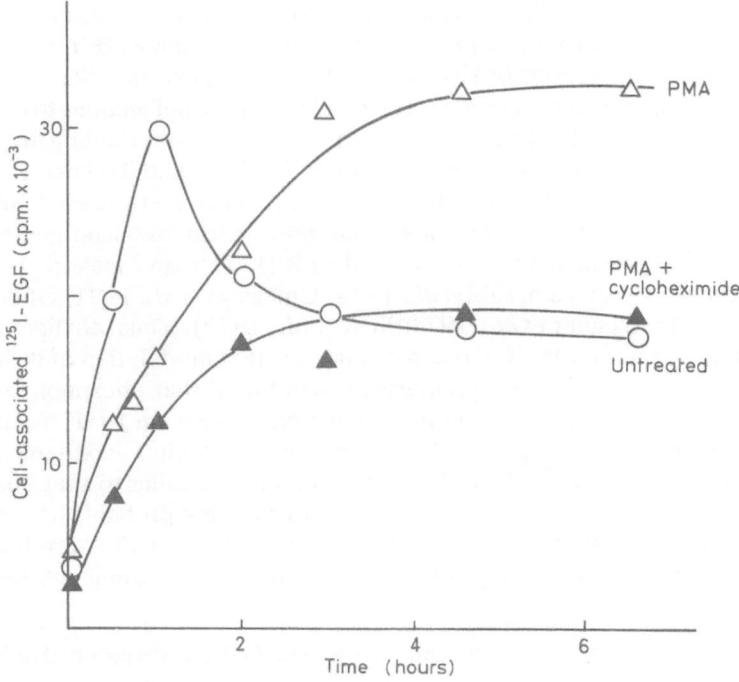

Fig. 3.8 Effect of phorbol myristate acetate (PMA) on the time course of the interaction of ^{125}I-EGF with KB cells. KB cells were harvested with 0.05% EDTA in phosphate-buffered saline and washed twice with Krebs' bicarbonate, pH 7.4, 0.1% BSA. Cells were resuspended into DMEM, 2% fetal calf serum (FCS) (10^6 cells/ml) and incubated at 37°C with 2.5 ng of ^{125}I-EGF/ml under three different incubation conditions: (○) untreated cells, (△) 10^{-8}M-PMA and (▲) 10^{-8}M-PMA with 10 µg of cycloheximide/ml. At different times, 200-µl aliquots of cells were vacuum-filtered through 1.0-µm-pore Millipore filters and washed twice with 5 ml of Krebs + 0.1% BSA. All values are corrected for non-specific binding by including 3 µg of unlabeled EGF/ml. Data from King and Cuatrecasas (1982b).

promoters on EGF binding is that they increase the sensitivity of cells to suboptimal concentrations of EGF and thereby lower the threshold of the cell's response (King and Cuatrecasas, 1982b). This might ensure fixation of carcinogen-induced DNA damage through DNA replication and cell proliferation.

Phorbol tumor promoters do not bind directly to EGF receptors, since EGF does not compete for binding to the phorbol receptor (Shoyab and Todaro, 1980). It is therefore believed that the inhibitory action of tumor promoters on EGF receptors depends on the interactions with their own specific receptors. Phorbol esters have no effect on EGF receptor binding at reduced temperatures (Brown *et al.*, 1979; Shoyab *et al.*, 1979; Fisher *et al.*, 1980), thus the phorbol receptor may be coupled to the EGF receptor to modulate the interactions of EGF with cells. To support the notion of membrane coupling, it was observed that the ability of phorbol analogs to compete with [^{3}H]dibutyryl TPA (phorbol 12-myristate 13-acetate) binding also correlates with their potency as tumor promoters (Shoyab and Todaro, 1980; Solanki and Slaga, 1981; Horowitz *et al.*, 1981). Phorbol esters are known to stimulate the turnover of arachidonic acid and phosphatidylcholine, and the production of prostaglandins in cultured cells (Levine and Hassid, 1977; Mufson *et al.*, 1979; Yamasaki *et al.*, 1979; Umezawa *et al.*, 1981; Ohuchi and Levine, 1978; Wenner *et al.*, 1974; Suss *et al.*, 1972). Thus, changes in the lipid environment of the EGF receptor may result in modulation of its binding capacity. In support of this hypothesis, it was found that phospholipase C, fillipin and millitin mimic the inhibitory action of TPA on EGF receptor binding (Cohen and Savage, 1974). Further, prostaglandins added to cultured cells directly decrease ^{125}I-EGF binding in a manner similar to that observed with phorbol esters (Lee, 1981). The effects of TPA are probably not caused by non-specific, chaotropic membrane perturbation, since TPA produces no inhibition of receptor binding for any other growth-promoting substance.

3.8.2 Phorbol esters cause transient reductions in EGF receptor affinity and number

Early studies aimed at the characterization of the inhibitory actions of phorbol esters on EGF–receptor binding demonstrated that there was a decrease in either the affinity and/or number of EGF receptors. Closer scrutiny revealed that the inhibitory effects of phorbol esters on EGF receptors involve both a decreased affinity and reduction in receptor number (Magun *et al.*, 1980; King and Cuatrecasas, 1982b; Shoyab *et al.*, 1979). The inhibition of EGF-binding capacity is transient and lasts only about 90 minutes (Magun *et al.*, 1980; King and Cuatrecasas, 1982b). After this time, the binding capacity of the cell is recovered and when low concentrations of serum are present, it is even enhanced above the levels of untreated cultures. TPA does not interfere with the ultimate loss of surface EGF receptors associated with the down regulation process (Magun *et al.*, 1980; King and Cuatrecasas, 1982b; Shoyab *et al.*, 1979). It may, however, delay the internalization of receptors, since much of the EGF remains accessible to trypsinization during early exposure periods (King and Cuatrecasas, 1982b). The phorbol tumor promoters also have little or no effect on the rate of degradation of ^{125}I-EGF that becomes transferred to the lysosomal compartment (Magun *et al.*, 1980; King and Cuatrecasas, 1982b). These results indicate that phorbol esters (and possibly other tumor-protecting agents) decrease EGF–receptor interactions only after very early exposure periods and explain why cells 'escaped' from the inhibitory effect on EGF–receptor interactions after prolonged exposures to TPA.

3.8.3 Resolution of two classes of EGF receptors

The inhibitory effects of phorbol esters and other tumor promoters on EGF receptors do not occur in plasma membranes of cells or in intact cells incubated at low temperatures (Brown *et al.*, 1979; Shoyab *et al.*, 1979; Fisher *et al.*, 1980). Thus, intact cells and physiological temperatures are required. This suggests that a metabolic event may be involved. We found that at 4°C, the affinity of cell surface EGF receptors is lower than that of cells incubated at 37°C. A temperature-dependent shift in receptor affinity explains why there is a marked temperature-dependence of EGF interactions in intact cells, plasma membranes and solubilized receptor preparations, and why phorbol esters have no effect on EGF–receptor interactions at low temperatures (King and Cuatrecasas, 1982b). No high-affinity binding sites are present on cells at reduced temperatures. This suggests that the coupling phenomenon between phorbol ester receptors and EGF receptors may be more complex than a simple membrane transduction effect. Phorbol esters have no effect on ^{125}I-EGF binding even at physiological temperatures when plasma membranes are prepared. We showed that unlike the intact cell, both high- and low-affinity receptor populations are present (King and Cuatrecasas, 1982b). These results suggest that a class of high-affinity EGF receptors are cryptic in the plasma membrane of intact cells. Their exposure at the cell surface is temperature-dependent, may require binding of EGF to the original complement of low-affinity sites, and is inhibited by the phorbol ester tumor promoters.

3.8.4 High-affinity EGF receptor interaction is sensitive to cycloheximide

We found that the appearance of these high-affinity EGF receptors in KB cells could be prevented by exposing cells to 10 μg of cycloheximide/ml for 2 to 4 hours (King and Cuatrecasas, 1982b). The same is true for the receptor found on diploid human fibroblasts (HF) (Fig. 3.9). Two possibilities may account for this. First, the subset of high-affinity EGF receptors may turnover at a more rapid rate than the larger number of low-affinity sites. Second, the expression of these sites from their 'masked' domain may require an effector molecule that turns over rapidly. Results with plasma membrane preparations cause us to favor the latter interpretation, since membranes prepared from cells penetrated with cycloheximide do not show a preferential loss of high-affinity binding sites (King and Cuatrecasas, 1982b).

These properties may represent a subset of high-affinity EGF receptors that have heretofore gone unrecognized. Phorbol esters may delay the

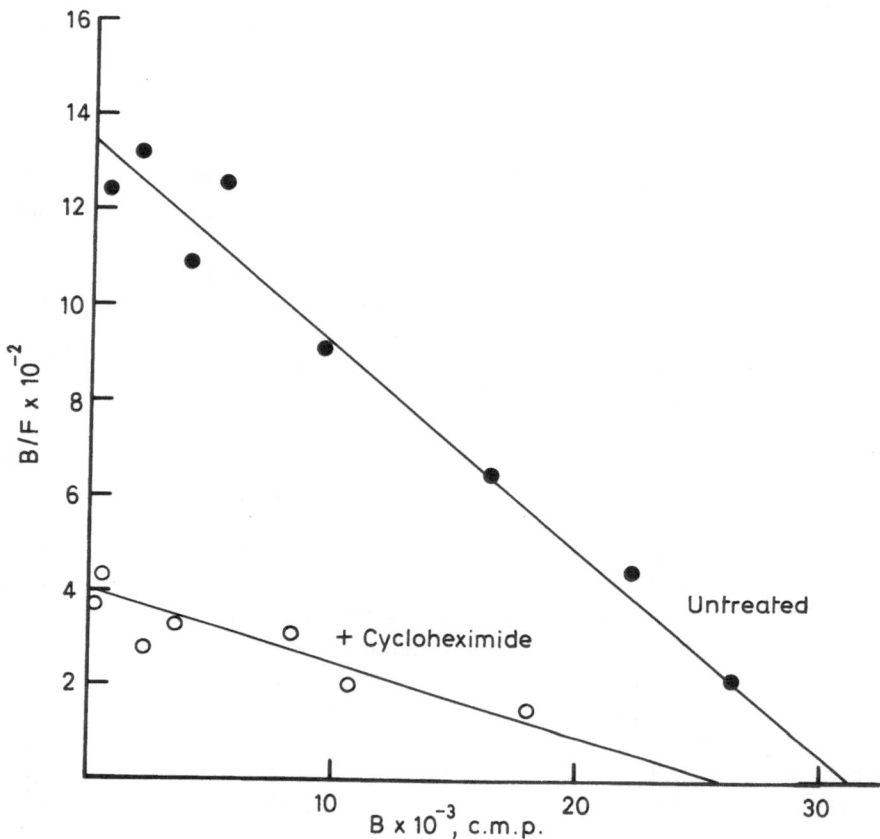

Fig. 3.9 Effect of cycloheximide on high-affinity interaction of ^{125}I-EGF with HF cells. HF cells were harvested and then incubated with (○) and without (●) 10 μg of cycloheximide/ml for 1 h at 37°C. Aliquots of cells were then vacuum-filtered and washed. The specific binding is plotted by the method of Scatchard. Cycloheximide inhibits the association of ^{125}I-EGF with HF cells at all concentrations tested when it was included throughout the incubation. Pretreatment of HF cells with cycloheximide for 4 h results in a greater degree of inhibition.

appearance at the plasma membrane of a class of cryptic, high-affinity EGF receptors. These results may not be compatible with the idea that there is an interconversion of EGF receptors from a low- to high-affinity state as has been described for nerve growth factor (NGF) and insulin receptors, since membrane preparations contain both classes of sites. However, lysis of intact cells may alter the membrane-lipid environment of the receptor in a way that artifactually converts a portion of the receptor population to a different conformational state.

3.9 EGF-RECEPTOR–TYROSINE PROTEIN KINASE COMPLEX

3.9.1 EGF enhances phosphorylation of the EGF receptor and other cellular proteins

When intact cells or membranes are treated with EGF, there is a 3–4-fold increase in the transfer of ^{32}P-labeled phosphate from $[\gamma\text{-}^{32}\text{P}]$ATP into membranes and cytoplasmic proteins (Carpenter *et al.*, 1979; Hunter and Cooper, 1981). Resolution of the phosphoproteins by sodium dodecyl sulphate (SDS)–polyacrylamide-gel electrophoresis showed that after a 1 min incubation with EGF at 0°C, the label is almost exclusively in a doublet of $M_r = 150\ 000$ and $170\ 000$ (King, L.E. *et al.*, 1980). After 5 minutes approximately ten other proteins are labeled. Cross-linking studies (Linsley *et al.*, 1979; Linsley and Fox, 1980a) and partial affinity purification of the EGF receptor (Cohen *et al.*, 1980) suggest that the 150 000–170 000-M_r doublet is the EGF receptor. In further support of this, these proteins can be cross-linked to EGF and immunoprecipitated with affinity purified antisera prepared against the mitogen (Hunter and Cooper, 1981; Linsley *et al.*, 1979; Cohen *et al.*, 1980). The 170 000-M_r species is thought to represent the holoreceptor, while the 150 000-M_r species is a candidate for a receptor fragment produced by an endogenous cell surface protease with trypsin-like specificity (Fox *et al.*, 1980; Wrann and Fox, 1979; Cohen *et al.*, 1982). It is not clear whether this represents part of the physiological pathway for receptor degradation or an artifact caused by cell lysis. Both species of the EGF receptor bind ^{125}I-EGF and are phosphorylated at tyrosine residues (Ushiro and Cohen, 1980), but the Triton X-100-solubilized 170 000-M_r species has a twofold higher affinity for ^{125}I-EGF (13 nM compared to 27 nM) and has a 5–10-fold greater capacity to be autophosphorylated than the 150 000-M_r component (Cohen *et al.*, 1982). These results suggest that two molecular forms of the EGF receptor may exist and are reflected by differences in affinity towards ^{125}I-EGF, molecular size, protease sensitivity and phosphorylation.

3.9.2 Does enzymatic modification of EGF receptors result in an altered affinity?

Thus, EGF receptors are susceptible to two different enzymatic modifications that take place at the plasma membrane prior to their internalization. Upon binding EGF, receptors are phosphorylated by a tyrosine-specific protein kinase (Cooper and Hunter, 1981b; Cohen *et al.*, 1980; Carpenter *et al.*, 1978; Ushiro and Cohen, 1980). In addition, EGF receptors identified by affinity cross-linking and SDS–polyacrylamide-gel electrophoresis are routinely represented by a doublet of M_r 170 and 150 (Linsley *et al.*, 1979; Linsley and Fox, 1980a,b). Both species are phosphorylated, and are thought to represent native receptor and a receptor fragment produced by an endogenous cell surface protease with trypsin-like specificity (Linsley and Fox, 1980a,b; Cohen *et al.*, 1982). Either of these enzymatic alterations may induce a conformational change in the receptor that alters its affinity for EGF. The affinities of insulin (Donner and Corin, 1980), nerve growth factor (Landreth and Shooter, 1980), estrogen (Weichman and Notides, 1977, 1979), growth hormone (Donner, 1980) and acetylcholine (Grunhagen and Changeux, 1976; Moore and Raftery, 1979) receptors are increased by their respective hormones. It has been proposed that transfer of the hormone receptor to a higher affinity state results in either activation and subsequent translocation to the nucleus (nerve growth factor and estrogen) or to a desensitized state where the hormone–receptor complex is no longer in equilibrium with free hormone (insulin, growth hormone and acetylcholine).

3.9.3 Is tyrosine phosphorylation relevant to growth control?

Currently, the significance of these modifications of EGF receptors is obscure. Studies with temperature-sensitive mutants show that transformation of cells by feline and avian sarcoma viruses is mediated by a protein kinase that preferentially phosphorylates tyrosine residues (Cooper and Hunter, 1981a; Erikson *et al.*, 1981). After both viral transformation and EGF treatment, cellular phosphotyrosine content increases approximately 100-fold (Hunter and Cooper, 1981; Sefton *et al.*, 1980; Pawson *et al.*, 1980). The finding that two different activators of cell growth enhance phosphorylation of tyrosine residues has led to the speculation that tyrosine-specific protein kinases may represent cellular controls that regulate normal and pathological growth. To date, the relationship of tyrosine phosphorylation to cell growth is not clear. Cells require exposure to EGF for 6 to 8 hours before they are committed to synthesis of DNA (Aharonov *et al.*, 1978b; Carpenter and Cohen, 1976b; Shechter *et al.*, 1978a). If EGF is removed prior to this critical 6 hour period, there is no response. EGF-induced phosphorylation reaches a maximum after 5 to 10 minutes (Carpenter and Cohen, 1979; Carpenter *et al.*, 1978). Thus, there is poor temporal correlation between EGF-enhanced tyrosine phosphorylation and the growth response. In addition, an analog of EGF that is non-mitogenic (cyanogen bromide-cleaved EGF) is capable of enhancing protein phosphorylation (Schreiber *et al.*, 1982), and EGF is not mitogenic for A_{431} cells which display the greatest EGF-dependent tyrosine protein kinase activity (Gill and Lazar, 1981; Barnes, 1982). Thus, receptor phosphorylation may be a necessary but insufficient signal for mediation of the growth response. There exists the possibility that receptor phosphorylation or proteolysis represents a means for regulation of receptor affinity. A conformational change in the receptor may induce alterations in affinity and lead to an activated state. This mechanism has been proposed for estrogen receptors where dimerization of receptors results in an increased affinity towards the ligand and activation of the receptor to a state capable of translocation to the nucleus (Weichman and Notides, 1977, 1979).

3.10 TRANSLOCATION OF EGF TO THE NUCLEUS

We reported previously that when cells are exposed to ^{125}I-EGF, homogenized, and fractionated by sucrose density gradient centrifugation, three peaks of radioactivity are recovered (King *et al.*, 1980a). It is well-known that EGF is internalized into cells, and with time, a large portion accumulates in lysosomes (Carpenter and Cohen, 1976b). Prior to lysosomal accumulation, EGF has been identified in a non-lysosomal vesicle (McKanna *et al.*, 1979; Maxfield *et al.*, 1979; Davies *et al.*, 1980). Lastly, when degradation of ^{125}I-EGF is prevented by a variety of pharmacological agents, EGF accumulates into a dense membrane-bound organelle (King *et al.*, 1980a) which has been tentatively identified as the nucleus (Johnson *et al.*, 1980; Savion *et al.*, 1980). This finding may be an extremely important consideration, but the biological consequences of nuclear accumulation of EGF are obscure. Drugs (primary and tertiary alkylamines) that enhance nuclear accumulation of EGF inhibit the mitogenic response (King *et al.*, 1981). Further, in morphological studies of the localization of EGF in cells treated with lysosomal inhibitors, there is no evidence for nuclear association (McKanna *et al.*, 1979; Cohen *et al.*, 1982). Although nuclear accumulation of other peptide hormones (insulin and nerve growth factor) has been reported (Goldfine, 1977; Yankner and Shooter, 1979), characterization of this putative translocation process and the compartment into which the hormone accumulates is fragmentary.

Identification of nuclear-associated EGF and NGF has relied largely on its resistance to detergent extraction (Johnson *et al.*, 1980; Savion *et al.*, 1980; Yanker and Shooter, 1979). Unlike NGF, no binding of ^{125}I-EGF to nuclei purified in the presence of detergents is observed (Savion *et al.*, 1980; King, unpublished results). Thus, it is possible that detergents interact with hydrophobic sites on the receptor to produce non-specific loss of binding activity or simply that the receptor is not involved in the translocation process. An alternative explanation for the transfer of ^{125}I-EGF to a detergent-resistant state is that it becomes associated with the cytoskeleton of cells. Further studies are required to clarify these issues.

REFERENCES

Aharonov, A., Passovoy, D.S. and Herschman, H.R. (1978b), *J. Supramol. Struct.*, **9**, 41–45.

Aharonov, A., Pruss, R.M. and Herschman, H.R. (1978a), *J. Biol. Chem.*, **253**, 3970–3977.

Anderson, R.G.W., Brown, M.S. and Goldstein, J.L. (1977), *Cell*, **10**, 351–364.

Anderson, R.G.W., Brown, M.S. and Goldstein, J.L. (1981), *J. Cell Biol.*, **88**, 441–452.

Barnes, D.W. (1982), *J. Cell Biol.*, **93**, 1–4.

Barnes, D. and Colowick, S.P. (1976), *J. Cell. Physiol.*, **89**, 633–640.

Blomberg, J., Reynolds, F.H., Van de Ven, W.J.M. and Stephenson, J.R. (1980), *Nature (London)*, **286**, 504–507.

Bradshaw, R.A. (1978), *Annu. Rev. Biochem.*, **47**, 191–216.

Brown, K.D., Dicker, P. and Rozengurt, E. (1979), *Biochem. Biophys. Res. Commun.*, **86**, 1037–1043.

Brown, K.D., Friedkin, M. and Rozengurt, E. (1980), *Proc. Natl. Acad. Sci. U.S.A.*, **77**, 480–484.

Brunk, U., Schellens, J. and Westermark, B. (1976), *Exp. Cell Res.*, **103**, 295–302.

Carpenter, G. and Cohen, S. (1976a), *J. Cell. Physiol.*, **88**, 227–237.

Carpenter, G. and Cohen, S. (1976b), *J. Cell Biol.*, **71**, 159–171.

Carpenter, G. and Cohen, S. (1979), *Annu. Rev. Biochem.*, **48**, 193–216.

Carpenter, G., King, L. and Cohen S. (1978), *Nature (London)*, **276**, 409–410.

Carpenter, G., King, L. and Cohen, S. (1979), *J. Biol. Chem.*, **254**, 4884–4891.

Carpentier, J-L., Gorden, P., Freychet, P. and Le Cam, A. (1979), *J. Clin. Invest.*, **63**, 1249–1261.

Chen, L.B. and Buchanan, J.M. (1975), *Proc. Natl. Acad. Sci. U.S.A.*, **72**, 131–135.

Chinkers, M., McKanna, J.A. and Cohen, S. (1979), *J. Cell. Biol.*, **83**, 260–265.

Chinkers, M., McKanna, J.A. and Cohen, S. (1981), *J. Cell. Biol.*, **83**, 260–265.

Cohen, S. and Savage, C.R. (1974), *Recent Prog. Hormone Res.*, **30**, 551.

Cohen, S. and Stastry, M. (1968), *Biochim. Biophys. Acta*, **166**, 427–437.

Cohen, S., Carpenter, G. and King, L. (1980), *J. Biol. Chem.*, **255**, 4834–4842.

Cohen, S., Ushiro, H., Stopscheck, C. and Chinkers, M. (1982), *J. Biol. Chem.*, **257**, 1523–1531.

Cooper, J.A. and Hunter, T. (1981a), *J. Cell Biol.*, **91**, 878–883.

Cooper, J. and Hunter, T. (1981b), *Mol. Cell Biol.*, **1**, 394–407.

Das, M. and Fox, C.F. (1978), *Proc. Natl. Acad. Sci. U.S.A.*, **75**, 2644–2648.

Davies, P.J.A., Davies, D.R., Levitzki, A., Maxfield, F.R., Milhaud, P., Willingham, M.C. and Pastan, I.H. (1980), *Nature (London)*, **283**, 162–167.

Diamond, I., Legg, A., Schneider, J.A. and Rozengurt, E. (1978), *J. Biol. Chem.*, **253**, 866–871.

Dicker, P. and Rozengurt, E. (1978), *Nature (London)*, **276**, 723–726.

DiPasquale, A., White, D. and McGuire, J. (1978), *Exp. Cell Res.*, **116**, 317–323.

Donner, D.B. (1980), *Biochemistry*, **19**, 3300–3306.

Donner, D.B. and Corin, R.E. (1980), *J. Biol. Chem.*, **255**, 9005–9008.

Erikson, E., Shealy, D.J. and Erikson, R.L. (1981), *J. Biol. Chem.*, **256**, 11381–11384.

Fernandez-Pol, J.A. (1981), *J. Biol. Chem.*, **256**, 9742–9749.

Fisher, P.B., Lee, L-S. and Weinstein, I.B. (1980), *Biochem. Biophys. Res. Commun.*, **93**, 1160–1169.

Fox, C.F. and Das, M. (1979), *J. Supramol. Struct.*, **10**, 199–214.

Fox, C.F., Wrann, M., Linsley, P. and Vale, R. (1980), *J. Supramol. Struct.*, **12**, 517, 535.

Gavin, J.R., Roth, J., Neville, D.M., DeMeyts, P. and Buell, D.N. (1974), *Proc. Natl. Acad. Sci. U.S.A.*, **71**, 84–88.

Gill, G.N. and Lazar, C.S. (1981), *Nature (London)*, **293**, 305–307.

Goldfine, I.D. (1977), *Diabetes*, **26**, 148–152.

Gorden, P., Carpentier, J-L., Cohen, S. and Orci, L. (1978), *Proc. Natl. Acad. Sci. U.S.A.*, **75**, 5025–5029.

Gospodarowicz, D. and Moran, J.S. (1976), *Annu. Rev. Biochem.*, **45**, 531–558.

Grunhagen, H.H. and Changeux, J-P. (1976), *J. Mol. Biol.*, **106**, 517–535.

Haigler, H.T., McKenna, J.A. and Cohen, S. (1979), *J. Cell. Biol.*, **81**, 382–395.

Heldin, C.H., Westermark, B. and Wasteson, A. (1979), *Nature (London)*, **282**, 419–420.

Hollenberg, M.D. and Cuatrecasas, P. (1973), *Proc. Natl. Acad. Sci. U.S.A.*, **70**, 2964–2968.

Hollenberg, M.D. and Cuatrecasas, P. (1975), *J. Biol. Chem.*, **250**, 3845–3853.

Hoober, J.K. and Cohen, S. (1967), *Biochim. Biophys. Acta*, **138**, 357–368.

Horowitz, A.D., Greenebaum, E. and Weinstein, I.B. (1981), *Proc. Natl. Acad. Sci. U.S.A.*, **78**, 2315–2319.

Hunter, T. and Cooper, J.A. (1981), *Cell*, **24**, 741–752.

Johnson, L.K., Vlodavsky, I., Baxter, J.D. and Gospodoarwicz, D. (1980), *Nature (London)*, **287**, 340–343.

Josefsberg, Z., Posner, B.I., Patel, B. and Bergeron, J.J.M. (1979), *J. Biol. Chem.*, **254**, 209–214.

King, A.C. and Cuatrecasas, P. (1981a), *N. Engl. J. Med.*, **305**, 77–88.

King, A.C. and Cuatrecasas, P. (1981b), *J. Supramol. Struct. Cell. Biochem.*, **17**, 377–387.

King, A.C. and Cuatrecasas, P. (1982a), *Biochem. Biophys. Res. Commun.*, **106**, 479–485.

King, A.C. and Cuatrecasas, P. (1982b), *J. Biol. Chem.*, **257**, 3053–3060.

King, L.E., Carpenter, G. and Cohen, S. (1980), *Biochemistry*, **19**, 1524–1528.

King, A.C., Davis, L.H. and Cuatrecasas, P. (1981), *Proc. Natl. Acad. Sci. U.S.A.*, **78**, 717–721.

King, A.C., Hernaez, L.J. and Cuatrecasas, P. (1980a), *Proc. Natl. Acad. Sci. U.S.A.*, **77**, 3283–3287.

King, A.C., Willis, R.A. and Cuatrecasas, P. (1980b), *Biochem. Biophys. Res. Commun.*, **97**, 840–845.

Kram, R., Mamont, P. and Tomkins, G.M. (1973), *Proc. Natl. Acad. Sci. U.S.A.*, **70**, 1432–1436.

Landreth, G.E. and Shooter, E.M. (1980), *Proc. Natl. Acad. Sci. U.S.A.*, **77**, 4751–4755.

Lee, L.S. (1981), *Proc. Natl. Acad. Sci. U.S.A.*, **78**, 1042–1046.

Levine, L. and Hassid, A. (1977), *Biochem. Biophys. Res. Commun.*, **79**, 477–483.

Linsley, P.S. and Fox, C.F. (1980a), *J. Supramol. Struct.*, **14**, 441–459.

Linsley, P.S. and Fox, C.F. (1980b), *J. Supramol. Struct.*, **14**, 461–471.

Linsley, P.S., Blifeld, C., Wrann, W. and Fox, C.F. (1979), *Nature (London)*, **278**, 745–748.

Magun, B.E., Matrisian, L.M. and Bowden, G.T. (1980), *J. Biol. Chem.*, **255**, 6373–6381.

Marchisio, P.C., Naldini, L. and Calissano, P. (1980), *Proc. Natl. Acad. Sci. U.S.A.*, **77**, 1656–1660.

Maxfield, F.R., Willingham, M.C., Davies, P.J.A. and Pastan, I. (1979), *Nature (London)*, **277**, 661–663.

McKanna, J.A., Haigler, H.T. and Cohen, S. (1979), *Proc. Natl. Acad. Sci. U.S.A.*, **76**, 5689–5693.

Moore, H-P. and Raftery, M.A. (1979), *Biochemistry*, **18**, 1907–1911.

Moriority, D., DiSorbo, S.M., Litwack, G. and Savage, C.R. (1981), *Proc. Natl. Acad. Sci. U.S.A.*, **78**, 2752–2756.

Mufson, R.A., DeFeo, D. and Weinstein, I.B. (1979), *Mol. Pharmacol.*, **16**, 569–578.

Ohkuma, S. and Poole, B. (1978), *Proc. Natl. Acad. Sci. U.S.A.*, **75**, 3327–3331.

Ohuchi, K. and Levine, L. (1978), *J. Biol. Chem.*, **253**, 4783–4790.

Otten, J., Johnson, G.S. and Pastan, I. (1971), *Biochem. Biophys. Res. Commun.*, **44**, 1192–1198.

Otten, J., Johnson, G.S. and Pastan, I. (1972), *J. Biol. Chem.*, **247**, 7082–7087.

Otto, A.M., Zumbe, A., Gibson, L., Kubler, A.M. and De Asua, L.J. (1979), *Proc. Natl. Acad. Sci. U.S.A.*, **76**, 6435–6438.

Pawson, T.J., Guyden, T.H., King, K., Radke, K., Gilmore, T. and Martin, G.S. (1980), *Cell*, **22**, 767–775.

Poole, B. and Ohkuma, S. (1981), *J. Cell Biol.*, **90**, 665–669.

Rose, S.P., Pruss, R.M. and Herschman, H.R. (1975), *J. Cell. Physiol.*, **86**, 593–598.

Rozengurt, E. and Heppel, L.A. (1975), *Proc. Natl. Acad. Sci. U.S.A.*, **72**, 4492–4495.

Salomon, D.S. (1981), *J. Biol. Chem.*, **256**, 7958–7966.

Savion, N., Vlodavsky, I. and Gospodarowicz, D. (1980), *Proc. Natl. Acad. Sci. U.S.A.*, **77**, 1466–1470.

Savion, N., Vlodavsky, I. and Gospodarowicz, D. (1981), *Proc. Natl. Acad. Sci. U.S.A.*, **71**, 1466–1470.

Sawyer, S.T. and Cohen, S. (1981), *Biochemistry*, **20**, 6280–6286.

Schlessinger, J., Shechter, Y., Willingham, M.C. and Pastan, I. (1978), *Proc. Natl. Acad. Sci. U.S.A.*, **75**, 2659–2663.

Schreiber, A.B., Yarden, Y. and Schlessinger, J. (1982), *Biochem. Biophys. Res. Commun.*, **101**, 517–523.

Sefton, B.M., Hunter, T., Beemon, K. and Eckhart, W. (1980), *Cell*, **20**, 807–816.

Seifert, W.E. and Rudland, P.S. (1974), *Nature (London)*, **248**, 138–140.

Shechter, Y., Hernaez, L. and Cuatrecasas, P. (1978a), *Proc. Natl. Acad. Sci. U.S.A.*, **75**, 5788–5791.

Shechter, Y., Schlessinger, J., Jacobs, S., Chang, K.-J. and Cuatrecasas, P. (1978b), *Proc. Natl. Acad. Sci. U.S.A.*, **75**, 2135–2139.

Shoyab, M. and Todaro, G.J. (1980), *J. Biol. Chem.*, **255**, 8735–8739.

Shoyab, M., DeLarco, J.E. and Todaro, G.J. (1979), *Nature (London)*, **279**, 387–391.

Smith, G.L. and Temin, H.M. (1974), *J. Cell. Physiol.*, **84**, 181–192.

Solanki, V. and Slaga, T.J. (1981), *Proc. Natl. Acad. Sci. U.S.A.*, **78**, 2549–2553.

Suss, R., Kreibich, G. and Kinzel, V. (1972), *Eur. J. Cancer*, **8**, 299–304.

Todaro, G.J., DeLarco, J.E. and Cohen, S. (1976), *Nature (London)*, **264**, 26–30.

Umezawa, K., Weinstein, I.B., Horowitz, A., Fujiki, H., Matsushima, T. and Sugimura, T. (1981), *Nature (London)*, **290**, 411–413.

Ushiro, H. and Cohen, S. (1980), *J. Biol. Chem.*, **255**, 8363–8365.

Vlodavsky, I., Brown K.D. and Gospodarowicz, D. (1978), *J. Biol. Chem.*, **253**, 3744–3750.

Weichman, B.B. and Notides, A.C. (1977), *J. Biol. Chem.*, **252**, 8856–8862.

Weichman, B.M. and Notides, A.C. (1979), *Biochemistry*, **18**, 220–225.

Wenner, E., Hackney, J., Kimelberg, H.K. and Mayhew, E. (1974), *Cancer Res.*, **34**, 1731–1737.

Wiley, H.S. and Cunningham, D.D. (1981), *Cell*, **25**, 433–440.

Wiley, H.S. and Cunningham, D. (1982), *J. Biol. Chem.*, **257**, 4222–4229.

Wrann, M.M. and Fox, C.F. (1979), *J. Biol. Chem.*, **254**, 8083–8086.

Yamasaki, H., Mufson, R.A. and Weinstein, I.B. (1979), *Biochem. Biophys. Res. Commun.*, **89**, 1018–1025.

Yankner, B.A. and Shooter, E.M. (1979), *Proc. Natl. Acad. Sci. U.S.A.*, **76**, 1269–1273.

Yarden, Y., Gabbay, M. and Schlessinger, J. (1981), *Biochim. Biophys. Acta*, **674**, 188–203.

4 The Binding and Internalization of Nerve Growth Factor

RONALD D. VALE, CHARLES E. CHANDLER, ARNE SUTTER and ERIC M. SHOOTER

Acknowledgements

This project was supported by grants from the NIH (NS 04270) and the American Cancer Society (BC 325A).

Receptor-Mediated Endocytosis
(*Receptors and Recognition*, Series B, Volume 15)
Edited by P. Cuatrecasas and T. F. Roth
Published in 1983 by Chapman and Hall, 11 New Fetter Lane, London EC4P 4EE
© 1983 Chapman and Hall

4.1 INTRODUCTION

The discovery of nerve growth factor (NGF) marked the beginning of a new era of developmental neurobiology. Since its discovery in the early 1950s, it has become apparent that the establishment of connections between neurons and target cells as well as the selective neuronal death which takes place during the development of the nervous system may be mediated, in part, by proteins secreted by the target tissues themselves (Varon and Bunge, 1978; Gottlieb and Glaser, 1980). Nerve growth factor, which serves an important role in the development of the sympathetic and sensory nervous systems, is very likely only one member of a class of proteins whose function it is to direct the differentiation and growth of the wide spectrum of neurons found both in the central and peripheral nervous systems. The discovery of an ample source of NGF in the mouse submaxillary gland (Cohen, 1960) along with its subsequent purification (Varon *et al.*, 1967a,b) have allowed the pursuit of the complex questions of the development of the nervous system at the molecular level.

NGF was first noted for its ability to stimulate the growth of sympathetic and sensory neurons (Levi-Montalcini and Angeletti, 1968). Increasing the level of circulating NGF in neonatal animals produced an increase in ganglion size along with an abundant outgrowth of sympathetic fibers (Levi-Montalcini, 1966). On the other hand, injection of an antibody to NGF resulted in a virtual destruction of the sympathetic nervous system (Levi-Montalcini and Booker, 1960; Levi-Montalcini and Angeletti, 1966) and a partial inhibition of the sensory nervous system (Gorin and Johnson, 1970) in the developing animal. NGF may also be required for the maintenance of the differentiated state of sympathetic neurons in mature animals (Thoenen and Barde, 1980; Hendry, 1976). Furthermore, sympathetic and sensory neurons demonstrate an absolute requirement for NGF for their growth and survival in tissue culture (Levi-Montalcini and Angeletti, 1963). In the presence of this hormone, these cells will extend extensive neuritic processes (Fig. 4.1). This factor may also play a critical role in the regeneration of sympathetic neurons, since such neurons will recover from trans-section only if NGF is available at the site of injury (Hendry, 1975b). There is also considerable evidence which implicates NGF not only as a differentiation signal but also as a chemotactic factor which directs the growth of axons to their appropriate target (Campenot, 1977; Menesini-Chen *et al.*, 1978; Gunderson and Barratt, 1979). The recent development of a clonal cell line, PC12, from a rat pheochromocytoma, a tumor derived from the adrenal medulla, has been important for understanding the biology and biochemical effects of NGF (Greene and Tischler, 1976; Tischler and Greene, 1975). This cell line does not require NGF for its growth and survival; however, in the presence of this factor, the

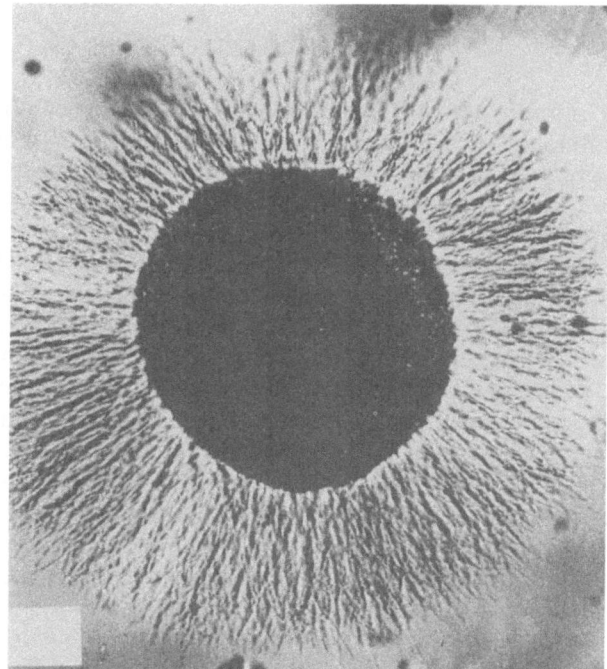

Fig. 4.1 Induction of neurites in a chick embryo sensory ganglion explant by NGF. NGF (30 ng/ml) was added for 24 h to an 8-day-old chick embryo sensory ganglion explant as described by Varon *et al.* (1972).

cells will extend neurites and become electrically excitable, and hence behave very much like sympathetic neurons. Such findings reinforce the importance of NGF as an important developmental signal capable of directing primitive neural crest-derived cells into becoming terminally differentiated sympathetic or sensory neurons.

Although the mechanism whereby NGF can evoke such important developmental decisions remains to be elucidated, a good deal is known of the protein itself. NGF is a small dimeric, basic protein (Serves and Shooter, 1977) whose amino acid sequence shows some homology to insulin (Frazier *et al.*, 1972). It is found in abundant amounts in one sympathetic target tissue, the mouse submaxillary gland, in a high molecular weight complex (7 S NGF) consisting of the NGF protein itself (the active β subunit), two γ subunits (a serine protease which processes a pro-NGF precursor (Berger and Shooter, 1977) analogous to the processing of pro-insulin) and two α subunits (a protein whose function in the complex is unknown). This complex protects the active β subunit from proteolytic degradation (Moore *et al.*, 1974; Mobley *et al.*, 1976), but since it is itself devoid of biological activity (Harris-Warrick *et al.*, 1980), the complex must dissociate at some point in order to release the active NGF protein.

The interactions of NGF with target cells are complex. It is clear that the binding of NGF to cell surface receptors is an obligatory first step towards the generation of subsequent biological responses. These receptors, along with the bound NGF, can also become internalized and be transported great distances along the axon from the neurite tip to the cell body. It is the purpose of this review to examine the interaction of NGF with its receptor and the possible fate(s) of the NGF–receptor complex subsequent to binding. Through the elucidation of these events, it should be possible to eventually understand the means by which NGF influences the differentiated state of responsive neurons.

4.2 NGF BINDS TO TWO CLASSES OF CELL SURFACE RECEPTORS

In the early 1970s, several laboratories developed an ^{125}I-labeled nerve growth factor derivative of high specific activity which retained full biological activity (Bannerjee *et al.*, 1973; Herrup and Shooter, 1973; Frazier *et al.*, 1974). This derivative allowed the characterization of high-affinity, specific cell surface receptors on sensory and sympathetic neurons. Bannerjee *et al.* (1973) demonstrated saturable binding of ^{125}I-NGF to rabbit cervical sympathetic ganglion membranes with an equilibrium dissociation constant (K_d) of 0.2 nM. A variety of other polypeptide hormones did not compete with the binding of NGF, and binding was not observed in tissues which do not respond to NGF. At the same time, Herrup and Shooter (1973) reported binding of ^{125}I-EGF to intact chick dorsal root ganglion cells with a similar K_d of 0.26 nM. NGF binding was developmentally regulated in these cells with maximal binding occurring at the time of maximal sensitivity to NGF (Herrup and Shooter, 1975).

In contrast to these studies which observed a homogeneous class of NGF receptors, Frazier *et al.* (1974) reported curvilinear Scatchard plots indicative of apparent affinities between 10^{-10} M and 10^{-6} M for NGF binding to membranes derived from chick embryonic sensory and sympathetic neurons. Furthermore, native NGF was observed to enhance the dissociation of labeled NGF bound to cell membranes by 30-fold compared to dilution condition alone. Biphasic Scatchard plots and enhanced dissociation kinetics with unlabeled ligand were previously observed by De Meyts and co-workers (De Meyts *et al.*, 1973, 1976) for insulin binding to lymphocytes. These inves- · tigators proposed that such results could be explained using a model of negative co-operativity in which increasing receptor occupancy results in decreased receptor affinity. Because of their similar results, Frazier *et al.* proposed that NGF receptors also interacted in a negatively co-operative fashion.

Several years later, Sutter *et al.* (1979a) re-examined the binding characteristics of NGF to intact chick embryo dorsal root ganglion cells. Two technical improvements were made in this study. An improved ^{125}I-NGF preparation with low non-specific binding allowed a better characterization of high-affinity sites, while a rapid centrifugation step to separate bound from free ligand provided a means to measure ^{125}I-NGF bound to receptors with rapid dissociation kinetics. The lengthy washing procedures used in the previous studies made it difficult to detect such sites. Equilibrium binding of ^{125}I-NGF to chick embryo sensory cells clearly revealed the presence of two saturable binding components as shown in the Scatchard plot in Figure 4.2. The two NGF binding sites differed by 100-fold in their affinity constants. The high-affinity, low-capacity site (designated as Site I) had a $K_d = 2.3 \times 10^{-11}$ M with 3000 sites per cell, while the low-affinity, high-capacity site (Site II) displayed a $K_d = 1.7 \times 10^{-9}$ M with 45 000 sites per cell. The two classes of binding

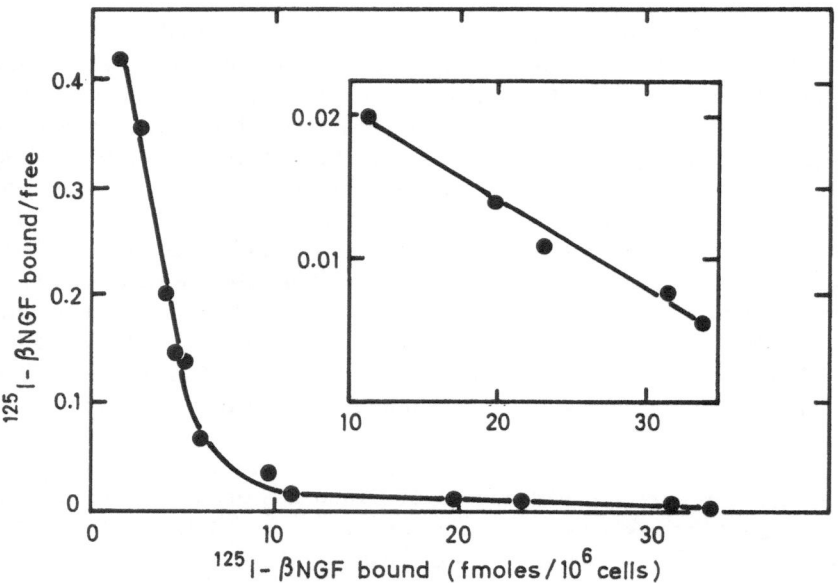

Fig. 4.2 Scatchard analysis of equilibrium binding of ^{125}I-NGF to 8-day-old chick embryo sensory ganglion cells. Cells (0.6×10^6/ml) were incubated at 37°C for 45 min with various concentrations of ^{125}-NGF (3 pM to 3.7 nM). Triplicate determinations of binding were made at each point, non-specific binding was subtracted, and the data obtained were transformed into a Scatchard plot. Binding data for the low-affinity region are expanded in the insert. Reproduced with permission from Sutter *et al.* (1979a).

sites were observed both at 37°C and 2°C. The affinity constants for the two receptor sites were unaltered by temperature; however, the number of Site I binding sites (but not Site II) was reduced three-fold at the lower temperature. The cell surface localization of the temperature-sensitive high affinity, low capacity binding site was shown in an NGF-dependent cytotoxicity assay (Zimmermann *et al.*, 1978). Under conditions where NGF was bound only to site I, sensory neurons were lysed upon addition of anti-NGF antiserum and complement.

The two classes of NGF receptors differed principally in their dissociation constants. Measurement of the association rates of Site I and II receptors showed rate constants (K_{+1}) of approximately 5×10^7 M^{-1}s^{-1} for both receptor types which suggests that the association rate is largely a diffusion-limited

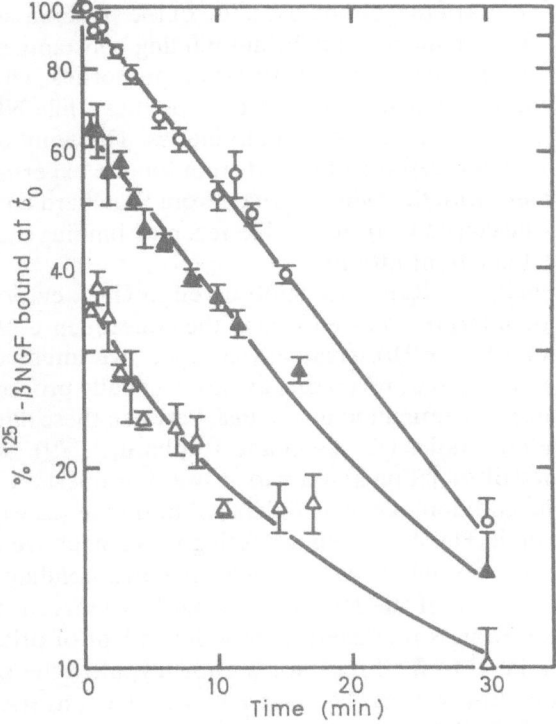

Fig. 4.3 Dissociation of ^{125}I-NGF after pre-equilibration with different ^{125}I-NGF concentrations at 37°C. Cells (3×10^6/ml) were preincubated for 30 min with 0.6×10^{-11} M (O), 1.6×10^{-10} M (▲) or 1.36×10^{-9} M (△) ^{125}I-NGF. The dissociation of ^{125}I-NGF was induced by the addition of 3.8×10^{-7} M unlabeled NGF. The specific binding at t_0 was measured in sextuplets and after different times of dissociation in triplicates of 100 µl each. The data are corrected for non-specific binding. Reproduced with permission from Sutter *et al.* (1979a).

process. On the other hand, the dissociation rates for ^{125}I-NGF bound to Site I and Site II receptors were quite distinct, as demonstrated by the experiment shown in Fig. 4.3. If cells were allowed to reach equilibrium with ^{125}I-NGF at concentrations below 10^{-11} M, such that greater than 90% of the binding was to Site I receptors, and then dissociation initiated by the addition of excess unlabeled NGF, monoexponential kinetics ($t_{\frac{1}{2}} = 10$ min) for the release of ^{125}I-NGF were observed. However, when increasing concentrations of ^{125}I-NGF were used to increase the occupancy of the low-affinity Site II receptors, biphasic dissociation curves were obtained. The first dissociation component displayed very rapid kinetics and corresponds to NGF being displaced from Site II receptors while the second component once again represented NGF released from Site I. Site II receptors were estimated to have a half-time of dissociation of about 3 s. Thus, the dissociation rates for the two receptors differed by two orders of magnitude. Because of the similar association rates, the 100-fold difference in the equilibrium binding constants of Site I and II receptors is a consequence of their dissociation properties. The high-affinity binding sites demonstrated slow dissociation kinetics while NGF bound to low-affinity sites was released with rapid kinetics. The equilibrium binding constants for the two receptors obtained from kinetic experiments were in excellent agreement with the values derived from Scatchard analysis, suggesting that the kinetic constants of reversible receptor binding are rate limiting for the measured apparent affinities.

The heterogeneity of NGF binding observed in chick embryonic sensory neurons could be interpreted as either (a) the consequence of multiple classes of binding sites, or (b) negatively co-operative interactions of a homogeneous population of receptors, as was originally proposed by Frazier *et al.* (1974). Since it is difficult to distinguish between these alternatives from equilibrium binding studies (De Lean and Rodbard, 1979), Sutter *et al.* (1979a) examined dissociation kinetics to answer this question. If dissociation is initiated by the addition of excess unlabeled hormone (as was true for the experiment shown in Fig. 4.3), then according to the negative co-operativity theory, a single exponential release of ^{125}I-NGF, corresponding to the kinetics of the low-affinity form of the receptor, should be observed. Under these conditions, dissociation is predicted to be independent of prior receptor occupancy but related to final receptor occupancy, after the addition of unlabeled ligand. However, as shown in Fig. 4.3, the converse situation is observed for NGF. With concentrations of ^{125}I-NGF greater than 10^{-11} M, two dissociation components were present. The pattern of dissociation was dependent upon the concentrations of ^{125}I-NGF used and hence upon the occupancy of Site I and II receptors prior to dissociation. In fact, given minimal incubation times to reach maximum binding the relative proportion of slow and rapid dissociation components could be predicted from the relative occupancy of Site I and II receptors as calculated using the data from the Scatchard analysis. A similar dependence of dissociation kinetics on the fractional receptor occupancy during the association phase has been observed for the interaction of ^{125}I-insulin with its receptor on adipocytes (Olefsky and Chang, 1979).

The enhancement of dissociation rates of cell-bound ^{125}I-NGF by the addition of unlabeled NGF was originally explained in terms of negative co-operativity as a result of increasing receptor occupancy during the dissociation phase (Frazier *et al.*, 1974). Sutter *et al.* (1979a) also observed this phenomenon for Site I receptors: however, the rate of dissociation from Site I receptors in the presence of excess unlabeled NGF did not approach the dissociation rate of Site II binding sites as the negative co-operativity hypothesis would have predicted and the dissociation rate was accelerated even at minimal concentrations of unlabeled NGF which allowed receptor occupancy to decrease rather than increase. Pollet and his co-workers (Pollet *et al.*, 1977) have performed similar experiments which indicate that negative co-operativity cannot adequately explain the dissociation of insulin from its receptor on lymphocytes.

These data support a two-site model for NGF binding to chick dorsal root ganglion cells. Binding studies of NGF to membranes from these same cells have given similar results (Riopelle *et al.*, 1980). Although the characteristics of the enhanced dissociation observed in the presence of unlabeled hormone are incompatible with a model of negative co-operativity, such an observation likewise is not reconcilable with a model of two independent receptor sites. However, it has been proposed that membrane components in the vicinity of the receptor could create a diffusion barrier which would retain ^{125}I-NGF at the surface long enough to allow some rebinding to occur. Addition of unlabeled NGF would dilute the specific activity of labeled hormone in the diffusion barrier and thereby prevent its rebinding. This rebinding phenomenon has been considered in detail by Silhavy *et al.* (1975). Some evidence for the existence of a diffusion barrier in sensory neuronal membranes, which could be partially disrupted by stirring membranes during dissociation experiments, has been obtained (Riopelle *et al.*, 1980). Nevertheless, concrete evidence for the existence of a diffusion barrier does not exist. Certain models of hormone–receptor interactions, besides negative co-operativity, are able to account for the enhancement of dissociation kinetics with unlabeled hormone without evoking diffusion barriers as an explanation.

The dissociation kinetics of ^{125}I-NGF bound to sensory neurons was recently re-examined by Tait *et al.* (1981). These investigators reported that dissociation kinetics at low ligand concentrations, where only Site I binding is expected, was monoexponential only for the first 60 min, after which time the dissociation no longer followed first-order kinetics. The enhancement of dissociation by unlabeled ligand, originally observed by Frazier *et al.* (1974), was also only observed in the first 30 to 40 min after the addition of unlabeled ligand. Neither of these results are compatible with either a two-site or a negatively co-operative receptor model. However, the long preincubations with ^{125}I-NGF (2 h at 27°C) used in the study allow the participation of the NGF–receptor complex in a number of events (such as internalization) subsequent to binding. These events make interpretations of the initial receptor binding interactions more difficult. This subject will be discussed further in a later section.

The presence of two classes of NGF receptors with different affinities may have important consequences for understanding the mode of action of this hormone. There is some evidence that distinct biological responses may be mediated through these two receptor classes. The dose–response curve of NGF-induced neurite outgrowth in sensory neurons is clearly correlated with occupancy of high-affinity Site I receptors (Sutter *et al.*, 1979b). Only neurons in the chick dorsal root ganglion possess Site I receptors, the non-neuronal cells having only Site II (Sutter *et al.*, 1979b). Moreover, the Site I receptors appear developmentally on the neuron at the embryonic age, day 6, when the cells become responsive to NGF (Zimmerman *et al.*, 1978). Similarly, neurite outgrowth in PC12 cells is initiated at low concentrations of NGF (half-maximal response at 10^{-11} M) (Tischler and Greene, 1975), while the stimulation of amino acid uptake by NGF in this same cell line requires much higher (half-maximal response at 4×10^{-10} M) concentrations (McGuire and Greene, 1979) and may be mediated by Site II receptors.

4.3 NGF-INDUCED RECEPTOR CONVERSION IN PC12 CELLS

As with chick embryo sympathetic and sensory neurons, two classes of NGF receptors have been documented on the rat pheochromocytoma cell line PC12. Studies performed first by Landreth and Shooter (1980) and then by Schechter and Bothwell (1981) revealed two receptor populations with rapid and slow dissociation kinetics, very similar to those reported for chick embryo sensory ganglion cells by Sutter *et al.* (1979a). At 37°C, in the presence of excess unlabeled NGF, ^{125}I-NGF dissociates from rapidly and slowly dissociating receptor subtypes with half-times of 30 s and 30 min respectively. However, at 4°C, ^{125}I-NGF is released completely from rapidly dissociating sites within 30 min while it remains persistently bound to slowly dissociating sites in the same time period. Thus, by incubating cells which have reached equilibrium with ^{125}I-NGF with an excess of unlabeled hormone for 30 min at 4°C, one can selectively dissociate NGF bound to rapidly dissociating receptors. By comparing total binding with such dissociation data, the amount of NGF associated with each receptor type can be determined.

The appearance of ^{125}I-NGF bound to rapidly and slowly dissociating receptors showed different time courses as opposed to the results obtained with sensory neurons. Binding to rapidly dissociating receptors reached a maximum level within 2 min, while binding to slowly dissociating receptors appeared after an apparent 30 s lag phase and then increased steadily, reaching a steady state by 30 min (Landreth and Shooter, 1980). Such results suggest that rapidly dissociating receptors occupied with ^{125}I-NGF can be converted to a high-affinity, slowly dissociating form.

An experiment to further probe this question is shown in Fig. 4.4. In this experiment, the fate of cell-bound ^{125}I-NGF was assessed after removing free labeled NGF from the medium. The majority of the NGF did not dissociate in

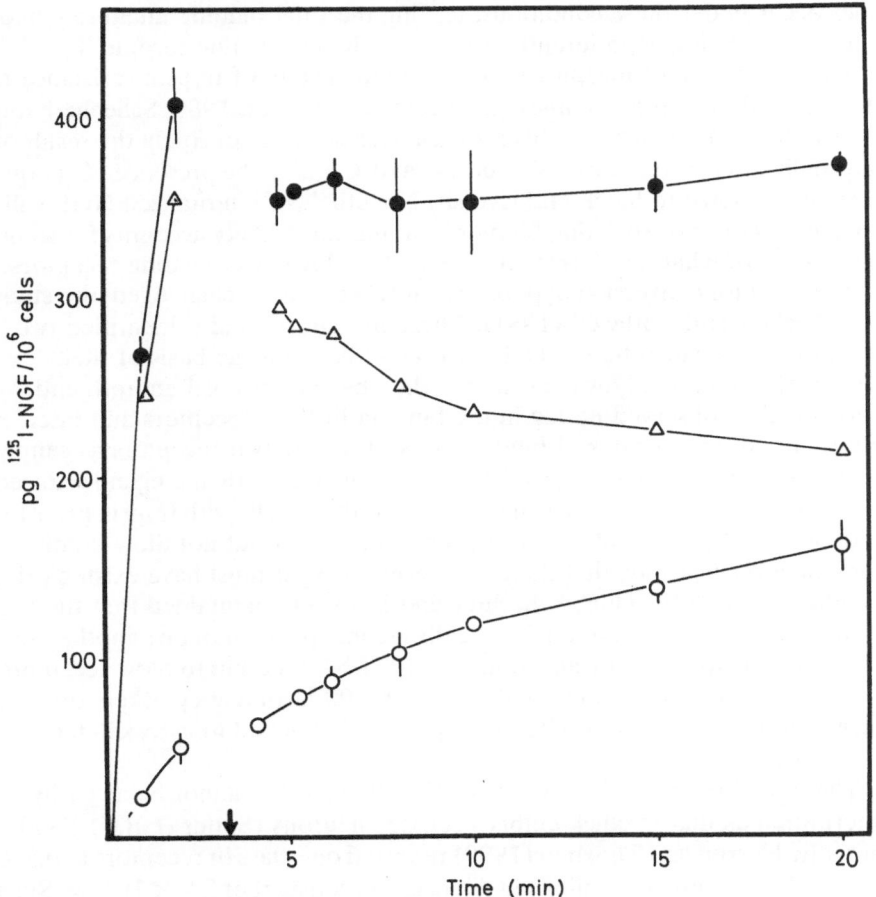

Fig. 4.4 Effect of removal of ^{125}I-NGF from the medium on NGF binding to PC12 cells: a demonstration of receptor conversion. PC12 cells were incubated with 35 pM-^{125}I-NGF for 2 min at 37°C. The cells were centrifuged for 1 min at 1000 g, the ^{125}I-NGF-containing medium was aspirated, and the cells were resuspended in the same volume of fresh medium (arrow). Total binding (●), slowly dissociating binding (○) and rapidly dissociating binding (△) were determined. Each point represents ∓SD of triplicate determinations. Reproduced with permission from Landreth and Shooter (1980).

this period of time, as the rate of dissociation in the absence of excess unlabeled ligand is relatively slow. After free ^{125}I-NGF was removed, a decrease in rapidly dissociating binding with a concomitant increase in slowly dissociating binding was observed. As total binding remained relatively unaltered, these results argue that slowly dissociating receptors may be generated as the result of NGF initially binding to rapidly dissociating receptors. An alternative explanation is that some dissociation and rebinding of ^{125}I-NGF does occur under these conditions, leaving the total binding unaltered, and that the rebinding is preferential to the slowly dissociating receptors.

The conversion of receptor affinity confers a state of trypsin resistance to the ^{125}I-NGF–receptor complex (Landreth and Shooter, 1980; Schechter and Bothwell, 1981). However, this protease resistance is not solely the result of internalization, as the conversion occurs at 0°C and in the presence of energy inhibitors known to block endocytosis. Nonetheless, internalized NGF will appear as slowly dissociating binding, and this most likely accounts for some but not all of what one terms NGF bound to slowly dissociating receptors.

The receptor conversion hypothesis has recently been challenged by a study of Schechter and Bothwell (1981). These investigators also identified two receptor types which they called 'Fast' and 'Slow' on the basis of their dissociation kinetics. However, unlike the observations of Landreth and Shooter, they observed no lag in the binding to Slow receptors and interpreted the different rates of binding of NGF to the two receptors as simply reflecting different kinetic association constants rather than a ligand-induced receptor conversion. Also, when they treated PC12 cells with trypsin prior to adding ^{125}I-NGF, they observed degradation of Fast but not Slow binding components suggesting that these two receptor types must have existed prior to addition of NGF. Thus, Schechter and Bothwell maintained that the two NGF receptor populations on PC12 cells are independent of one another and do not interconvert. They also found that ^{125}I-NGF bound to Slow receptors was preferentially associated with Triton X-100-insoluble cytoskeletons and suggested that this type of NGF–receptor was attached to cytoskeletal elements.

The equilibrium binding of ^{125}I-NGF to PC12 cells cannot be as easily interpreted as that in chick embryo sensory neurons (Sutter *et al.*, 1979a). Initially, Herrup and Thoenen (1979) reported one class of receptors at 0.5°C on PC12 cells with an equilibrium dissociation constant of 2.9×10^{-9} M. Such receptors correspond to the low-affinity receptor class of the sensory neurons. Cohen *et al.* (1980) also noted, in competition assays, a single class of sites on PC12 cells with a similar affinity. However, equilibrium binding studies at 37°C (G. E. Landreth and E. M. Shooter, unpublished observations) showed a slight curvilinearity of the Scatchard plots, indicating the presence of a higher-affinity class of receptors. Furthermore, Scatchard analysis of the slowly dissociating binding component clearly indicated it was of higher affinity than rapidly dissociating binding, especially when conditions which block endocytosis were used. On the other hand, Scatchard analysis by Schechter and Bothwell (1981) showed that slowly and rapidly dissociating receptors had similar affinities, although 10% of the slowly dissociating binding may have been of somewhat higher affinity. They suggested that Slow and Fast receptors have similar equilibrium binding constants because of cancelling effects of their association and dissociating rates. However, their kinetic experiments apparently do not agree with their equilibrium binding studies. At 37°C, the ratio of the dissociation rate constants for Slow to Fast receptors is 40:1 while the ratio of the respective association rates is 1:4. These data predict that the slowly dissociating receptors should have a 10-fold higher affinity if true equilibrium binding is present. An involvement of the ^{125}I-NGF–receptor complex in events which produce non-equilibrium binding conditions must be considered in such a situation.

4.4 MOLECULAR MODELS OF RECEPTOR CONVERSION

Some aspects of hormone interaction with their receptors can be explained by sequential events which alter the affinity of the hormone–receptor complex. Under conditions of equilibrium binding, interactions of this nature can be predicted according to the mobile receptor hypothesis of Jacobs and Cuatrecasas (1976) and Boeynaems and Dumont (1975). This hypothesis proposes that hormone receptors interact reversibly with effector molecules in the plane of the membrane to alter receptor activity and affinity. Receptor affinity for the effector protein is assumed to be enhanced when occupied with its hormone. This situation could produce an effector–receptor complex with a greater affinity for the ligand than the receptor alone. The binding of hormone to two populations of receptors will be observed if either limiting effector concentrations or a low receptor–effector affinity is present. This model has been applied to insulin binding and can successfully account for the curvilinear Scatchard plots and enhanced dissociation kinetics with unlabeled hormone which have been observed in this system.

This model is attractive in terms of the receptor conversion hypothesis. It supposes that a ligand induces a conformational change in the receptor which enables the complex to recognize the effector protein. In the case of insulin, hormone binding, in fact, does produce a conformational change in its receptor which is detectable, as it makes the receptor more accessible to degradation by trypsin (Pilch and Czech, 1980). NGF binding may also produce a receptor conformational change, but it seems to have an opposite effect, making its receptors more resistant to trypsin digestion.

Biochemical and pharmacological evidence for the existence of effector proteins as proposed by the mobile receptor hypothesis have been obtained. The best example of this is the interaction of the β-adrenergic receptor with the guanine nucleotide regulatory protein (G site), a protein which plays an essential role in the activation of adenylate cyclase by adrenergic agonists (Spiegel and Downs, 1981). A high-affinity ternary complex involving the hormone-occupied receptor and the G site appears to be an obligatory intermediate for enzyme activation (De Lean et al., 1980). Interaction of the hormone–receptor complex with the G site converts the low-affinity complex to a high-affinity form which can be readily reversed by adding guanine nucleotides to dissociate the ternary complex (Siegel and Downs, 1981). Antagonists, on the other hand, only bind to one class of receptors, as they are unable to induce receptor coupling to the G site. Biochemical evidence supporting an agonist-induced interaction with the G site has also been obtained (Limbird and Lefkowitz, 1978). A glycoprotein which can alter the affinity of the insulin receptor also has been described (Maturo and Hollenberg, 1978). This protein, when added to a preparation of partially purified and solubilized insulin receptors, can convert them into a higher-affinity form.

If an endogenous membrane protein does play a role in modulating the affinity and dissociating properties of the NGF receptor, it may be possible to inhibit, enhance, or mimic its activity by adding an exogenous agent. In fact, Vale and Shooter (1982) discovered that the lectin wheatgerm agglutinin (WGA) substantially altered the ratio of NGF bound to slowly and rapidly dissociating receptors. As shown in Fig. 4.5, at 500 pM-^{125}I-NGF, about 25% of the hormone was bound to slowly dissociating receptors; however, in the presence of WGA, 90% of the binding became slowly dissociating. As total binding remained unchanged, it appeared as if WGA converted rapidly dissociating receptors occupied with NGF into slowly dissociating receptors. Out of nine lectins tested, only WGA produced this effect. The WGA-induced receptor conversion occurred very rapidly, reaching completion within 2 min at 37°C or 4°C. Internalization, therefore, cannot account for this sudden induction of slowly dissociating receptors. It is unclear whether normal and WGA-induced slowly dissociating receptors represent identical forms of the receptor; however, similar trypsin sensitivities and association and dissociation rates argue that they are at least similar.

As Schechter and Bothwell (1981) proposed that slowly dissociating receptors are associated with Triton X-100-insoluble cytoskeletons, Vale and Shooter (1982) investigated whether the cytoskeleton may play a role in the WGA-mediated receptor conversion. As shown in Fig. 4.5, only 10–15% of NGF binding is normally associated with Triton X-100-insoluble cytoskeletons. However, in the presence of WGA, 90% of the total binding becomes associated with the cytoskeleton. These data indicate that WGA can induce a physical interaction of the NGF receptor with the cytoskeleton or a cytoskeleton-associated protein.

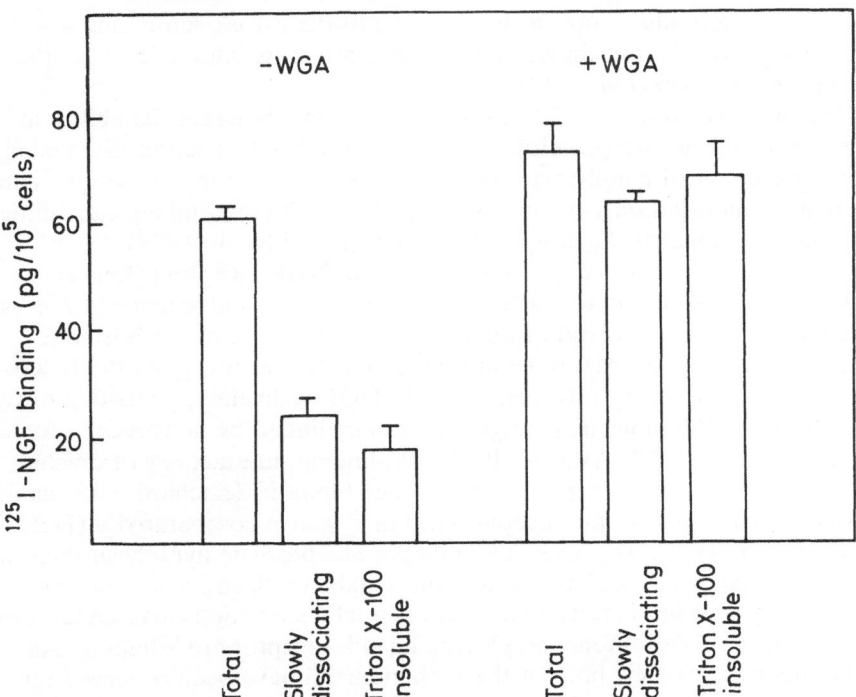

Fig. 4.5 Effects of wheatgerm agglutinin on ^{125}I-NGF binding to PC12 cells. Cells (10^6/ml) were incubated for 30 min at 37°C with 5×10^{-10} M-^{125}I-NGF followed by a 30 min incubation in the absence or presence of 50 μg of wheatgerm agglutinin (WGA)/ml. Total, slowly dissociating and Triton X-100-insoluble binding were determined at this time, and non-specific binding was subtracted from these values. Triton X-100-insoluble binding was determined by centrifuging 100 μl aliquots of ^{125}I-NGF and PC12 cells through 200 μl of 0.5% Triton in 0.3 M-sucrose/3 mM-MgCl$_2$/20 mM-Tris/HCl, pH 7.4. ^{125}I-NGF which pelleted to the bottom of the tube was considered to be Triton X-100-insoluble binding. Values represent the mean and standard deviations of triplicate determinations.

The conversion of a heterogeneous population of NGF receptors to a slowly dissociating receptor population could be the result of WGA either stabilizing a receptor–effector complex or mimicking the effector activity itself. Since receptor conversion and cytoskeletal association induced by WGA exhibited similar quantitative relationships and time courses (Vale and Shooter, 1982), the cytoskeletal association may be important for the conversion process. Therefore, it is possible that either a cytoskeleton-associated membrane protein or the cytoskeleton itself could play the role of an effector molecule for the NGF receptor on PC12 cells. The G site of the β-adrenergic receptor system also seems to be attached to the cytoskeleton, and agents which regulate the attachment process also affect its interaction with the receptor (Sahyoan *et al.*, 1981).

It is not known how WGA produces its effects; however, its ability to cross-link cell surface proteins may be important for its action. By analogy, the attachment of lymphocyte surface immunoglobulin to cytoskeletal components is induced only by anti-Ig antibodies with cross-linking capabilities and not by univalent F_{ab} fragments (Flanagan and Koch, 1978).

A number of hormone receptors, including NGF receptors (Levi *et al.*, 1980) have been found to cluster on the cell surface, and in some instances, clustering may be involved in the mechanism of action of the hormone (Schechter *et al.*, 1979a). Receptor clustering may also explain the heterogeneous binding of ligands such as NGF. NGF is divalent, consisting of two identical, 13 000-molecular-weight monomers linked by noncovalent forces (Frazier *et al.*, 1972). As a result of cross-linking, interactions of divalent ligands with divalent receptors can produce biphasic Scatchard plots, as do models involving multiple receptor sites or negative co-operativity (DeLisi and Chabay, 1979). However, the multiple-receptor-site hypothesis does not predict enhanced dissociation rates with unlabeled ligand, and negative co-operativity does not predict a multiexponential dissociation curve under these conditions as does a model employing ligand–receptor cross-linking. As mentioned previously, both of these phenomena have been observed for NGF binding.

The cross-linking hypothesis has been employed to model the binding of insulin to lymphocytes (DeLisi and Chabay, 1979). Although insulin is normally a monomer in solution, it may dimerize at the cell surface. A divalent receptor was utilized for modeling binding data according to the cross-linking hypothesis, and recent work has indicated that the insulin receptor contains two binding subunits (Pollet *et al.*, 1982; Jacobs *et al.*, 1980). Furthermore, when the cross-linking of occupied insulin receptors is enhanced by anti-insulin antibodies, a conversion of receptors from a low- to a high-affinity state is observed (Schechter *et al.*, 1979b). Monovalent F_{ab} fragments, without cross-linking ability, do not produce this effect.

As has been observed for insulin and EGF (Schlessinger *et al.*, 1978), a fluorescent-labeled derivative of NGF was found to cluster on cell surfaces (Levi *et al.*, 1980). Cross-linking of receptors should also be considered when interpreting NGF-binding data. The clustering of EGF receptors on A431 carinoma cells was observed after 30 s (Haigler *et al.*, 1979), which is similar to the time lag observed for the appearance of slowly dissociating receptors on PC12 cells (Landreth and Shooter, 1980). In accordance with the cross-linking hypothesis, WGA could enhance slowly dissociating binding (Fig. 4.5) by clustering NGF receptors, and thereby increasing the probability of NGF interacting with two receptors. Similar, NGF antibodies have also been found to increase slowly dissociating binding, perhaps by cross-linking NGF-receptor complexes (R. Vale and E.M. Shooter, unpublished observations). Therefore, while receptor clustering cannot explain heterogeneous receptor populations for a monovalent ligand, it must be considered for a divalent molecule like NGF.

4.5 BIOCHEMICAL IDENTIFICATION OF THE NGF RECEPTOR

Banerjee *et al.* (1976) demonstrated that NGF receptors solubilized from rabbit superior cervical ganglion with the non-ionic detergent Triton X-100 retain their specific binding activity. The K_d observed for this solubilized receptor preparation was 2×10^{-10} M, which is in close agreement with the K_d obtained using intact membranes (Banerjee *et al.*, 1973). Further characterization of the solubilized NGF receptor was performed by Costrini and co-workers (Costrini and Bradshaw, 1979; Costrini *et al.*, 1979). Their studies indicated that the receptor is a minimally hydrophobic, highly asymmetric membrane protein with an apparent molecular weight of 135 000 and a Stokes radius of 7.1 nm. These results are in good agreement with a recent report of the cross-linking of ^{125}I-NGF to its receptors in rabbit superior cervical ganglion membranes (Massague *et al.*, 1981b). Using a photoreactive compound as the cross-linking reagent, ^{125}I-NGF was found covalently attached to two molecules, forming complexes with molecular weights of 143 000 and 112 000. Proteolytic peptide maps of the 143 K and 112 K-mol.wt. proteins were similar, indicating a precursor–product relationship between these two species, conceivably involving proteolytic degradation. Transformations of the epidermal growth factor (Linsley and Fox, 1980) and insulin (Massague *et al.*, 1981a) receptors by a proteolytic mechanism have also been proposed.

The structure of the insulin receptor is known in some detail (Jacobs and Cuatrecasas, 1981). Physical studies of the insulin and NGF receptors indicate that the gross structures of the two receptors are quite different, although the hormones themselves are very homologous and probably arose from a common primordial gene (Frazier *et al.*, 1972). However, like the insulin receptors (Cuatrecasas and Tell, 1973), recent work has revealed that the NGF receptor is a glycoprotein. Costrini and Kogan (1981) observed that concanavalin A and wheatgerm agglutinin decreased NGF binding to sympathetic ganglion cells. These two lectins also impaired NGF binding to a solubilized receptor preparation indicating that the lectins interacted directly with the receptor. Vale and Shooter (1982) also found that wheatgerm agglutinin produced dramatic changes in NGF binding to PC12 cells. Unlike sympathetic ganglion cells, concanavalin A produced only a small alteration in receptor-binding properties in PC12 cells, indicating that the NGF receptor may undergo different patterns of glycosylation in these two cell types. The ability of wheatgerm agglutinin to interact with the NGF receptor should make it possible to partially purify the receptor by lectin affinity chromatography.

4.6 CELLULAR EVENTS INFLUENCING NGF BINDING

Steady-state binding may not be satisfactorily described by a reversible equilibrium between a ligand and its receptor. This is a particularly apparent after long incubation with the hormone. It is known that hormone–receptor complexes participate in a number of events subsequent to their initial binding interaction. Such complexes can become internalized, compartmentalized on the cell surface, covalently attached, or undergo affinity changes, all of which will affect binding measurements. Changes in the binding properties of the ^{125}I-NGF–receptor complex indicate that it may participate in one, if not more, of these pathways. Fig. 4.6 shows that the dissociation properties of ^{125}I-NGF from chick sensory cells are affected by the time of incubation of the ligand (Sutter *et al.*, 1979a). After 15 min at 37°C, ^{125}I-NGF at 4.3×10^{-10} M reaches a steady state with cells, and its dissociating properties after such an incubation are displayed in the lower curve. At longer times of incubation (60 min), even though an apparent steady state was reached 45 min earlier, the proportion of slowly dissociating to rapidly dissociating binding increased. Furthermore, the dissociation curve for the slower component became noticeably curved. Estimates of site numbers from equilibrium and kinetic experiments are in closer agreement at shorter preincubation times. Since an increase in the proportion of slowly dissociating binding is observed after steady-state conditions have been reached in sensory neuron membranes as well, internalization cannot completely explain this phenomenon (Riopelle *et al.*, 1980).

An accumulation of slowly dissociating (chase-stable) binding as a function of time of ^{125}I-NGF incubation has been observed in studies involving PC12 cells (Calisano and Shelanski, 1980) and chick embryo sympathetic (Olender and Stach, 1980) and sensory (Olender *et al.*, 1981) neurons. Some of this cell-bound ^{125}I-NGF, in fact, is even resistant to a 90 min dissociation at 37°C with unlabeled ligand (Olender and Stach, 1980). Olender and Stach have termed this phenomenon 'sequestration' and have found this process to be independent of temperature but dependent upon energy (Olender and Stach, 1980; Olender *et al.*, 1981). They have therefore suggested that such events might be cell-surface-mediated rather than simply reflecting internalization. As a result of certain kinetic experiments, high-affinity Site I receptors were implicated in the sequestration process. Olender and Stach and others (Landreth and Shooter, 1980) have considered a two-step model for the interactions of NGF with a single receptor species. Similar to a model

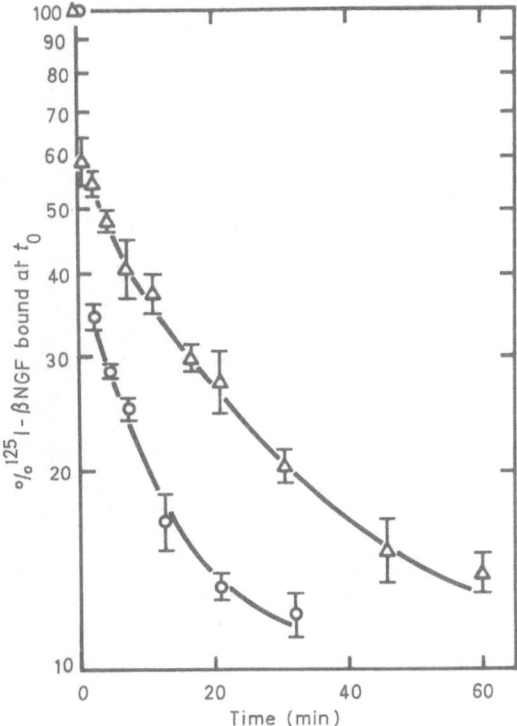

Fig. 4.6 Effect of time of incubation on the dissociation kinetics of ^{125}I-NGF. Cells (1.2×10^6/ml) were incubated at 37°C with 4.3×10^{-10} M-^{125}I-NGF for 15 min (○) and 60 min (△). Dissociation was initiated after those times by the addition of 3.8×10^{-7} M unlabeled NGF. The specific binding at t_0 was measured in sextuplets, and at all other times in triplicates. Data are corrected for non-specific binding. Reproduced with permission from Sutter *et al.* (1979a).

proposed by Corin and Donner (1982) for the insulin receptor, binding of NGF can be portrayed in the following manner, with NGF–R* signifying either an 'activated' or 'sequestered' hormone–receptor complex.

$$\text{NGF} + \text{R} \underset{k_{-1}}{\overset{k_{+1}}{\rightleftharpoons}} \text{NGF–R} \underset{k_{-2}}{\overset{k_{+2}}{\rightleftharpoons}} \text{NGF–R}^* \overset{k_{+3}}{\longrightarrow} \text{NGF} + \text{R}^*$$

Such sequential receptor–ligand interactions could explain how biphasic dissociation kinetics could co-exist with linear Scatchard plots, as has been observed with human growth hormone (Donner, 1980). If $k_{-2} \gg k_{+3}$, the activated complex (NGF–R*) will not dissociate directly into the medium but must dissociate indirectly through the non-activated complex. Furthermore, if

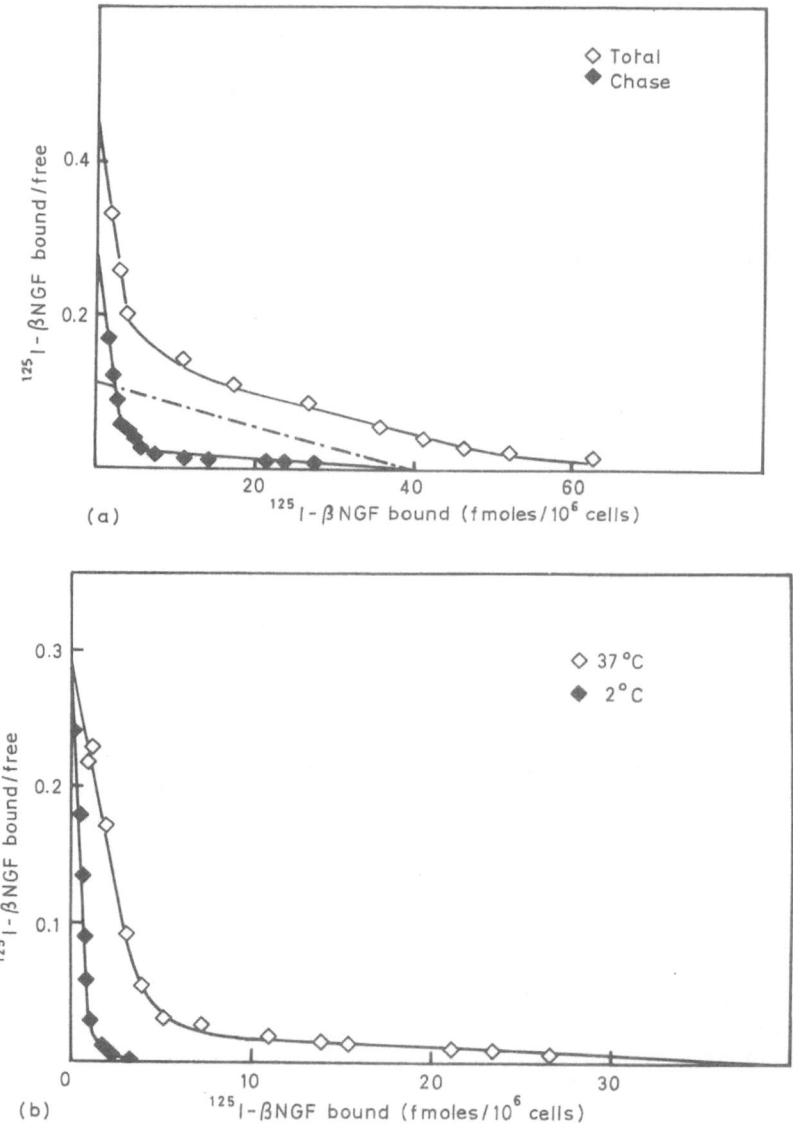

Fig. 4.7 Heterogeneity of the low-affinity binding component shown in a 'steady state chase' experiment. (a) Sensory cells were incubated at 37°C with 5×10^{-12}–2×10^{-9} M-^{125}I-NGF. After 45 min, total specific binding ($\lozenge$) as well as chase-stable specific binding ($\blacklozenge$) (binding resistant to a 10 min incubation with 5×10^{-7} M-unlabeled NGF) were determined and transformed into Scatchard plots. (b) Chase-stable binding was determined after 45 min incubation at 37°C ($\lozenge$) and 2°C ($\blacklozenge$) in the manner described above.

$k_{-1} \gg k_{-2}$, biphasic dissociation curves will be observed while only equilibrium binding, reflecting the low-affinity site, will be seen. Such a model should be explored in the instance of NGF binding to PC12 cells in which some studies revealed only one site by Scatchard analysis yet two sites by dissocation kinetics (Schechter and Bothwell, 1981; Buxser and Johnson, 1982).

While Olender and Stach (1980) suggested the sequestration phenomenon to be a property of Site I receptors, Sutter (unpublished observations) found that such secondary binding events were the result of NGF first being bound to Site II receptors. At short incubation times, as discussed earlier, NGF dissociates from Site II receptors with rapid kinetics at both 37°C and 0°C. As shown in Fig. 4.6, a third binding component develops over time secondary to initial Site I and II binding. Kinetic analysis revealed that the third binding component has slow dissociation kinetics ($t_{\frac{1}{2}}$ at 2°C of > 180 min in the presence of an unlabeled NGF chase), yet Scatchard analysis demonstrated that its affinity was comparable to that of Site II (Fig. 4.7a). Thus, at longer incubation times, Site II equilibrium binding is heterogeneous, consisting of chase-stable and chase-labile components.

Unlike initial Site I or Site II binding, the chase-stable Site II binding component did not appear at 0°C (Fig. 4.7b) or in the presence of inhibitors of metabolic energy, suggesting that this third binding component does indeed arise secondarily to ^{125}I-NGF initially bound to Site II receptors. This third binding component can be further distinguished by its slow rate of appearance, consistent with its low affinity and slow rate of dissociation, which stands in contrast to the rapid appearance of binding to both Site I and II receptors on sensory neurons (Sutter *et al.*, 1979a). In contrast to the results of Olender and Stach, the sensitivity to temperature of the chase-stable Site II binding argues that internalization may be involved in its generation. The influence of internalization on equilibrium binding should be considered for any hormone–receptor system with the potential for endocytosis and will be examined in further detail for NGF in the next section.

4.7 INTERNALIZATION OF NGF BY CULTURED CELLS

4.7.1 The chase-stable NGF-binding component

As discussed in the previous section, a number of investigators have reported that a portion of the cell-associated ^{125}I-NGF is resistant to dissociation with excess unlabeled hormone. As internalized NGF will behave as chase-stable binding, it is important to determine whether chase-stable NGF is internalized or present at the cell surface. Norr and Varon (1975) found that ^{125}I-NGF associated with both whole sensory ganglia and dissociated cells in two processes: a time-saturable uptake due to binding of NGF to cell surface receptors, and a time-linear, temperature sensitive uptake which included a non-dissociable binding component. Similarly, in addition to a chase-stable high affinity cell surface binding component Sutter and Shooter (unpublished observations) observed a temperature- and energy-sensitive accumulation of chase-stable low affinity binding in dorsal root ganglion cells. It is likely that internalization contributes to chase-stable ^{125}I-NGF binding in both these studies.

Olender and Stach observed that as much as 50% of cell-associated NGF in sympathetic and sensory neurons was not released after a 90 min incubation with excess unlabeled NGF at 37°C. In contrast to the above-mentioned studies, this sequestration of NGF was insensitive to temperature but could be blocked by energy inhibitors. Since this process occurred at 4°C, they argued that the sequestration of NGF is not due to internalization. Instead, they proposed that sequestration may represent covalent linkage of NGF to its receptor or modification of the receptor by phosphorylation, two processes which have been observed in the interaction of EGF with its receptor (Linsley *et al.*, 1979; Carpenter *et al.*, 1979).

Several groups have also documented slowly dissociating (chase-stable at 4°C) ^{125}I-NGF associated with PC12 cells, as discussed previously. This cell-associated NGF is resistant both to dissociation with excess unlabeled NGF at 4°C as well as trypsin degradation at the same temperature. Both Landreth and Shooter (1980) and Schechter and Bothwell (1981) contend that a significant fraction of this chase-stable, trypsin-resistant NGF is present at the cell surface, as its formation could not be blocked entirely by low temperature or energy poisons. Furthermore, when incubated at 37°C instead of 0.5°C in the presence of unlabeled NGF, this cell-associated NGF dissociated with first-order kinetics, and the released NGF was intact, as judged by precipitability in trichloroacetic acid (TCA). It is not clear how this low-temperature, chase-stable NGF binding to PC12 cells is related to the sequestered NGF of chick sensory ganglion cells.

Chandler and Herschman (1980) further examined the contribution of internalization to the chase-stable NGF pool in PC12 cells. These investigators found that 25–45% of the total cell-associated NGF was sequestered (not removed by a 60 min incubation at 0.5°C with excess unlabeled NGF) as was documented in earlier reports. This chase-stable NGF was then incubated for a further 2 h only at 37°C instead of 0.5°C. During this period of time, the cells released essentially all of their ^{125}I label as intact, TCA-precipitable NGF (50%), or as degraded TCA-soluble radioactivity (50%). The time of the initial ^{125}I-NGF incubation (5–75 min) did not alter these percentages. When the same experiment was performed in the presence of ammonium chloride to inhibit lysosomal function, the same amount of NGF was chase-stable (25–45%), but upon subsequent incubation at 37°C, only 50% of this cell-associated ^{125}I-NGF was released into the medium, all as intact, TCA-precipitable counts. The remaining 50% of the chase-stable ^{125}I-NGF was still cell-associated, presumably undigested inside the cell. Thus, this study indicates that the chase-stable NGF-binding component of PC12 cells is comprised of both cell surface and internalized components in approximately equal amounts.

4.7.2 NGF receptors are down regulated and internalized

A number of polypeptide hormones, including NGF, are capable of being internalized through a process of receptor-mediated endocytosis (Goldstein *et al.*, 1979; Gordon *et al.*, 1980). This pathway involves the clustering of receptors in regions of the cell surface known as coated pits with the subsequent budding off of clathrin-coated vesicles. These vesicles, in turn, may deliver the hormone and its receptor to appropriate intracellular sites. The fate of the internalized hormone is currently the subject of extensive work. It appears as though the majority of internalized polypeptide hormones are degraded in lysosomes; however, association of the internalized hormone with other intracellular organelles has also been reported (Willingham and Pastan, 1982). Also, it is now apparent that receptor down regulation, the decrease in receptor number seen with long incubations with ligand, is mediated by receptor endocytosis (Baldwin *et al.*, 1980). The endocytosis of hormone–receptor complexes may be important in mediating some of the delayed cellular responses to hormones such as the regulation of DNA synthesis and gene transcription.

Internalization of NGF in PC12 cells and embryonic dorsal root ganglia was directly visualized by Levi *et al.* (1980) using a fluorescent rhodamine–NGF conjugate in a manner analogous to their previous studies with insulin and epidermal growth factor (Schlessinger *et al.*, 1978). At 4°C, NGF was bound diffusely to the cell surface of both cell types. Upon subsequent warming to 37°C, the diffuse labeling rapidly coalesced into large patches and become endocytosed. An internal location was ascribed to the fluorescent signal as it focused between the planes of the upper and lower cell membranes and could not be removed with excess unlabeled NGF. In the presence of the energy poison sodium azide, patching still occurred at 37°C; however, endocytosis was blocked, as demonstrated by the ability of unlabeled NGF to displace the cell-bound fluorescent NGF. In dorsal root ganglion cells treated for 20 h with fluorescent NGF, trypsin-resistant clusters of fluorescence (presumed to be internalized NGF) were observed at the neurite tips, along the neurite shafts, and in the cell body. Furthermore, fluoresent clusters were found to move in a retrograde fashion from neurite tip to cell body.

As has been demonstrated for the low-density lipoprotein (LDL) receptor (Goldstein *et al.*, 1979), the NGF–receptor complex may become internalized via coated pits. Using scanning as well as transmission electron microscopy, Connolly *et al*. (1981) observed a three-fold increase in the number of 60–130 nM pits on the cell surface within 0.5 to 3 min after addition of NGF to both sympathetic neurons as well as PC12 cells. By transmission electron microscopy, these pits were found to be of the coated variety. The formation of other coated regions on the membrane not associated with pits was also induced by NGF. It is tempting to speculate that these NGF-induced coated pits may be the means by which this hormone undergoes endocytosis.

NGF receptors are also down regulated in a manner similar to the process described for other hormone receptors. Binding of ^{125}I-NGF to PC12 cells reaches a maximum after 90 min and then steadily declines to 25% of maximal binding by 10 h (Layer and Shooter, 1983). The decrease in total binding, due to a decrease in receptor number, was accompanied by the degradation of ^{125}I-NGF and the appearance of acid-soluble [^{125}I]-monoiodotyrosine in the medium. The decrease in binding and the degradation of NGF could be blocked either by low temperature or by adding the lysosomal inhibitor, chloroquine. In the presence of chloroquine, the proportion of chase-stable ^{125}I-NGF increased significantly, presumably owing to trapping intracellular NGF which cannot be degraded by lysosomes. EGF (King *et al*., 1980) and insulin (Marshall and Olefsky, 1979) also accumulate within cells in an intact form when lysosomal processing is inhibited. Since neurite outgrowth in response to NGF occurs normally and perhaps at an accelerated rate in the presence of chloroquine (Shooter *et al*., 1981), lysosomal processing of either the hormone or receptor is unlikely to be involved in generating a signal needed for the biological response.

4.8 CELLULAR SITES OF ACTION

Although NGF becomes internalized by all target cells examined, it has not been established whether this process is required for biological activity. It appears, at any rate, that the hormone must be present at the cell surface for at least 4 h to elicit neurite outgrowth (Shooter *et al*., 1981). If an antibody to NGF is added within 4 h after addition of NGF to PC12 cells, neurite outgrowth is inhibited. During the time prior to antibody addition, considerable internalization has already occurred; however, a critical pool size of internalized NGF may be needed to initiate a biological response. Similar to these results, a continuous 6 h incubation with EGF is required to elicit a mitogenic response in 3T3 cells (Shechter *et al*., 1978).

In order to determine if internalization is required for NGF action, Frazier *et al*. (1973) covalently linked NGF to cyanogen-activated Sepharose beads, thereby preventing the hormone from becoming internalized. They found that this NGF–Sepharose complex produced an identical neurite outgrowth response to native NGF, albeit at a 100-fold higher concentration. On the basis of this experiment, it was proposed that NGF produced its biological action exclusively by acting at the cell surface. However, leakage of 1% of the NGF from the beads would be sufficient to generate a maximal biological response. Although control experiments were conducted to test for spontaneous leakage of NGF from the beads during the course of the assay, they may not have been rigorous enough to exclude the possibility that proteolytically generated active NGF fragments or small bead fragments with the potential for internalization could have been formed during the incubation with cells. In the control experiment, NGF–Sepharose was added to a clot culture of several ganglia to see if active fragments were generated from the beads which could stimulate an adjacent ganglion not directly exposed to the NGF–Sepharose beads. It was assumed that many ganglia would be more destructive to the conjugate than one; however, many ganglia would also bind more of the released NGF which would be therefore unavailable to stimulate the adjacent control ganglion.

Recently, Heumann *et al.* (1981) reported that NGF when microinjected into PC12 cells or delivered by fusion with NGF-loaded erythrocyte ghosts does not induce neurite outgrowth. Also, NGF antibodies introduced into the cytoplasm do not inhibit the response of cells to NGF added in the medium. Negative results such as these do not necessarily preclude a biological role for internalized NGF, as intracellular NGF may only be active when internalized as part of a receptor-complex and not as free hormone present in the cytosol. Furthermore, compartmentalized intracellular NGF may not be accessible to NGF antibodies in the cytoplasm and NGF compartmentalized in specialized vesicles might be delivered to internal cellular sites which are not reached by free, cytoplasmic NGF.

Where are the potential target sites of internalized NGF? As mentioned above, biochemical studies indicate that the majority of intracellular NGF is degraded in lysosomes. This conclusion is supported by EM radioautography experiments (Claude *et al.*, 1982). However, several studies have documented that NGF can associate with the nucleus. Andres *et al.* (1977) demonstrated that a portion of ^{125}I-NGF bound to chick embryonic sensory neurons was resistant to solubilization with the detergent Triton X-100. Subcellular fractionation revealed that the nuclear fraction accounted for the majority of the Triton X-100-resistant NGF binding. Isolated nuclei from these cells also demonstrated specific and saturable NGF-binding sites with an apparent K_d of 2×10^{-10} M. Further experiments indicated that these NGF receptors were contained within the chromatin portion of the nucleus.

Yankner and Shooter (1979) found that isolated nuclei from PC12 cells demonstrated two distinct receptor types with K_d values of 10^{-10} and 10^{-8} M; however, unlike the study on chick sensory neurons, these binding sites were localized to the nuclear membrane as opposed to chromatin. NGF appeared in the nuclear fraction of PC12 cells grown in monolayer after a lag of a few hours, indicating that nuclear high-affinity receptors may arise from a translocation and insertion of high-affinity surface receptors to the nuclear membrane. This translocation could be enhanced by inhibiting lysosomal degradation with chloroquine (Shooter *et al.*, 1981). Moreover, the proportion of PC12 cells which extended neurites on initial exposure to NGF correlated with the amount of nuclear bound NGF, and the ability of a cell to rapidly grow neurites after shearing off the original neurites again was proportional to nuclear-bound NGF.

Marchisio *et al.* (1980) showed perinuclear and intranuclear binding of NGF in PC12 cells by indirect immunofluorescence and radioautographic techniques. Within 1 h of addition, NGF was primarily localized to the cell membrane. With a further incubation (6–24 h), NGF was found in a perinuclear location, and at even longer incubations (24–48 h), NGF was seen in the nucleus itself. When cells were incubated with ^{125}I-NGF and subsequently solubilized with Triton X-100, radioautographs also showed grains directly over the Triton X-100-insoluble nucleus.

In a recent radioautographic analysis of the binding and internalization of ^{125}I-NGF to PC12 cells in monolayers on collagen (P. Bernd and L.A. Green, unpublished observations) it was emphasized that the internalization process is rapid and that after only 15 min, a significant fraction of the bound NGF was inside the cell. Over time periods from 2 min to 1 week, internalized ^{125}I-NGF was found to be associated with lysosomes and the nuclear membrane. Interestingly, the amount of ^{125}I-NGF on the nuclear membrane paralleled that bound to the plasma membrane, including a decrease over the first 6 h and then a subsequent slow increase. Primed PC12 cells (i.e., those exposed to unlabeled NGF for 1 week and the neurites removed before exposure to ^{125}I-NGF) also showed ^{125}I-NGF bound to the nuclear membrane. The reports of nuclear-bound NGF have, however, been challenged. Schechter and Bothwell (1981) argued that nuclei prepared from PC12 cells by Triton X-100 or mechanical treatment alone are contaminated with associated cytoskeletal elements to which cell-surface NGF receptors may be attached. More complete homogenization of these cytoskeleton/nuclear structures in their hands produced a cleaner nuclear preparation and concomitantly reduced the amount of associated ^{125}I-NGF. In contrast, extensive shearing of PC12 nuclei, to which ^{125}I-NGF was bound by internalization in whole cells, reduced only the non-specific binding and not the specific binding (B.A. Yankner and E.M. Shooter, unpublished observations), arguing that nuclear binding is not a result of cytoskeleton contamination. The intracellular distribution of ^{125}I-NGF in PC12 cells has also been followed by quantitative EM radioautography in an independent study (Rohrer *et al.*, 1982). No evidence for nuclear or nuclear membrane accumulation was observed in cell monolayers incubated with ^{125}I-NGF for between 1 h to 8 days. Instead, the labeled NGF appeared in membrane-bound cytoplasmic compartments including lysosomes. These authors also showed that the accumulation of ^{125}I-NGF in the nuclear fraction of PC12 cells incubated with the labeled growth factor in suspension was an artefact resulting from damaged cells. It should be explained that in all the studies with PC12 cells in monolayer (Shooter *et al.*, 1981; Rohrer *et al.*, 1982; P. Bernd and L.A. Greene, unpublished data), the cells remained viable and produced extensive neurites. The reasons for the differing results are not clear.

The fate of internalized ^{125}I-NGF after retrograde transport in autonomic neurons (Claude *et al.*, 1982) also failed to provide evidence for nuclear accumulation. Instead, the ^{125}I-NGF was located in lysosomes and multivesicular bodies. Thus, while it is clear that internalization and intracellular transport of NGF (e.g. by retrograde transport) is physiologically significant (see the next section), the final location from which the intracellular signals emanate has not yet been resolved.

4.9 RETROGRADE TRANSPORT OF NGF

4.9.1 Specific and saturable uptake of NGF from nerve terminals and its transport to the cell body

In order for NGF to function as a messenger between effector organs and innervating neurons, some signal must be carried to the distant cell body in order that gene expression can be appropriately regulated. The possibility that NGF itself may be used to provide information to the cell body was suggested by the experiments of Hendry *et al*. (1974a) who first observed the retrograde axonal transport of NGF. In their study, ^{125}I-NGF was injected into the anterior chamber of one eye of a mouse. Radioactive label was then found to accumulate in the superior cervical ganglion (SCG), which provides sympathetic innervation to the iris, reaching a maximum 6 h after injection. Although radioactivity was found in both SCG, the ganglion on the side of the injected eye accumulated three-fold more label by 16 h than the ganglion on the other side. This preferential accumulation of radioactivity on the side of the injection could be blocked by severing the postganglionic nerves or inhibiting axonal microtubules with colchicine, suggesting that the preferential accumulation of label in the SCG on the side of injection was axon-mediated and not due to diffusion.

The presence of ^{125}I-NGF in the SCG opposite the side of an intraocular injection suggested that delivery via the general circulation could also occur. This could be demonstrated by a subcutaneous injection of ^{125}I-NGF which produced an equal labeling of both SCG (Hendry *et al.*, 1974a). A study by Paravicini *et al.* (1975) indicated that radioactivity present in the SCG within

4 h after an ^{125}I-NGF injection reached this ganglion through the circulation while after that time, accumulation of ^{125}I-NGF in the ganglion was primarily due to retrograde axonal transport. The spillover of ^{125}I-NGF into the general circulation became less of a concern with the development of ^{125}I-NGF with high specific activity which allowed the injection of 10–100-fold lower concentrations of the hormone. Using such a preparation, Johnson *et al.* (1978) found no label in either SCG within the first 5 h after intraocular injection, and less radioactivity was subsequently observed in the ganglion on the opposite side of injection than had been seen in previous studies. The rate of transport of ^{125}I-NGF was estimated from this study to be 3 mm/h.

The uptake of NGF by adrenergic nerve endings is highly specific. A variety of proteins including some with similar isoelectric points to NGF, such as cytochrome *c*, or with a structural homology, such as insulin, were not transported to the ganglion (Stockel *et al.*, 1974). Neither were the α- and γ-subunits which, in addition to the biologically active β-subunit, comprise the circulating 7 S NGF complex (Hendry *et al.*, 1974b). Furthermore, oxidation of tryptophan residues on NGF with *N*-bromosuccinimide, which results in a loss of receptor-binding activity (Cohen *et al.*, 1980), also results in the loss of its ability to to transported (Stockel *et al.*, 1974).

The uptake and transport of NGF is saturable at concentrations which characterize its binding to receptors on sympathetic neurons. Hendry *et al.* (1974b) and Johnson *et al.* (1978) found that transport of NGF could be saturated. Half-maximal transport was seen by Johnson *et al.* (1978) at 15 ng/eye with a maximal accumulation of 40 pg of NGF in the ganglion. Receptors for NGF rather than capacity of the transport system appear to be the rate-limiting factor. Such a conclusion can be reached because even when the retrograde transport of lectins is fully saturated, retrograde transport of NGF still occurs at its normal rate, unaffected by the simultaneous transport of the lectins (Dumas *et al.*, 1979). While Hendry *et al.* (1974b) and Johnson *et al.* (1978) only documented one affinity for transport of NGF, Dumas *et al.* (1979) found that NGF was transported through two receptor-mediated processes with affinities which differed by twenty-fold. In addition to the higher-affinity transport described above, a lower-affinity, higher-capacity (half-maximal transport at 260 ng/eye; maximal accumulation of 130 pg/ganglion) was also observed. Thus, the two receptor populations which mediate the uptake of NGF from nerve terminals appear similar to the two classes of NGF receptors on sympathetic neurons defined by binding studies.

Radioautography of the SCG after an intraocular injection of ^{125}I-NGF revealed that only 10% of the cells in the ganglion were labeled (Hendry *et al.*, 1974b), most likely representing the neurons which innervate the iris. When ^{125}I-NGF was injected both into the anterior chamber of the eye and submandibular gland, 15–20% of the SCG neurons were labeled due to the additional axonal transport from neurons supplying the salivary gland (Iverson *et al.*,

1975). Schwab and Thoenen (1977) reported that no radioactivity was detected in the extracellular space or in the glial component of the SCG. They also found labeling exclusively in postganglionic neurons with little or no transsynaptic transfer to preganglionic fibers. The radioactivity which accumulates in the SCG after retrograde transport is primarily intact NGF as judged by recognition by specific antibodies as well as by gel electrophoresis under denaturing conditions (Hendry *et al.*, 1974a; Dumas *et al.*, 1979; Stockel *et al.*, 1976).

A number of studies have been conducted to determine the subcellular location of transported NGF. In the nerve terminals and postganglionic axons, label was primarily localized to smooth vesicles and tubules of the smooth endoplasmic reticulum (Schwab and Thoenen, 1977). Crude cell fractionation of SCG showed ^{125}I equally distributed between nuclei, mitochondria, microsomes and supernatant fractions (Stockel *et al.*, 1976). However, with more precise radioautographic and cytochemical (using horseradish peroxidase-conjugated NGF) techniques, label was distributed within the perikaryon in smooth vesicles, smooth endoplasmic reticulum, multivesicular bodies and secondary lysosomes but not within Golgi, nuclei, or free in the cytoplasm (Schwab and Thoenen, 1977; Schwab, 1977). With increasing times of incubation, a transfer of radioactivity was seen from the smooth endoplasmic reticulum to multivesicular bodies. The results of EM radioautography on PC12 cells and autonomic neurons which confirm and extend these data have been mentioned earlier.

Retrograde transport of NGF is largely independent of neuronal activity, occurring at the same rate regardless of the frequency of neuron firing (Thoenen and Barde, 1980). Transport is also not influenced by the release of neurotransmitter granules through pharmacological means (Lees *et al.*, 1981). Furthermore, animals which were decentralized (preganglionic fibers severed) prior to ^{125}I-NGF injection (Stockel *et al.*, 1978; Johnson *et al.*, 1979b) or which received chlorisondamine (a ganglionic-blocking agent) (Stockel *et al.*, 1978) were not impaired in their retrograde axonal transport of NGF. Curiously, however, reserpine, a drug which causes the intracellular release of norepinephrine from secretory storage granules, decreased the accumulation of NGF in the SCG by 40%, whether the ^{125}I-NGF was injected intraocularly or intravenously (Johnson *et al.*, 1976b). Nonetheless, despite the curious findings with reserpine, the majority of evidence indicates that retrograde transport of NGF is independent of the state of amine storage granules in the nerve terminal.

Boegman and Riopelle (1980a) studied how two drugs, tetrodotoxin and batrachytoxin, which block nerve impulse conduction, affected retrograde transport. Tetrodotoxin, a sodium channel blocker, had no effect on NGF transport; however, after batrachytoxin treatment, an agent which blocks orthograde axonal transport as well as impulse conduction, NGF was only transported up to the point of the batrachytoxin injection, where it subsequently accumulated. In a later study (Boegman and Riopelle, 1980b), batrachytoxin or tetrodotoxin was injected 12–36 h before the ^{125}I-NGF instead of simultaneously. Similar results were obtained in these experiments; retrograde transport was completely blocked in the batrachytoxin-treated nerves. Also, ligating the nerve many hours prior to ^{125}I-NGF injection resulted in the inability of the distal nerve segment to transport the hormone. These studies concluded that while retrograde transport does not require nerve electrical activity, it is dependent upon an intact orthograde transport system, which can be disrupted by batrachytoxin or ligating the nerve. Orthograde transport may be needed to deliver necessary proteins such as NGF receptors to the nerve terminals.

4.9.2 Physiological implications of retrograde transport

While sympathetic neurons require NGF to maintain their differentiated state, sensory neurons, which are responsive to NGF during ontogenesis, are not dependent upon this factor thereafter (Levi-Montalcini and Angeletti, 1968). This change in responsiveness to NGF is correlated with the loss of NGF receptors (Herrup and Shooter, 1975). It was therefore of interest to see whether such mature neurons, which are unresponsive to NGF, maintain their ability to transport the hormone from their nerve terminals. Stockel *et al.* (1975a) indeed showed that the dorsal root ganglion of the rat did accumulate immunologically intact ^{125}I-NGF injected into the periphery. Like sympathetic retrograde transport, transport to the dorsal root ganglion could be abolished by severing the nerve or treating with colchicine, but the calculated rate of transport was five times faster than its sympathetic counterpart. Only a portion of the neurons in the dorsal root ganglion were intensely labeled and more label could be seen in the cytoplasm than the nucleus. In contrast to these results, motor neurons in the spinal cord, which are capable of retrogradely transporting toxins and lectins, are unable to transport and accumulate NGF (Stockel *et al.*, 1975b). Thus retrograde axonal transport of NGF does not seem to be a property of all neurons, but rather a property of a subset of neurons which are at one time or another responsive to NGF.

Several exceptions to this generalization, however, have been noted in the literature. Max *et al.* (1978) found evidence for NGF transport by cholinergic parasympathetic neurons in the ciliary ganglion which innervate the iris. While parasympathetic neurons are neural crest derivatives, they do not require NGF for survival (Helfand *et al.*, 1978). However, this same study showed that not all parasympathetic neurons of this ganglion can transport NGF, as ^{125}I-NGF injected into the submandibular gland was not transported by the parasympathetic axons which innervate it. Another study reported the transport of NGF injected into the cerebellar cortex to cells located in the medulla and pons (Ebbot and Hendry, 1978). It is not clear from these results whether such cells are responsive to NGF at some point in their development or whether NGF can be retrogradely transported by some neurons which never respond to this hormone.

A characteristic response of sympathetic neurons to NGF is the induction of the neurotransmitter enzyme, tyrosine hydroxylase (TH). The observation that target tissues containing NGF influence the neuronal levels of this enzyme was made by Hendry and Iverson (1973). When they removed the submaxillary glands of mice, a decrease in the TH activity of the SCG was subsequently seen. This decrease could be prevented by exogenous NGF supplied either systemically or at the site of gland removal. It has also been shown that a unilateral injection of NGF into the eye and submaxillary gland causes a preferential 160% increase in the TH activity on the side of injection (Paravicini *et al.*, 1975). However, the control ganglion opposite the side of injection also had increased TH activity. To circumvent the problem of leakage of NGF into the general circulation so that a better specific induction of TH could be seen on the side of injection, Hendry (1977) attached NGF to cellulose in a manner in which it was released slowly from the injection site. Using such a technique, an increase in TH activity (140% higher than controls) was only observed on the SCG on the side of injection. The increase in TH activity seen in the SCG after NGF administration is relatively small because only 10% of the SCG neurons innervate the iris. Although retrograde transport of NGF seems to be required for the induction of TH, it is not known whether NGF acts directly in the cell body or whether second messengers formed at the nerve endings are also required. In a similar study to those conducted with tyrosine hydroxylase, NGF was shown to increase the levels of ornithine decarboxylase in SCG neurons (Hendry and Bonyhady, 1980).

Retrograde transport of NGF also results in an increase in the size of the neuronal soma (Hendry, 1977). Hendry identified neurons in the SCG innervating the iris by injecting ^{125}I-tetanus toxin or ^{125}I-NGF into the eye and found that only those neurons which were radiolabeled responded to a 7-day previous intraocular injection with NGF with an increase in their cell diameter corresponding to a two-fold increase in volume. Thus it appears that NGF affects macromolecular synthesis only in cells which transport it from their nerve terminals.

Campenot (1977) devised an ingenious series of experiments which demonstrated that retrograde transport of NGF from the periphery can provide all the NGF required for the survival of the cell body. A three-chambered apparatus was developed which could separate the neurites from the neuronal cell body and maintain individual fluid compartments. In this fashion, the neurite tips and cell bodies could be exposed to different local environments. Neurites would only grow into a side chamber if NGF were present in its medium. The neurites and cell bodies could then be maintained if NGF were present in all chambers, but cell death resulted from the withdrawal of NGF from all of these chambers. However, if NGF was removed from only the central chamber containing the cell bodies but was present still in the medium exposed to the neurites, cell viability was maintained. Such results indicate that the cell body need not be directly exposed to NGF in order to survive as long as retrograde transport of NGF from peripheral neurites takes place. Thus, retrograde transport of NGF appears to play an essential biological role for cell survival.

The influence of retrograde transport on neuronal survival can be further demonstrated by the destructive effects on neurons of 6-hydroxydopamine, an agent which destroys adrenergic nerve terminals, and vinblastine, an inhibitor of ortho- and retro-grade transport (Johnson *et al.*, 1979a). These drugs were also demonstrated to inhibit the transport of ^{125}I-NGF from the anterior chamber of the eye to the SCG. In early postnatal animals, the destruction of nerve terminals by these drugs was eventually followed by the degeneration of the cell body, but this degeneration could be prevented by administering large amounts of NGF. These results suggested that vinblastine and 6-hydroxydopamine sympathectomize immature animals by inhibiting the retrograde transport of NGF released by target tissues. In mature animals, interference with axonal transport results in chromatolysis rather than cell death, and administered NGF can prevent these events as well (Purves and Nja, 1976). NGF also influences preganglionic nerve fibers, with decreased levels of this hormone resulting in a reduction in the number of such fibers (Hendry, 1975; Aguayo *et al.*, 1976). However, NGF appears to produce its effects on preganglionic nerve fibers by some indirect mechanism.

REFERENCES

Aguayo, A.J., Peyronnard, J.M., Terry, L.C., Romine, J.S. and Bray, G.M. (1976), *J. Neurocytol.*, **5**, 137–155.

Andres, R.Y., Jeng, I. and Bradshaw, R.A. (1977), *Proc. Natl. Acad. Sci. U.S.A.*, **74**, 2785–2789.

Baldwin, D., Jr., Prince, M., Marshall, S., Davies, P. and Olefsky, J.M. (1980), *Proc. Natl. Acad. Sci. U.S.A.*, **77**, 5975–5978.

Banerjee, S.P., Cuatrecasas, P. and Snyder, S.H. (1976), *J. Biol. Chem.*, **251**, 5680–5685.

Banerjee, S.P., Snyder, S.H., Cuatrecasas, P. and Greene, L.A. (1973), *Proc. Natl. Acad. Sci. U.S.A.*, **70**, 2519–2523.

Berger, E.A. and Shooter, E.M. (1977), *Proc. Natl. Acad. Sci. U.S.A.*, **74**, 3647–3651.

Boegman, R.J. and Riopelle, R.J. (1980a), *Neurosci. Lett.*, **18**, 143–147.

Boegman, R.J. and Riopelle, R.J. (1980b), *J. Neurobiol.*, **11**, 497–501.

Boeynaems, J.M. and Dumont, J.E. (1975), *J. Cyclic Nucleotide Res.*, **1**, 123–142.

Buxser, S.F. and Johnson, G.L. (1982), *J. Cell. Biochem. Supplement* **6**, 165.

Calissano, P. and Shelanski, M.L. (1980), *Neuroscience*, **5**, 1033–1039.

Campenot, R.B. (1977), *Proc. Natl. Acad. Sci. U.S.A.*, **74**, 4516–4519.

Carpenter, G., King, I. and Cohen, S. (1979), *J. Biol. Chem.*, **254**, 4884–4891.

Chandler, C. and Herschman, H. (1980), Ph.D. Thesis, Los Angeles: University of California.

Claude, P., Hawrot, E., Dunis, D.A. and Campenot, R.B. (1982), *J. Neurosci.*, **2**, 431–442.

Cohen, P., Sutter, A., Landreth, G., Zimmermann, A. and Shooter, E.M. (1980), *J. Biol. Chem.*, **255**, 2949–2954.

Cohen, S. (1960), *Proc. Natl. Acad. Sci. U.S.A.*, **46**, 302–311.

Connolly, J.L., Green, S.A. and Green, L.A. (1981), *J. Cell Biol.*, **90**, 176–180.

Corin, R.E. and Donner, D.B. (1982), *J. Biol. Chem.*, **257**, 104–110.

Costrini, N.V. and Bradshaw, R.A. (1979), *Proc. Natl. Acad. Sci. U.S.A.*, **76**, 3242–3245.

Costrini, N.V. and Kogan, M. (1981), *J. Neurochem.*, **36**, 1175–1180.

Costrini, N.V., Kogan, M., Kukreja, K. and Bradshaw, R.A. (1979), *J. Biol. Chem.*, **254**, 11242–11246.

Cuatrecasas, P. and Tell, G.P.E. (1973), *Proc. Natl. Acad. Sci. U.S.A.*, **70**, 485–489.

De Lean, A. and Rodbard, D. (1979), in *The Receptors* (R.D. O'Brien, ed.) Plenum, New York, vol. 1, pp. 143–192.

De Lean, A., Stadel, J.M. and Lefkowitz, R.J. (1980), *J. Biol. Chem.*, **255**, 7108–7117.

DeLisi, C. and Chabay, R. (1979), *Cell Biophys.*, **1**, 117–131.

De Meyts, P., Bianco, A.R. and Roth, J. (1976), *J. Biol. Chem.*, **251**, 1877–1888.

De Meyts, P., Roth, J., Neville, D.M., Jr., Gavin, J.R., III and Lesniak, M.A. (1973), *Biochem. Biophys. Res. Commun.*, **55**, 154–161.

Donner, D.B. (1980), *Biochemistry*, **19**, 3300–3306.

Dumas, M., Schwab, M.E. and Thoenen, H. (1979), *J. Neurobiol.*, **10**, 179–197.

Ebbott, S. and Hendry, I. (1978), *Brain Res.*, **139**, 160–163.

Flanagan, J. and Koch, G.L.E. (1978), *Nature (London)*, **273**, 278–281.

Frazier, W.A., Angeletti, R.H. and Brandshaw, R.A. (1972), *Science*, **176**, 482–488.

Frazier, W.A., Boyd, L.F. and Bradshaw, R.A. (1973), *Proc. Natl. Acad. Sci. U.S.A.*, **70**, 2931–2935.

Frazier, W.A., Boyd, L.F. and Bradshaw, R.A. (1974), *J. Biol. Chem.*, **249**, 5513–5519.

Goldstein, J.L., Andersen, R.G.W. and Brown, M.S. (1979), *Nature (London)*, **279**, 679–685.

Gordon, P., Carpentier, J.L., Freychet, P. and Orci, L. (1980), *Diabetologia*, **18**, 263–274.

Gorin, P. and Johnson, E.M. (1979), *Proc. Natl. Acad. Sci. U.S.A.*, **76**, 5382–5386.

Gottlieb, D.I. and Glaser, L. (1980), *Annu. Rev. Neurosci.*, **3**, 303–318.

Greene, L.A. and Tischler, A.S. (1976), *Proc. Natl. Acad. Sci. U.S.A.*, **73**, 2424–2428.

Gunderson, R.W. and Barrett, J.N. (1979), *Science*, **206**, 1079–1080.

Haigler, H.T., McKanna, J.A. and Cohen, S. (1979), *J. Cell Biol.*, **81**, 382–395.

Harris-Warrick, R.M., Bothwell, M.A. and Shooter, E.M. (1980), *J. Biol. Chem.*, **255**, 11284–11289.

Helfand, St.L., Riopelle, R.J. and Wessells, N.K. (1978), *Exp. Cell Res.*, **113**, 39–45.

Hendry, I.A. (1975a), *Brain Res.*, **86**, 483–487.

Hendry, I.A. (1975b), *Brain Res.*, **94**, 87–97.

Hendry, I.A. (1976), in *Review of Neuroscience* (S. Ephrenreis and I.J. Kopin, eds), Raven, New York, vol. 2, pp. 149–194.

Hendry, I.A. (1977), *Brain Res.*, **134**, 213–223.

Hendry, I.A. and Bonyhady, R. (1980), *Brain Res.*, **200**, 39–45.

Hendry, I.A. and Iversen, L.L. (1973), *Nature (London)*, **243**, 500–504.

Hendry, I.A., Stockel, K., Theonen, H. and Iversen, L.L. (1974a), *Brain Res.*, **68**, 103–121.

Hendry, I.A., Stach, R. and Herrup, K. (1974b), *Brain Res.*, **82**, 117–128.

Herrup, K. and Shooter, E.M. (1973), *Proc. Natl. Acad. Sci. U.S.A.*, **70**, 3884–3888.

Herrup, K. and Shooter, E.M. (1975), *J. Cell Biol.*, **67**, 118–125.

Herrup, K. and Thoenen, H. (1979), *Exp. Cell Res.*, **121**, 71–78.

Heumann, R., Schwab, M. and Thoenen, H. (1981), *Nature (London)*, **292**, 838–840.

Iversen, L.L., Stockel, K. and Thoenen, H. (1975), *Brain Res.*, **88**, 37–43.

Jacobs, S. and Cuatrecasas, P. (1976), *Biochim. Biophys. Acta*, **433**, 482–495.

Jacobs, S. and Cuatrecasas, P. (1981), *Endocrine Rev.*, **2**, 251–263.

Jacobs, S., Hazum, E. and Cuatrecasas, P. (1980), *J. Biol. Chem.*, **255**, 6937–6940.

Johnson, E.M., Andres, R.Y. and Bradshaw, R.A. (1978), *Brain Res.*, **150**, 319–331.

Johnson, E.M., Macia, R.A., Andres, R.Y. and Bradshaw, R.A. (1979a), *Brain Res.*, **171**, 461–472.

Johnson, E.M., Blumberg, H.M., Costrini, N.V. and Bradshaw, R.A. (1979b), *Brain Res.*, **178**, 389–401.

King, A.C., Davis-Hernaez, L. and Cuatrecasas, P. (1980), *Proc. Natl. Acad. Sci. U.S.A.*, **77**, 3283–3287.

Landreth, G.E. and Shooter, E.M. (1980), *Proc. Natl. Acad. Sci. U.S.A.*, **77**, 4751–4755.

Layer, P.G. and Shooter, E.M. (1983), *J. Biol. Chem.*, **258**, 3012–3018.

Lees, G., Chubb, I., Freeman, C., Geffin, L. and Rush, R. (1981), *Brain Res.*, **214**, 186–189.

Levi, A., Shechter, Y., Neufeld, E.J. and Schlessinger, J. (1980), *Proc. Natl. Acad. Sci. U.S.A.*, **77**, 3469–3473.

Levi-Montalcini, R. (1966), *Harvey Lect.*, **60**, 217–259.

Levi-Montalcini, R. and Angeletti, P.U. (1963), *Dev. Biol.*, **7**, 653–657.

Levi-Montalcini, R. and Angeletti, P.U. (1966), *Pharmacol. Rev.*, **48**, 534–569.

Levi-Montalcini, R. and Angeletti, P.U. (1968), *Physiol. Rev.*, **48**, 534–569.

Levi-Montalcini, R. and Booker, B. (1960), *Proc. Natl. Acad. Sci. U.S.A.*, **46**, 384–391.

Limbird, L.E. and Lefkowitz, R.J. (1978), *Proc. Natl. Acad. Sci. U.S.A.*, **75**, 228–232.

Linsley, P. and Fox. C.F. (1980), *J. Supramol. Struct.*, **14**, 461–471.

Linsley, P., Blifeld, C., Wrann, M. and Fox, C.F. (1979), *Nature (London)*, **278**, 745–748.

Marchisio, P.C., Naldini, L. and Calissano, P. (1980), *Proc. Natl. Acad. Sci. U.S.A.*, **77**, 1656–1660.

Marshall, S. and Olefsky, J.M. (1979), *J. Biol. Chem.*, **254**, 10153–10160.

Massague, J., Pilch, P.F. and Czech, M.P. (1981a), *J. Biol. Chem.*, **256**, 3182–3190.

Massague, J., Guillette, B.J., Czech, M.P., Morgan, C.J. and Bradshaw, R.A. (1981b), *J. Biol. Chem.*, **256**, 9419–9424.

Maturo, J.M., III and Hollenberg, M.D. (1978), *Proc. Natl. Acad. Sci. U.S.A.*, **75**, 3070–3074.

Max, S.R., Schwab, M., Dumas, M. and Thoenen, H. (1978), *Brain Res.*, **159**, 411–415.

McGuire, J.C. and Greene, L.A. (1979), *J. Biol. Chem.*, **254**, 3362–3367.

Menesini-Chen, M.G., Chen, J.S. and Levi-Montalcini, R. (1978), *Arch. Ital. Biol.*, **116**, 53–84.

Mobley, W.C., Schenker, A. and Shooter, E.M. (1976), *Biochemistry*, **15**, 5543–5551.

Moore, J.B., Jr., Mobley, W.C. and Shooter, E.M. (1974), *Biochemistry*, **13**, 833–840.

Norr, S.C. and Varon, S. (1975), *Neurobiology*, **5**, 101–118.

Olefsky, J.M. and Chang, H. (1979), *Endocrinology*, **104**, 462–466.

Olender, E.J. and Stach, R.W. (1980), *J. Biol. Chem.*, **255**, 9338–9343.

Olender, E.J., Wagner, B.J. and Stach, R.W. (1981), *J. Neurochem.*, **37**, 436–442.

Paravicini, V., Stockel, K. and Theonen, H. (1975), *Brain Res.*, **84**, 279–291.

Pilch, P.F. and Czech, M.P. (1980), *Science*, **210**, 1152–1153.

Pollet, R.J., Standaert, M.L. and Hasse, B.A. (1977), *J. Biol. Chem.*, **252**, 5828–5834.

Pollet, R.J., Kempner, E.S., Standaert, M.L. and Haase, B.A. (1982), *J. Biol. Chem.*, **257**, 894–898.

Purves, D. and Nja, A. (1976), *Nature (London)*, **260**, 535–536.

Riopelle, R.J., Klearman, M. and Sutter, A. (1980), *Brain Res.*, **199**, 63–77.

Rohrer, H., Schafer, T., Korsching, S. and Thoenen, H. (1982), *J. Neurosci.*, **2**, 687–697.

Sahyoan, N.E., Le Vine, H., III, Davis, J., Hedon, G.M. and Cuatrecasas, P. (1981), *Proc. Natl. Acad. Sci, U.S.A.*, **78**, 6158–6162.

Schechter, A.L. and Bothwell, M.A. (1981), *Cell*, **24**, 867–874.

Schlessinger, J., Shechter, Y., Willingham, M.C. and Pastan, I. (1978), *Proc. Natl. Acad. Sci. U.S.A.*, **75**, 2659–2663.

Schwab, M. (1977), *Brain Res.*, **130**, 190–196.

Schwab, M. and Thoenen, H. (1977), *Brain Res.*, **122**, 459–474.

Server, A.C. and Shooter, E.M. (1977), *Adv. Protein Chem.*, **31**, 339–409.

Shechter, Y., Chang, K.-J., Jacobs, S. and Cuatrecasas, P. (1978), *Proc. Natl. Acad. Sci. U.S.A.*, **75**, 5788–5791.

Shechter, Y., Hernaez, L., Schlessinger, J. and Cuatrecasas, P. (1979a), *Nature (London)*, **278**, 835–838.

Shechter, Y., Chang, K.-J., Jacobs, S. and Cuatrecasas, P. (1979b), *Proc. Natl. Acad. Sci. U.S.A.*, **76**, 2720–2724.

Shooter, E.M., Yankner, B.A., Landreth, G.E. and Sutter, A. (1981), *Recent Prog. Horm. Res.*, **37**, 417–466.

Silhavy, T.J., Smekman, S., Boos, W. and Schwartz, M. (1975), *Proc. Natl. Acad. Sci. U.S.A.*, **72**, 2120–2124.

Spiegel, A.M. and Downs, R.W., Jr. (1981), *Endocrine Rev.*, **2**, 275–305.

Stockel, K., Dumas, M., and Theonen, H. (1978), *Neurosci. Lett.*, **10**, 61–64.

Stockel, K., Guroff, G., Schwab, M.E. and Theonen, H. (1976), *Brain Res.*, **109**, 271–284.

Stockel, K., Paravicini, V. and Thoenen, H. (1974), *Brain Res.*, **76**, 413–421.

Stockel, K., Schwab, M. and Theonen, H. (1975a), *Brain Res.*, **89**, 1–14.

Stockel, K., Schwab, M. and Thoenen, H. (1975b), *Brain Res.*, **99**, 1–16.

Sutter, A., Riopelle, R.J., Harris-Warrick, R.M. and Shooter, E.M. (1979a), *J. Biol. Chem.*, **254**, 1516–1523.

Sutter, A., Riopelle, R.J., Harris-Warrick, R.M. and Shooter, E.M. (1979b), *Transmembrane Signalling* (M. Bilensky, R.J. Collier, D.F. Steiner and C.F. Fox, eds), Alan R. Liss Inc., New York, pp. 659–667.

Tait, J.F., Weinman, S.A. and Bradshaw, R.A. (1981), *J. Biol. Chem.*, **256**, 11086–11092.

Thoenen, H. and Barde, Y.-A. (1980), *Physiol. Rev.*, **60**, 1284–1335.

Tischler, A.S. and Greene, L.A. (1975), *Nature (London)*, **258**, 341–342.

Vale, R.D. and Shooter, E.M. (1982), *J. Cell Biol.* (in press).

Varon, S.S. and Bunge, R.P. (1978), *Annu. Rev. Neurosci.*, **1**, 327–361.

Varon, S., Nomura, J., Perez-Polo, J.R. and Shooter, E.M. (1972), in *Methods of Neurochemistry* (R. Fried, ed.), Marcel Dekker, Inc., New York, vol. 3, pp. 203–229.

Varon, S., Nomura, J. and Shooter, E.M. (1967a), *Proc. Natl. Acad. Sci. U.S.A.*, **57**, 1782–1789.

Varon, S., Nomura, J. and Shooter, E.M. (1967b), *Biochemistry*, **6**, 2202–2209.

Willingham, M.C. and Pastan, I.H. (1982), *J. Cell. Biol.*, **94**, 207–212.

Yankner, B.A. and Shooter, E.M. (1979), *Proc. Natl. Acad. Sci. U.S.A.*, **76**, 1269–1273.

Zimmerman, A., Sutter, A., Samuelson, J. and Shooter, E.M. (1978), *J. Supramolecular Struct.*, **9**, 351–361.

5 Entry of Enveloped Viruses into Cells

JOHN LENARD and DOUGLAS K. MILLER

Receptor-Mediated Endocytosis
(*Receptors and Recognition*, Series B, Volume 15)
Edited by P. Cuatrecasas and T. F. Roth
Published in 1983 by Chapman and Hall, 11 New Fetter Lane, London EC4P 4EE
© 1983 Chapman and Hall

5.1 INTRODUCTION

In this chapter we will describe our current understanding of the mode of penetration and uncoating of four commonly studied enveloped viruses – Semliki Forest virus (SFV), vesicular stomatitis virus (VSV), influenza A virus and Sendai virus. In common with other enveloped viruses, all four of these possess the following properties: (1) a small size (*ca*. 70–200 nm in diameter); (2) RNA genomes of limited coding capacity; (3) a limited number of different virus-coded proteins (generally between 5 and 12); (4) a continuous lipid bilayer envelope derived from the host cell membrane (usually plasma membrane) from which the virus buds; (5) an envelope containing one to three different transmembrane glycoproteins. The structural features are typified by VSV (Fig. 5.1).

Despite their general similarity, the enveloped viruses show considerable variation in their shape and size, the sense and intactness of their genome, the geometry of their nucleocapsids, the geometry and disposition of their surface glycoproteins, and the presence or absence of an internal, non-glycosylated membrane protein (designated M) (Table 5.1). The structure and assembly of enveloped viruses has been reviewed in greater detail elsewhere (Lenard and Compans, 1974; Lenard, 1978).

VSV, SFV, influenza and Sendai are among the most widely studied of the diverse group of lipid-enveloped viruses. Their widespread use as virus models reflects a combination of their high yield and ease of growth in cell culture,

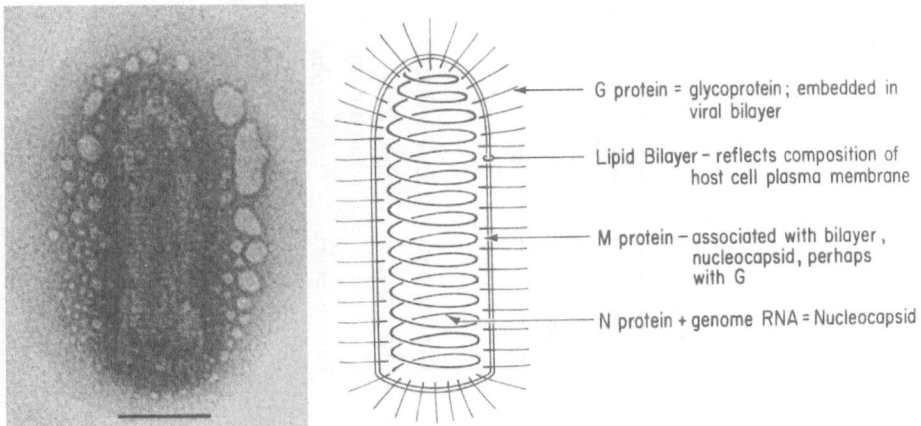

Fig. 5.1 Structure of vesicular stomatitis virus. Left: VSV negatively stained with 1% phosphotungstic acid (bar indicates 50 nm); Right: schematic drawing of VSV indicating the RNA and the three major structural viral proteins.

Table 5.1 Structural differences between four enveloped viruses

Virus	Class	Shape	Size (nm)	Genome	Nucleocapsid geometry	M Protein	Glycoprotein geometry
SFV	Alphavirus	Spherical	70	+Sense, integral	Icosahedral	No	E_1, E_2 and E_3 heterotrimer
VSV	Rhabdovirus	Bullet-shaped	80 × 170	−Sense, integral	Helical	Yes	G-monomer
Influenza A	Myxovirus	Spherical–pleiomorphic	100	−Sense, segmented	Helical	Yes	HA-homodimer NA-homotetramer
Sendai	Paramyxovirus	Pleiomorphic	120–200	−Sense, integral	Helical	Yes	HN–? F–?

their epidemiological importance and their relative lack of pathogenicity. While they represent quite different classes of enveloped viruses, substantial similarities are apparent in their entry processes. Although the processes by which other enveloped viruses enter cells have not been studied in detail, it may be anticipated that the generalizations that emerge from the consideration of these four viruses will apply to the entry strategies of others as well.

5.2 PENETRATION OF ENVELOPED VIRUSES

5.2.1 Fusion versus endocytosis

The geonome of the attacking enveloped virus is separated from the cell cytoplasm by two barriers: the cell plasma membrane and the viral envelope. It is now generally accepted that viral transcription, the first virus-directed synthetic event in infected cells, is initiated by a viral ribonucleoprotein complex, the nucleocapsid, in the cytoplasmic or nuclear compartments of the cell. The complete entry of these viruses into cells in such a way as to initiate an infection necessitates, therefore, the crossing of both the plasma and viral membranes by the viral genome. Passage of the genome across the plasma membrane is termed 'penetration', and passage across the viral membrane into the cytoplasm, 'uncoating'. Penetration is thus necessary, but is not sufficient, to complete the process of uncoating.

Two mechanisms have historically been proposed to accomplish viral penetration and uncoating: endocytosis and fusion (reviewed by Dales, 1973; Lonberg-Holm and Philipson, 1974). These proposals have generally been regarded as being mutually exclusive, and each has suffered from a major limitation. The fusion proposal states that infection is initiated by fusion of the viral envelope with the plasma membrane, thus accomplishing penetration and uncoating simultaneously in a single step. Of the commonly studied enveloped viruses, however, only the paramyxoviruses seem to possess the requisite fusion ability with the cell surface and even within this group, the fusibility of different strains varies greatly (Choppin and Compans, 1975). The proposal that viruses enter the cell by what is now known as adsorptive endocytosis ('viropexis') was originally made by Fazekas de St. Groth (1948). The endocytosis mechanism, however, merely internalizes the membrane transit problem; two membranes still separate the viral genome from the cell cytoplasm. Until recently, very little serious attention was given to the problem of how uncoating might occur following penetration by viropexis. As detailed below, our current understanding represents an elegant synthesis of what once seemed to be two disparate ideas.

Early electron microscope studies indicated that most viruses could enter cells by endocytosis (reviewed by Dales, 1973; Lonberg-Holm and Philipson, 1974), but several contemporaneous reports also showed micrographs suggestive of fusion between the viral envelope and the plasma membrane.

Endocytosis of viral particles is intrinsically better observed than fusion using electron microscopy. Endocytosed particles maintain their morphology within recognizable organelles for an extended period of time, while recognition of fusion depends upon the identification of a membrane continuity between a cell membrane and a still recognizable viral membrane. Some technical difficulties involved in recognizing fused viral particles in electron micrographs have been discussed by Dales (1973). These electron microscopic observations, however, led to the differing views concerning the nature of the viral penetration process.

Several quantitative electron microscopic studies have concluded that adsorptive endocytosis represents the overwhelmingly preferred route for penetration by VSV, SFV and influenza (Simpson *et al.*, 1969; Dahlberg, 1974; Helenius *et al.*, 1980; Dourmashkin and Tyrrell, 1974). In only one of these studies was fusion observed between the viral particle and cell membranes under normal conditions of infection, and then only as an extremely rare event (Dahlberg, 1974). Studies with radioactively labeled SFV and VSV showed that the proportion of added viral particles that bind and penetrate cells is independent of multiplicity from very low to very high particle:cell ratios (Marsh and Helenius, 1980; Miller and Lenard, 1980; see below). Consequently, the electron microscope observations, necessarily made at very high particle:cell ratios, most likely reflect the penetration process occurring at much lower multiplicities during normal infection. In agreement with this conclusion, Fan and Sefton (1978), using antibody-mediated, complement-dependent cell lysis as an assay to detect the presence of viral antigens fused into the cell surface, found no evidence of fusion of Sindbis virus (a close relative of SFV) or VSV with the cell surface during virus penetration. In contrast, infection of cells by Sendai virus, which can fuse at neutral pH, did elicit cell lysis in this assay (Fan and Sefton, 1978). The qualitative results obtained from this assay, however, do not provide a quantitative measure of the fraction of Sendai particles that penetrate the cell by fusion with the cell surface. Electron microscopic studies have shown that Sendai, as well as other paramyxoviruses, also penetrate cells by endocytosis (Dales, 1973; Choppin and Compans, 1975).

5.2.2 Receptors for enveloped viruses

The infecting virus must first stably associate with a 'receptor' on the outer surface of the cell. The word 'receptor' has been used in cell biology to express two quite different concepts. When applied to such entities as receptors for hormones, asialoglycoproteins, or low-density lipoprotein (LDL) (see other chapters in this volume), the word refers to a molecule or molecular complex in a precisely controlled autoregulatory system that has evolved to serve the physiological needs of the multicellular organism. The same word has been applied, however, to those cell surface molecules to which viruses initially attach en route to a productive infection (e.g. see Lonberg-Holm, 1981). Since a virus must evolve so as to maximize its opportunities to overcome the cell's defenses, it might be expected that many viruses display a broad range of potential receptors. In the case of penetration of viruses by endocytosis, viral molecules must constitute subsets of those cell surface molecules that can be internalized into endocytic vacuoles by the ongoing process of adsorptive endocytosis. Many, but not all, of the cell surface protein and lipid molecules are internalized in this way (Schneider *et al.*, 1979a,b; Bretscher *et al.*, 1980; Goldstein *et al.*, 1979). Potential virus receptors thus constitute a large population of the total cell surface molecules, and indeed, the viruses under discussion all possess many binding sites on a wide variety of different cells. Each BHK cell possesses several thousand binding sites for SFV (Marsh and Helenius, 1980) and VSV (Miller and Lenard, 1980), several orders of magnitude more than are required to initiate an infection. Penetration, as measured by the acquisition by the virus of resistance to external protease, represents a constant fraction of bound virus (usually around one-third) up to very high multiplicities (Miller and Lenard, 1980; Marsh and Helenius, 1980). The particles that remain on the surface are presumably bound to molecules that do not undergo endocytosis, and represent dead-end particles, incapable of initiating infection.

Within the limitation imposed by the requirement for endocytosis, considerable diversity of virus receptors exists. The histocompatibility antigen apparently serves as the major receptor for SFV on the surface of several human and mouse cell lines (Helenius *et al.*, 1978). SFV can, however, also infect cells possessing no such antigens on their surface, and such infection is obviously mediated by different molecules (Oldstone *et al.*, 1980). Influenza and Sendai viruses both attach to sialic acid-containing proteins and lipids on the cell surface, but the specific sialyl linkage required for optimum affinity differs (Paulson *et al.*, 1979; Holmgren *et al.*, 1980; Markwell and Paulson, 1980). Different sialyl linkages are even required for binding of two very closely related strains of influenza (Carroll *et al.*, 1981). This indicates that subpopulations of sialic acid residues may serve as receptors for different viruses that appear to penetrate the cell, uncoat and replicate by very similar mechanisms. Unique subsets of endocytosed cell surface molecules can thus serve in an equivalent way for the penetration of different enveloped viruses.

The widespread use of DEAE-dextran to enhance virus infectivity by increasing binding underscores the lack of a requirement for specific surface molecules as receptors. Under one set of conditions, DEAE-dextran was found to increase BHK cell-associated VSV particles about 3–4-fold (Fig. 5.2(a)), and to increase the amount of penetration, uncoating and primary RNA synthesis (the first measurable post-uncoating event) by the same amount (Fig. 5.2(b)) without altering the time course of these events (Fig.

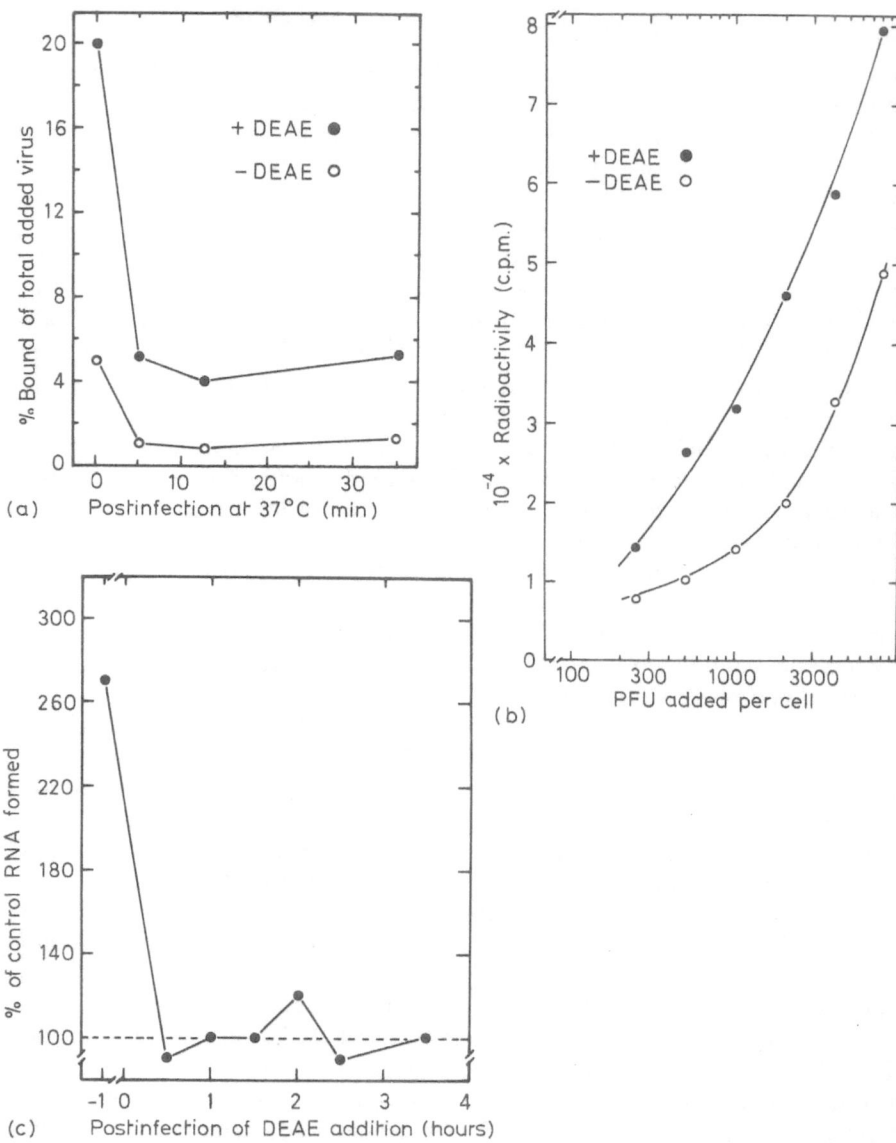

Fig. 5.2 Effect of DEAE-dextran on VSV binding and infection. (a) Proportion of a radioactive VSV inoculum bound to cells in the presence or absence of 10 μg of DEAE-dextran/ml. After VSV (10 pfu/cell) was bound to the cells for 1 hour at 5°C, the cells were washed and then allowed to warm up for the indicated times. (b) Primary VSV RNA formed by 5 hours postinfection in the presence or absence of 10 μg of DEAE-dextran/ml. DEAE-dextran was present only during the time of binding of virus to cells (5°C, 1 hour) at the indicated multiplicity of infection. (c) Primary VSV RNA formed at 5 hours postinfection when 10 μg of DEAE-dextran/ml is added at the time of binding (−1 hour; 500 pfu/cell) or at the indicated time postinfection. After its addition, DEAE-dextran was maintained throughout the time of the assay.

5.2(a)(c)). Since both VSV envelopes and cell surfaces are negatively charged at neutral pH, the DEAE-dextran-mediated binding is probably electrostatic. It seems unlikely that this co-ordinate increase in binding, penetration and transcription requires the presence of any specific receptor protein or lipid.

5.2.3 Sequence of events in viral endocytosis

Endocytosis of virus bound to the cell surface follows a rather precise sequence of events, which are closely similar for SFV, VSV and influenza. Attachment of viruses to the cell surface was often preferential for certain recognizable morphological features, especially microvilli (Bächi, 1970; Helenius *et al.*, 1980; Matlin *et al.*, 1981). If viruses were allowed to bind to cells in the cold (4°C), they remained on the cell surface and were accessible to added proteases.

Within a few minutes of the start of incubation at 37°C, individual virus particles were seen associated with coated pits on the cell surface, within coated and uncoated vesicles inside the cell, and within larger and uncoated intracellular vacuoles containing several viral particles (Fig. 5.3). Viruses were seen at later times, although less reproducibly, in secondary lysosomes (Dahlberg, 1974; Dourmashkin and Tyrell, 1974; Helenius *et al.*, 1980; Matlin *et al.*, 1981). Except for the rate of entry into lysosomes, the timing of these events are closely similar to those reported for internalization of α_2-macrogobulin, low-density lipoprotein, and a variety of hormones that enter cells exclusively through the coated pits by receptor-mediated endocytosis (Goldstein *et al.*, 1979). Indeed, one recent study has shown that α_2-macroglobulin and VSV are found in the same coated pits on the cell surface and in the same intracellular vacuoles following internalization (Dickson *et al.*, 1981).

These results suggest that viral penetration does not require specific modification of the cell surface by the invading virus, or of the virus by the cell surface (cf. discussion of poliovirus by Boulanger and Lonberg-Holm, 1981). Instead, the virus must simply become attached to a cell surface molecule that is capable of being internalized through coated pits; the normal cellular processes then assure the arrival of the virus particle at the correct destination (Marsh and Helenius, 1980). The use of normal cellular processes by invading viruses is, of course, a classical strategy by which viruses survive with only a limited genome.

5.3 pH AND VIRAL UNCOATING

5.3.1 Intralysosomal and intravacuolar pH

The key to viral uncoating following penetration is the low pH that characterizes both the lysosomes and the intracellular vacuoles of the endocytic path-

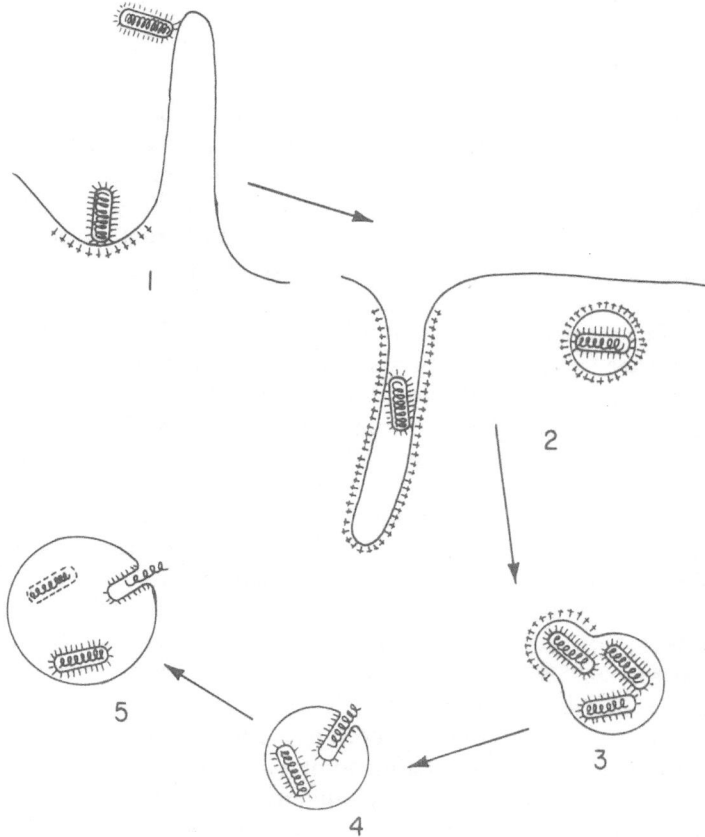

Fig. 5.3 Postulated sequence of events in VSV penetration and uncoating (adapted from Helenius *et al.*, 1980 and Wehland *et al.*, 1981). 1. VSV binds to the cell plasma membrane at the microvilli and eventually at clathrin-coated regions. 2. The virus becomes associated with coated regions beneath the plane of the membrane in either deep coated pits (seen in longitudinal section at the left or in cross section at the right) and/or in coated regions which are true coated vesicles (also shown by the structure at the right). 3. VSV is next found in non-coated vesicles formed either from coated pits or by fusion with coated vesicles bearing virus. 4. These vesicles acidify very quickly (Maxfield and Tycko, 1982) probably allowing some VSV to fuse with the vesicular membrane and uncoat. 5. Within 15–20 min the endocytic vesicles fuse with lysosomes. The acid pH of the lysosome may allow viral uncoating to continue. At the same time proteolytic degradation of the virus particle may occur (as indicated by the disintegrating particle in the upper left).

way. Ohkuma and Poole (1978) measured the intralysosomal pH directly in mouse peritoneal macrophages. They allowed fluoresceinated dextran to accumulate in lysosomes and then measured the pH-dependent fluorescence excitation spectrum of the intralysosomal fluorescein. With this technique they measured the intralysosomal pH to be 4.75 ± 0.06. Tycko and Maxfield (1982) attached fluorescein to α_2-macroglobulin, and measured its excitation spectrum by video intensification microscopy within a few minutes after endocytosis, when the fluoresceinated complex was still present in endocytic vacuoles prior to fusion with lysosomes. They measured a pH of 5.0 ± 0.2 in endocytic vacuoles of mouse 3T3 fibroblasts, while the lysosomes of the same cells had a pH of 4.6 ± 0.2, in close agreement with the measurements of Ohkuma and Poole (1978) in a different cell type.

5.3.2 pH-dependence of viral fusion

The importance of low pH to viral uncoating was discovered by Helenius and co-workers, who showed that SFV and other enveloped viruses acquired the capacity of undergoing fusion with a variety of other membranes upon exposure to pH's between 5.0 and 6.5 (Fig. 5.4). Several methods have been used to measure this low-pH-induced viral fusion. Electron microscopic examination showed fusion of SFV to cell membranes that occurred only at low pH (White *et al.*, 1980, 1981). Such viral fusion in confluent monolayers led to syncytia formation. BHK cells could be directly infected at low pH by SFV bound to the cell surface, evidence indirectly indicating fusion between the viral envelope and the cell membrane (Fig. 5.4(a); White *et al.*, 1980). Fusion of SFV bound to protein-free liposomes occurred quantitatively at pH 5.0 as measured by the digestion of the SFV nucleocapsids by nucleases or proteases trapped in the lysosomes (Fig. 5.4(a); White and Helenius, 1980).

Virus-induced hemolysis has proved to be a versatile and simple experimental approach to measure the fusibility of viral envelopes. One consequence of fusion of viruses with membranes is the leakage of molecules across the membrane (Poste and Pasternak, 1978). When viruses fuse with erythrocytes, hemoglobin leaks into the supernatant, providing a convenient assay. Although definitive evidence that virus-induced hemolysis is associated with virus–erythrocyte fusion has so far been obtained only for Sendai (Lyles, 1979), fusion is generally regarded as necessary for viral hemolysis. Consistent with this, erythrocyte–erythrocyte fusion has frequently been observed in association with viral hemolysis (Knutton, 1977; Väänänen and Kääriäinen, 1980; Maeda and Ohnishi, 1970).

The pH-dependence of hemolysis by SFV was similar to that for fusion of SFV with liposomes (Fig. 5.4), supporting the contention that hemolysis depends upon virus–erythrocyte fusion. A similar pH-dependence of hemolysis

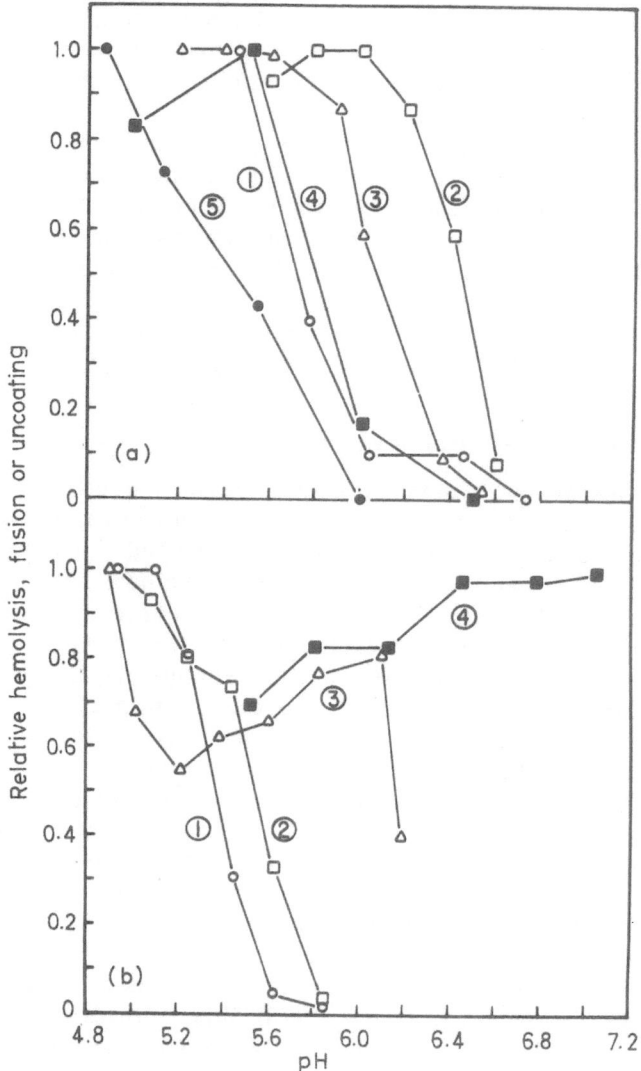

Fig. 5.4 (a) pH-dependence of hemolysis, fusion or uncoating by SFV and VSV. 1. Hemolysis by SFV (Lenard and Miller, 1981). 2. Hemolysis by SFV (Väänänen and Kääriäninen, 1979). 3. Fusion of SFV with liposomes (White and Helenius, 1980). 4. Uncoating of SFV by BHK cells in the presence of 20 mM-methylamine, determined by measurement of pH-induced viral transcription (White et al., 1980). 5. Hemolysis by VSV (Bailey et al., 1981). (b) pH-dependence of hemolysis by influenza virus and Sendai virus. 1. Influenza strain A/HK/8/68 (H3N2, Lenard et al., 1982). 2. Influenza strain A/PR/8/34 (HONI, Lenard et al., 1982). 3. Strain A/WSN/33, HONI, (unpublished). 4. Sendai virus (Lenard and Miller, 1981).

has been found also for VSV (Fig. 5.4(a)), and for influenza, although there were differences between strains examined under identical conditions (Fig. 5.4(b); cf. Maeda and Ohnishi, 1980; Huang *et al.*, 1981; Lenard and Miller, 1981). Hemolysis by Sendai was high at pH 7.0 and dropped to about 70% of this value at pH 5.5 (Fig. 5.4(b); cf. Hosaka, 1958; Neurath, 1965; Huang *et al.*, 1981).

The results presented by all these different techniques were similar (Fig. 5.4); viral fusion was maximal around pH 5.0 and dropped off sharply to negligible values above pH 6.5. Since fusion of the SFV envelope occurred readily with liposomes (White and Helenius, 1980), fusion was intrinsic to the viral envelope, and required no cellular proteins to activate or mediate the process. Cholesterol was an essential constituent of the liposomes, but no specific phospholipid was required (White and Helenius, 1980). In contrast, intact viral glycoproteins were required for fusion. Spikeless particles of SFV or VSV could not fuse with liposomes (White and Helenius, 1980; Bailey *et al.*, 1981). Moreover, VSV lipid vesicles containing only the spike glycoproteins and none of the other viral proteins had a pH profile of hemolysis identical with that of intact VSV (Bailey *et al.*, 1981).

5.3.3 Inhibition of viral infection by amines

Substances that can raise lysosomal pH when added to cells have been shown to be effective inhibitors of infection by enveloped viruses. Ammonia, simple aliphatic amines, and organic amines of diverse chemical structure all have the capacity to raise lysosomal (and presumably endosomal) pH in a reversible fashion. These lysosomotropic agents (de Duve *et al.*, 1974) diffuse across lipid membranes in their uncharged (basic) form, and become protonated (acidified) and trapped in the low pH of the lysosomal or vacuolar compartments, where they may be concentrated up to several-hundred-fold. Lysosomotropic agents raise lysosomal pH within 2 minutes after addition to cells. The characteristic lysosomal pH is re-established by removing the amine from the cells, although recovery times vary somewhat for different amines (Ohkuma and Poole, 1978).

It was reported very early that any of several ammonium salts inhibited influenza virus infection at low concentration (<1 mM), both in tissue culture cells (Higo *et al.*, 1957; Jensen *et al.*, 1961; Jensen and Liu, 1961; Eaton and Scala, 1961; Oxford and Schild, 1968) and in live mice (Oxford and Schild, 1967). Ammonium salts affected the cells rather than the viruses themselves, since incubation of the viruses with ammonia did not inactivate them. The ammonium salts acted early in the infectious cycle; addition even 1 hour after infection greatly decreased their inhibitory effect. Cell binding and internalization were unaffected by the presence of ammonium salts.

Simple aliphatic amines were soon discovered to act very similarly to ammonium salts (Jensen and Liu, 1963; Oxford and Schild, 1968). Chloroquine, an antimalarial drug with lysosomotropic activity, was found to inhibit SFV, VSV and influenza infections in a manner very similar to that of aliphatic amines (Helenius *et al.*, 1980; Miller and Lenard, 1980, 1981; Matlin *et al.*, 1981).

Of particular interest was the aliphatic amine amantadine (Davies *et al.*, 1964; Hoffman *et al.*, 1965; Oxford and Schild, 1967, 1968) and its methylated analog rimantadine. Amantadine is the only licensed pharmacological agent against certain strains of influenza in the United States. Although very low concentrations of amantadine may act more specifically on highly sensitive strains of influenza (Zvonarjev and Ghendon, 1980; Lenard *et al.*, 1982), this agent does raise lysosomal, and presumably vacuolar, pH at the concentrations generally used to inhibit viral infections. This lysosomotropic action therefore most likely accounts for its inhibitory action against SFV, VSV, Sendai and the more resistant strains of influenza (Hoffman *et al.*, 1965; Oxford and Schild, 1968; Kato and Eggers, 1969; Shehel *et al.*, 1978; Lubeck *et al.*, 1978; Koff and Knight, 1970; Helenius *et al.*, 1980).

Several other amines of complex structure, commonly used as antihistamines or local anaesthetics, have similar effects to chloroquine and methylamine on VSV infection (Miller and Lenard, 1981). Several of these compounds are known to have lysosomotropic activity and all have the combination of hydrophobic structure and ionizable amino group required for such activity. All of these agents as well as the aliphatic amines, chloroquine, and amantadine discussed above were found to act very early in the infectious process. The amines did not affect virus binding to the cell, virus penetration, or virus-induced fusion provided the pH was properly controlled. In the case of VSV infection, inhibition by methylamine and chloroquine was found to be reversible. Within 1 hour of removal of methylamine from infected cells, or 2 hours of chloroquine removal, a burst of viral RNA transcription was observed, with total RNA synthesis rising rapidly to the level found in uninfected cells. This suggested that intact virus particles were accumulating intracellularly, and becoming uncoated after recovery of the low pH in the vacuoles and lysosomes (Miller and Lenard, 1981).

A second requirement, in addition to endocytosis, for virus–receptor interactions now becomes apparent. The virus must be able to associate with the receptor at the vacuolar pH, since close contact between the membrane and the viral envelope would seem to be necessary for fusion. Consistent with this requirement, binding to cells of SFV (Fries and Helenius, 1979), VSV (unpublished observations) and influenza (Matlin *et al.*, 1981; unpublished observations) is comparable at extracellular and vacuolar pH. Virus receptors differ in this regard from some physiological receptors, such as those for asialoglycoprotein and LDL, which have been observed to dissociate from their ligands inside endocytic vesicles (Wall *et al.*, 1980; Anderson *et al.*, 1977).

The use by these enveloped viruses of the endocytic pathway and the low intravesicular and intralysosomal pH is a specific variation of a general theme that is used by other invasive agents to enter cells. In direct analogy with enveloped viruses, diphtheria toxin has recently been found to enter cells by adsorptive endocytosis and to undergo at low pH a conformational change to a membrane-active form that permits passage of the toxic fragment into the cytoplasm. The entry of diphtheria toxin into cells is thus inhibited by chloroquine or methylamine, and this inhibition can be overcome by incubation of the cells at low pH (Chapter 8; Sandvig and Olsnes, 1980; Draper and Simon, 1980). Reovirus undergoes a specific proteolytic activation in lysosomes to the active form required for virus replication (Silverstein and Dales, 1968; Silverstein *et al.*, 1972). While it is not known how these activated particles leave the lysosome, it seems likely that the proteolysis generates a membrane-active protein in the particle's outer shell. Vaccinia virus, mouse hepatitis virus, and adenovirus are all inhibited by chloroquine, although it remains to be determined whether this inhibition arises from the drug's lysosomotropic effect (Mallucci, 1966; Allison, 1967).

5.3.4 Is low-pH-mediated fusion sufficient for viral uncoating?

As has been described above, brief treatment at low pH of cells that have virus adsorbed to their plasma membranes leads to massive fusion between the cell membranes and the viral envelopes (White *et al.*, 1980, 1981). Since the fusion reaction itself is unaffected by chloroquine or any of the other lysosomotropic agents, it may be predicted that such fusion will initiate infection without inhibition by chloroquine, provided that this fusion event is the *only* requirement for uncoating. This prediction was successfully tested using SFV. Low-pH-induced fusion between the viral envelope and the cell membrane in the presence of chloroquine initiated an infection at least as effectively as occurred in the absence of chloroquine at neutral pH (Helenius *et al.*, 1980; White *et al.*, 1980).

Similar experiments were not successful, however, in inducing infection by either VSV (unpublished observations) or influenza (Matlin *et al.*, 1981). This negative result suggested that, at least for these viruses, the low-pH-mediated fusion step, while necessary for uncoating, is not sufficient; an additional uncoating step could be inferred.

Observations on Sendai virus also point to a further requirement for uncoating. Sendai fuses readily with cell membranes during infection at neutral pH (Fan and Sefton, 1978). Yet, surprisingly, Sendai infection is inhibited by all the lysosomotropic agents tested, at similar concentrations to those required for inhibition of SFV, VSV and influenza infection, and at a similar early stage in infection (Miller and Lenard, 1981). While this clearly indicates a low pH requirement for initiation of Sendai infection, it is not at all clear what such a requirement might be. Apparently, fusion of Sendai virus with the cell membrane is not sufficient in itself to uncoat the nucleocapsid and initiate infection.

The obvious difference between SFV, on the one hand, and VSV, influenza and Sendai on the other, is the existence of M protein in the latter. In all three of these viruses, M protein is the smallest and most abundant viral protein, generally accounting for about 30–40% of the total (Lenard and Compans, 1974). It is tightly associated with both the viral envelope and the nucleocapsid, and is generally considered necessary to organize the nascent viral particle during budding. Its association with nucleocapsids causes inhibition of RNA synthesis in both VSV (Clinton *et al.*, 1978; Martinet *et al.*, 1979; Carroll and Wagner, 1979; Combard and Printz-Ané, 1979) and influenza (Zvonarjev and Ghendon, 1980). It is also the protein chiefly responsible for conferring amantadine sensitivity on highly sensitive strains of influenza (Lubeck *et al.*, 1978; Hay *et al.*, 1979).

It seems likely that the viral nucleocapsid must free itself from M protein after fusion, before it can start transcribing viral RNA. In view of the association of M protein with the viral envelope, this process may be envisioned as occurring on the cytoplasmic side of the vacuolar or lysosomal membrane, and may well constitute a second step in viral uncoating. It has been suggested that amantadine acts in highly sensitive strains of influenza by potentiating the M-protein-mediated inhibition of viral RNA polymerase (Zvonarjev and Ghendon, 1980; Lenard *et al.*, 1982) and may thus act on the second rather than the first uncoating step at the low concentrations that effectively inhibit these strains. With the development of *in vitro* fusion reactions between enveloped viruses and liposomes or erythrocyte membranes, it should be possible to define precisely the requirements for functional uncoating of each enveloped virus *in vitro*.

5.4 VIRUS UNCOATING AND CELLULAR CYTOPATHIC EFFECTS

The association of virus with lysosomes during the process of infection may be the cause of some of the cellular cytopathic effects (CPE) induced by many different viruses (see Enders, 1954; Allison, 1967; Bablanian, 1975 for reviews). CPE are observed as effects on cell morphology (producing cell rounding) and inhibition of macromolecular synthesis. The virus-induced CPE can be initiated in two ways. First, CPE can be produced by specific, preformed viral components in the absence of viral transcription and translation. The CPE occur rapidly after addition of virus to cells and require high concentrations of the viral effectors. The second type of CPE produced by viruses occurs in cells several hours after infection. This effect requires primary transcription and translation of some viral protein and, consequently, can be seen after infection at low multiplicity (Bablanian, 1972, 1975).

These two causes of CPE have been extensively studied with paramyxoviruses, and correlate with the phenomena of fusion from without (FFWO) and fusion from within (FFWI) (reviewed in Poste, 1972; Poste and Pasternak, 1978). FFWO results from addition to cells of very high concentrations of either infectious or ultraviolet-light-treated Sendai virus or Newcastle Disease Virus (NDV), and occurs as rapidly as 7 min after viral adsorption (Lyles and Landsberger, 1979). FFWI requires only a low multiplicity of infectious virus, and occurs several hours after the first addition of virus. Common to, and occurring simultaneously with, both types of fusion is a transient increased permeability to cations and a depolarization of the membrane (Fuchs *et al.*, 1978; Poste and Pasternak, 1978). This permeability change, presumably caused by the presence of viral envelope proteins in the plasma membrane, has been suggested as the general cause of all the CPE induced by viruses (Carrasco and Smith, 1976; Carrasco, 1977). Prevention of the association of those viral membrane components with cellular membranes should prevent the permeability changes and resultant CPE. Such a prediction was supported by the existence of a conditional lethal mutant of Sendai lacking the HN (hemagglutinin/neuraminidase) glycoprotein (Portner *et al.*, 1975). At the non-permissive temperature complete virus lacking only the HN proteins could be produced in normal amounts but the CPE of inhibited cellular protein synthesis was largely absent.

The above example suggests the importance both of membrane permeability changes and viral glycoproteins in the etiology of the CPE. The specific importance of lysosomal permeability to the onset of CPE was emphasized by Allison and co-workers. Prior to the onset of noticeable CPE in cells infected by viruses, the lysosomes became reversibly permeable to low-molecular-weight molecules (Allison and Mallucci, 1965). Soon thereafter the lysosomes became permeable to larger molecules including lysosomal hydrolases. These latter permeability changes occurred coincidentally with observed morphological CPE. The CPE could also be elicited in the absence of virus by permeability changes induced specifically in lysosomal membranes (Allison *et al.*, 1966). In the case of hepatitis virus, the permeability changes, the production of virus, and the subsequent polykaryocytosis induced by the virus could all be prevented by addition of lysosomotropic amines such as chloroquine (Allison, 1967).

These earlier observations fit neatly into what we now know about the behavior of viral glycoproteins at low pH; that is, these proteins induce permeability changes in membranes to which they fuse. It seems entirely reasonable to conclude that the CPE, particularly that form induced by large concentrations of irradiated non-infectious virus (cf. Miller and Lenard, 1982) could be elicited by the leakage of lysosomal hydrolases consequent to the low-pH-induced fusion of the virus with lysosomes. The specific action then of lysosomotropic amines in preventing those CPE would be due to prevention of that intralysosomal fusion.

5.5 CONCLUSION AND SUMMARY

In this review we have tried to document the assertion that the basic processes involved in uncoating of enveloped viruses are now known, and to explore the implications of this knowledge. Uncoating consists of adsorptive endocytosis of virus particles attached to the cell surface, followed by fusion with vacuolar or lysosomal membranes potentiated by the low pH present in those organelles. This process implies that only these molecules that are internalized by the endocytic process can serve as viral receptors, and that viruses will associate with these receptors at low pH. Within these limitations, a variety of different surface molecules seem to be capable of serving as virus receptors for each enveloped virus, with different specific requirements being evident for different virus types. The available evidence suggests that an additional uncoating step, beyond fusion, is required to initiate infection by those enveloped viruses possessing an M protein. This step may involve the release of the viral nucleocapsid from the M-protein-containing patch of cytoplasmic membrane. Finally, the low-pH-induced viral fusion may have implications for the development of cellular cytopathic effects.

REFERENCES

Allison, A.C. (1967), *Perspect Virol.*, **5**, 29–61.

Allison, A.C. and Mallucci, L. (1965), *J. Exp. Med.*, **121**, 463–485.

Allison, A.C., Magnus, I.A. and Young, M.R. (1966), *Nature (London)*, **209**, 874–878.

Anderson, R.G., Goldstein, J.L. and Brown, M.S. (1977), *Nature (London)*, **270**, 695–699.

Bablanian, R. (1972), *Symp. Soc. Gen. Microbiol.*, **22**, 359–381.

Bablanian, R. (1975), *Prog. Med. Virol.*, **19**, 40–83.

Bächi, T. (1970), *Pathol. Microbiol.*, **36**, 81–107.

Bailey, C.A., Miller, D.K. and Lenard, J. (1981), *J. Cell Biol.*, **91**, 111a.

Bretscher, M.S., Thomson, J.H. and Pearse, B.M.F. (1980), *Proc. Natl. Acad. Sci. U.S.A.*, **77**, 4156–4159.

Carrasco, L. (1977), *FEBS Lett.*, **76**, 11–15.

Carrasco, L. and Smith, A.E. (1976), *Nature (London)*, **264**, 807–809.

Carroll, A.R. and Wagner, R.R. (1979), *J. Virol.*, **29**, 134–142.

Carroll, S.M., Higa, H.M. and Paulson, J.C. (1981), *J. Biol. Chem.*, **256**, 8357–8363.

Choppin, P.W. and Compans, R.W. (1975), in *Comprehensive Virology* (H. Fraenkel-Conrat and R.R. Wagner, eds), Plenum Press, New York, vol. 5, pp. 95–178.

Clinton, G.M., Little, S.P., Hagen, F.S. and Huang, A.S. (1978), *Cell*, **15**, 1455–1462.

Combard, A. and Printz-Ané, C. (1979), *Biochem. Biophys. Res. Commun.*, **88**, 117–123.

Dahlberg, J.E. (1974), *Virology*, **58**, 250–262.

Entry of Enveloped Viruses into Cells 137

Dales, S. (1973), *Bacteriol. Rev.*, **37**, 103–135.

Davies, W.L., Grunert, R.R., Haff, R.F., McGahen, J.W., Neumayer, E.M., Paulshock, M., Watts, J.C., Wood, R.T., Herrmann, E.C. and Hoffmann, C.E. (1964), *Science*, **144**, 862–863.

de Duve, C., De Barsy, T., Poole, B., Trouet, A., Tulkens, P. and van Hoof, F. (1974), *Biochem. Pharmacol.*, **23**, 2495–2531.

Dickson, R.B., Willingham, M.C. and Pastan, I. (1981), *J. Cell. Biol.*, **89**, 29–34.

Dourmashkin, R.R. and Tyrrell (1974), *J. Gen. Virol.*, **24**, 129–141.

Draper, R.K. and Simon, M.I. (1980), *J. Cell. Biol.*, **87**, 849–854.

Eaton, M.D. and Scala, A.R. (1961), *Virology*, **13**, 300–307.

Enders, J.F. (1954), *Annu. Rev. Microbiol.*, **8**, 473–502.

Fan, D.P. and Sefton, B.M. (1978), *Cell*, **15**, 985–992.

Fazekas de St. Groth, S. (1948), *Nature (London)*, **162**, 294–295.

Fries, E. and Helenius, A. (1979), *Eur. J. Biochem.*, **97**, 213–220.

Fuchs, P., Spiegelstein, M., Haimsohn, M., Gitelman, J. and Kohn, A. (1978), *J. Cell Physiol.*, **95**, 223–234.

Goldstein, J.L., Anderson, R.G.W. and Brown, M.S. (1979), *Nature (London)*, **279**, 679–685.

Hay, A.J., Kennedy, N.C.T., Skehel, J.J. and Appleyard, G. (1979), *J. Gen. Virol.*, **42**, 189–191.

Helenius, A., Kartenbeck, J., Simons, K. and Fries, E. (1980), *J. Cell Biol.*, **84**, 404–419.

Helenius, A., Morein, B., Fries, E., Simons, K., Robinson, P., Schirrmacher, V., Terhorst, L. and Stromiger, J. (1978), *Proc. Natl. Acad. Sci. U.S.A.*, **75**, 3846–3850.

Higo, N., Kaneko, T., Kabayashi, N., Nakajima, S. and Sasaki, Y. (1957), *Virus (Osaka)*, **7**, 381–387.

Hoffman, C.E., Neumayer, E.M., Haff, R.F. and Goldsby, R.A. (1965), *J. Bacteriol.*, **90**, 623–628.

Holmgren, J., Svennerholm, L., Elwing, H., Fredman, P. and Strannegärd, Ö. (1980), *Proc. Natl. Acad. Sci. U.S.A.*, **77**, 1947–1950.

Hosaka, Y. (1958), *Biken's J.*, **1**, 70–89.

Huang, R.T.C., Rott, R. and Klenk, H.-D. (1981), *Virology*, **110**, 243–247.

Jensen, E.M. and Liu, O.C. (1961), *Proc. Soc. Exp. Biol. Med.*, **107**, 834–838.

Jensen, E.M. and Liu, O.C. (1963), *Proc. Soc. Exp. Biol. Med.*, **112**, 456–459.

Jensen, E.M., Force, E.E. and Unger, J.B. (1961), *Proc. Soc. Exp. Biol. Med.*, **107**, 447–451.

Kato, N. and Eggers, H.J. (1969), *Virology*, **37**, 632–641.

Knutton, S. (1977), *J. Cell Sci.*, **28**, 189–210.

Koff, W.C. and Knight, V. (1970), *J. Virol.*, **31**, 261–263.

Lenard, J. (1978), *Annu. Rev. Biophys. Bioeng.*, **7**, 139–166.

Lenard, J. and Compans, R.W. (1974), *Biochim. Biophys. Acta*, **344**, 51–94.

Lenard, J. and Miller, D.K. (1981), *Virology*, **110**, 479–482.

Lenard, J., Bailey, C.A. and Miller, D.K. (1982), *J. Gen. Virol.*, **62**, 353–355.

Lonberg-Holm, K. (1981) in *Virus Receptors, Part 2, Animal Viruses* (K. Lonberg-Holm and L. Philipson, eds), Chapman and Hall, London, pp. 3–20.

Lonberg-Holm, K. and Philipson (1974), *Monogr. Virol.*, **9**, 1–149.

138   *Receptor-Mediated Endocytosis*

Lubeck, M.D., Schulman, J.L. and Palese, P. (1978), *J. Virol.*, **28**, 710–716.
Lyles, D.S. (1979), *Proc. Natl. Acad. Sci. U.S.A.*, **76**, 5621–5625.
Lyles, D.S. and Landsberger, F.R. (1979), *Biochemistry*, **18**, 5088–5095.
Maeda, T. and Ohnishi, S. (1980), *FEBS Lett.*, **122**, 283–287.
Mallucci, L. (1966), *Virology*, **28**, 355–362.
Markwell, M.A.K. and Paulson, J.C. (1980), *Proc. Natl. Acad. Sci. U.S.A.*, **77**, 5693–5697.
Marsh, M. and Helenius, A. (1980), *J. Mol. Biol.*, **142**, 439–454.
Martinet, C., Combard, A., Printz-Ané, C. and Printz, P. (1979), *J. Virol.*, **29**, 123–133.
Matlin, K.S., Reggio, H., Helenius, A. and Simons, K. (1981), *J. Cell Biol.*, **91**, 601–613.
Maxfield, F.R. and Tycko, B. (1982), *Cell* (in press).
Miller, D.K. and Lenard, J. (1980), *J. Cell. Biol.*, **84**, 430–437.
Miller, D.K. and Lenard, J. (1981), *Proc. Natl. Acad. Sci. U.S.A.*, **78**, 3605–3609.
Miller, D.K. and Lenard, J. (1982), *J. Gen. Virol.*, **60**, 327–333.
Neurath, A.R. (1965), *Acta Virol.*, **9**, 34–46.
Ohkuma, S. and Poole, B. (1978), *Proc. Natl. Acad. Sci. U.S.A.*, **75**, 3327–3331.
Oldstone, M.B.A., Tishon, A., Dutko, F.J., Kennedy, I.T., Holland, J.J. and Lampert, P.W. (1980), *J. Virol.*, **34**, 256–265.
Oxford, J.S. and Schild, G.C. (1967), *Br. J. Exp. Pathol.*, **48**, 235–243.
Oxford, J.S. and Schild, G.C. (1968), *J. Gen. Virol.*, **2**, 377–384.
Paulson, J.C., Sadler, J.E. and Hill, R.L. (1979), *J. Biol. Chem.*, **254**, 2120–2124.
Portner, A., Scroggs, R.A., Marx, P.A. and Kingsbury, D.W. (1975), *Virology*, **67**, 179–187.
Poste, G. (1972), *Int. Rev. Cytol.*, **33**, 157–252.
Poste, G. and Pasternak, C.A. (1978), *Cell Surface Rev.*, **5**, 306–367.
Sandvig, K. and Olsnes, S. (1980), *J. Cell. Biol.*, **87**, 828–832.
Schneider, Y.-J., Tulkens, P., de Duve, C. and Trouet, A. (1979a), *J. Cell Biol.*, **82**, 449–465.
Schneider, Y.-J., Tulkens, P., de Duve, C. and Trouet, A. (1979b), *J. Cell Biol.*, **82**, 466–474.
Silverstein, S.C., Astell, C., Levin, D.H., Schonberg, M. and Acs, G., (1972), *Virology*, **47**, 797–806.
Silverstein, S.C. and Dales, S. (1968), *J. Cell. Biol.*, **36**, 197–230.
Simpson, R.W., Hauser, R.F. and Dales, S. (1969), *Virology*, **37**, 285–290.
Skehel, J.J., Hay, A.J. and Armstrong, J.A. (1978), *J. Gen. Virol.*, **38**, 97–110.
Tycko, B. and Maxfield, F.R. (1982), *Cell*, **28**, 643–651.
Väänänen, P. and Kääriäinen, L. (1979), *J. Gen. Virol.*, **43**, 593–601.
Väänänen, P. and Kääriäinen, L. (1980), *J. Gen. Virol.*, **46**, 467–475.
Wall, D.A., Wilson, G. and Hubbard, A.L. (1980), *Cell*, **21**, 79–93.
Wehland, J., Willingham, M.C., Dickson, R. and Pastan, I. (1981), *Cell*, **25**, 105–119.
White, J. and Helenius, A. (1980), *Proc. Natl. Acad. Sci. U.S.A.*, **77**, 3273–3277.
White, J., Kartenbeck, J. and Helenius, A. (1980), *J. Cell Biol.*, **87**, 264–272.
White, J., Matlin, K. and Helenius, A. (1981), *J. Cell Biol.*, **89**, 674–679.
Zvonarjev, A.Y. and Ghendon, Y.Z. (1980), *J. Virol.*, **33**, 583–586.

6 Lysosomes and Mononuclear Phagocytes

PHILIP STAHL

Acknowledgements

The contributions of Virgina Shepherd, Paul Schlesinger, Chris Tietze, Christian De Schryver, Ritan Boshans, Siamon Gordon, Alan Ezekowitz and Marilyn Konish to the experimental work presented here and by way of many fruitful discussions is gratefully acknowledged.

Receptor-Mediated Endocytosis
(*Receptors and Recognition*, Series B, Volume 15)
Edited by P. Cuatrecasas and T. F. Roth
Published in 1983 by Chapman and Hall, 11 New Fetter Lane, London EC4P 4EE
© 1983 Chapman and Hall

6.1 INTRODUCTION

Lysosomes are multifaceted intracellular organelles positioned at the cross-roads of the cell biosynthetic and catabolic pathways. For this reason their structure and function have received considerable attention during the past decade. The progress of this effort is summarized in a recent review by Bainton (1981). Morphologically, lysosomes are a collection of diverse intracellular structures which have in common the presence of acid hydrolase activity. The pleiomorphic nature of the lysosomal system has made structure–function assignments difficult. The morphological variation may, in fact, reflect stages in the lysosomal aging process. Following their biosynthesis, it seems that lysosomal enzymes first appear in dense bodies after which they are transferred to autophagic vacuoles, multivesicular bodies and ultimately, residual bodies. Physiologically, lysosomes are characterized by the presence of a low pH relative to the cytoplasm. The low pH provides an optimal environment in which the acid hydrolases operate; it is generated by the action of a proton pump or transport system (Schneider, 1981). The proton translocation mechanism in lysosomes and lysosomal membrane precursors may play an important role in the intracellular transport of enzymes and substrates to the lysosomal system (i.e. proton pumping may be the driving force for some forms of receptor-mediated endocytosis). While all eucaryotic cells have lysosomes, mononuclear phagocytes are unusually rich in lysosomal hydrolase activity, presumably because lysosomes play an important role in those pathophysiological mechanisms which are macrophage-dependent. These include inflammation, antigen processing and presentation, phagocytosis and digestion of bacteria, as well as tumor cell lysis and tissue remodelling (Nathan *et al.*, 1980). The lysosomal system of the mononuclear phagocyte must therefore be highly regulated and for this reason it has been a good system for experimental study.

6.2 LYSOSOME BIOGENESIS

Lysosomal enzyme levels in cultured macrophages are elevated by any number of stimuli and often enzyme secretion is likewise elevated *pari pasu*. However, the details of how lysosomal enzymes are packaged in mononuclear phago-cytes have not been revealed. In fibroblasts, on the other hand, the intracellular transport and packaging of lysosomal enzymes has been the subject of intense research activity over the past decade. Largely based on studies with fibroblasts from patients with lysosomal storage diseases and I-cell disease, a thumb-nail sketch of the packaging pathway and the mechanisms involved has emerged. The recognition marker mannose 6-phosphate (Kaplan *et al.*, 1977;

Sando and Neufeld, 1977) is transferred to the high-mannose oligosaccharide chain of newly synthesized lysosomal enzymes (Hasilik and Neufeld, 1980). The presence of the mannose 6-phosphate marker on newly formed enzyme mediates the transfer to lysosomes. The recognition marker is apparently removed once the enzyme has been transported to lysosomes probably via the action of various lysosomal phosphatases. The biochemical details of the transfer mechanism leading to the formation of the recognition marker have been elucidated by the studies of Kornfeld and associates (Tabas and Kornfeld, 1980; Reitman and Kornfeld, 1981) and by Waheed *et al.* (1981). GlcNAc phosphate is transferred from UDP-GlcNAc to one or more mannose units of the high-mannose chain on the lysosomal enzyme. The resulting phosphodiester-linked GlcNAc is then clipped, resulting in the exposure of a terminal mannose 6-phosphate moiety. The two key enzymes in the assembly of the mannose phosphate recognition marker, a phosphotransferase and a phosphodiesterase, have been partially purified and studied in some detail. Studies with fibroblasts from patients with I-cell disease, a very rare lysosomal storage disease initially studied by Leroy *et al.* (1972), provided the first clue that a common recognition marker may accommodate the intracellular transport of many lysosomal enzymes. I-cell fibroblasts are unable to package many lysosomal enzymes and as a result, they accumulate in the extracellular compartment (Hickman and Neufeld, 1972). Reitman and Kornfeld (1981) have recently shown that certain forms of I-cell fibroblasts lack the phosphotransferase necessary for the assembly of the recognition marker. This appears to be the solution to the enigma of I-cell disease.

The generality of the recognition marker has been demonstrated in a paper by Fischer *et al.* (1980) where membranes from many tissues have been shown to have receptors for mannose phosphate-containing lysosomal enzymes. There is, however, an observation or two which suggests that other recognition mechanisms may be operative in lysosomal biogenesis. First, the distribution of some lysosomal enzymes (e.g. α-L-fucosidase) is strongly affected in I-cell disease whereas other enzymes are unaffected (e.g. acid phosphatase) (Miller *et al.*, 1981). Second, I-cell disease affects some tissues much more than others. In this regard, the mechanism of intracellular recognition and transport of lysosomal enzymes in macrophages has recently come under careful study because macrophages regulate their lysosmal activities closely and because a second receptor, specific for mannose-terminal oligosaccharides including many lysosomal hydrolases, is expressed by macrophages.

6.3 LYSOSOMAL ENZYME SECRETION BY MACROPHAGES

Macrophages are well-known secretors of lysosomal enzymes. Resident and elicited macrophages continuously secrete lysosomal hydrolases into the culture media by a process which appears to require protein synthesis. This would suggest that secreted enzymes are diverted from the normal intracellular biosynthetic pathway into the extracellular compartment rather than arising from the fusion of secondary lysosomes with the plasma membrane (Schnyder and Baggiolini, 1978). The mode of release by macrophages contrasts sharply with induced secretion of lysosomal hydrolases by polymorphonuclear leukocytes where release is rapid (Smolen *et al.*, 1981). Consistent with the conclusion that macrophages divert enzymes is the observation that particles, such as zymosan and asbestos, trigger secretion of lysosomal hydrolases over and above normal secretory levels but the release carries on for several days (Davies and Bonney, 1979; Dean *et al.*, 1979). In fact, the stimulator can be removed from the media beforehand and enzyme release will still ensue. Amines are well known to trigger lysosomal enzyme release in fibroblasts (Gonzalez-Noriega *et al.*, 1980) presumably by by-passing the intracellular packaging machinery. Macrophages appear to respond to amines similarly (Riches and Stanworth, 1980). Agents such as C5a (McCarthy and Henson, 1979) also stimulate secretion; however, experimental data are not available as to whether protein synthesis is required.

The biosynthesis of two lysosomal enzymes, β-glucuronidase and β-galactosidase, have been studied in primary macrophage cultures (Skudlarek and Swank, 1979, 1981). These two hydrolases are synthesized as precursors with higher molecular weights than the mature enzyme. β-Glucuronidase is processed from a precursor of 75 000 daltons to a product of 73 000 daltons. β-Galactosidase is processed from 82K to 63K. The processing is rapid ($t_{1/2} = 1$ hour). Macrophage β-glucuronidase is unusual in this regard since almost all other lysosomal enzymes (e.g. in fibroblasts) are synthesized as large precursors (Hasilik and Neufeld, 1980). The turnover of biosynthetically labelled cellular β-glucuronidase and β-galactosidase is rapid ($t_{1/2} = 1.8$ days and 3.5 days respectively) and secretion accounts for at least half of the total cellular loss for both enzymes. The nature of the secretory product in terms of whether it has the mannose phosphate recognition marker is unknown. It is known that macrophage cell lines P388 and J774 secrete large amounts of lysosomal hydrolases (Jessup and Dean, 1980) and that secreted enzyme contains the mannose phosphate marker (Goldberg and Kornfeld, 1981; Shepherd, Miller and Stahl, unpublished observations). The question which arises then is how lysosomal enzyme secretion is managed by the macrophage. Several possible mechanisms are as follows: (i) The transferase which catalyzes the formation of the GlcNAc-phosphomannose oligosaccharide may be blocked. This, in turn, may lead to the processing of the high-mannose oligosaccharides into complex types (i.e. containing galactose and sialic acid). In fact, this may be occurring in I-cell disease (Miller *et al.*, 1981). Alternatively, (ii) it is possible that some of the GlcNAc cover on the mannose phosphate recognition marker is not removed by the Golgi phosphodiesterase. This may lead to a poorly recognized enzyme which makes its way into the extracellular compartment. (iii) Yet another possibility is that some portion of the newly synthesized enzyme is secreted quickly and recaptured by cell surface receptors specific for mannose phosphate. Primary macrophages do have mannose phosphate receptors on their surfaces (Shepherd, Stahl, Miller and Freeze, manuscript in preparation). The macrophage might regulate the flow of enzyme through this pathway by either immobilizing receptors on the cell surface or by shifting them into a futile-cycling pathway which is unable to deliver ligand to the lysosome. Finally, if another intracellular mechanism were present in cells (e.g. a system where receptors recognize some aspect of the protein structure of the lysosomal enzyme or some sugar other than mannose phosphate) it may well be that regulation of secretion occurs at this level.

6.4 FLUID-PHASE AND RECEPTOR-MEDIATED PINOCYTOSIS BY MACROPHAGES

Pinocytosis is characteristic of most, if not all, eucaryotic cells (Pratten *et al.*, 1980). The macrophage specializes in endocytic activity and, just as the biological and immunological activities of these cells are under close humoral control (Ezekowitz *et al.*, 1981), pinocytosis in macrophages appears to be regulated as well. Regulation of pinocytosis in macrophages is achieved in two ways: (i) regulation of fluid-phase pinocytosis; and (ii) regulation of receptor-mediated pinocytosis. Fluid-phase pinocytosis begins with a pinching off of the plasma membrane followed by translocation of the 'pinosome' toward the perinuclear region where fusion with lysosomes occurs. Following the progress of fluid-phase pinocytosis requires a soluble, membrane-impermeable marker which does not absorb to plasma membrane. The movement of the marker into the cells under study must then reflect the movement of a volume of pinocytosed fluid in which the marker is dissolved. Horseradish peroxidase (HRP) has been commonly used as a fluid-phase marker. Using these techniques, along with quantitative stereological measurements, Steinman and colleagues (Steinman *et al.*, 1976) in pioneering experiments demonstrated that surface membrane is continuously internalized by macrophages at a rate which far exceeds the ability of the cell to produce membrane *de novo*. The marker (HRP) used to identify the pinocytic vesicles was efficiently transported to lysosomes. However, during the period of observation the volume of the lysosomal system remained constant. From their data, Steinman *et al.* (1976) concluded that internalized membrane must be recycled back to the cell surface where it is reutilized. More recently, Mellman *et al.* (1980) have asked whether there is something peculiar about the patch of membrane which pinches off from the surface to form the pino-some. A technique was developed whereby newly internalized membrane was radiolabeled selectively. Lactoperoxidase (LPO) and glucose oxidase were allowed to be taken up by macrophages at 37°C after which the cells were cooled to block further intracellular movement. $Na^{125}I$ was then added and the incubation was carried out on ice allowing the LPO within pinocytic vesicles to catalyze membrane iodination. Sodium dodecyl sulphate (SDS)–polyacrylamide gel electrophoresis (PAGE) analysis of the radiolabeled membranes indicated that pinosome membranes and plasma membranes are very similar. These results argue against the presence of plasma membrane domains which mediate fluid-phase pinocytosis. Fluid-phase pinocytosis does appear to be under some form of regulatory control since there are humoral factors [epidermal growth factor (EGF), macrophage-activation factor] which increase pinocytic activity. How such regulation is achieved is not known.

Receptor-mediated endocytosis is another closely regulated cellular pro-cess found in many eucaryotic cells. The process involves receptors on the cell surface which bind soluble ligands with high affinity. Some receptors [e.g. low-density lipoprotein (LDL)] appear to be located in highly differentiated regions of the plasma membrane referred to as coated pits. [The coat derives from the presence of clathrin and related proteins (reviewed in Chapter 2).] This coated region may also be peculiar in having an altered cholesterol content (Montesano *et al.*, 1981). Other receptors may have a more random distribution over the plasma membrane from which they diffuse toward coated pits (Zidovetski *et al.*, 1981). Receptor–ligand complexes are rapidly internalized as the coated pit forms a coated vesicle. Coated pits are thought to depart inwardly from the surface every few minutes. It is not clear whether the coated pit is a pre-existing structure to which receptors migrate or whether receptors first cluster followed by the formation of a coated pit around them. Nevertheless, the coated pit appears to be the major structure through which pinocytosis receptors enter the cell.

The macrophage has a large repertoire of cell surface receptors whose synthesis and turnover are closely regulated and modulated. These include hormone/effector receptors [colony-stimulating factor (Stanley, 1979), macrophage-activation factor (Schubert *et al.*, 1980), etc.], pinocytosis recep-tors [elastase receptor (Campbell *et al.*, 1979), mannose receptor (Stahl *et al.*, 1978), α_2-macroglobulin–protease receptor (Kaplan and Nielsen, 1979), acetylated-LDL receptor (Traber *et al.*, 1981), C5a receptor (McCarthy and Henson, 1979)] and phagocytosis receptors [F_c receptor, C_3b receptor and a so-called non-specific receptor]. The pinocytosis receptors all appear to operate via a similar mechanism. One of these pinocytic mechanisms, the mannose/fucose receptor, will be discussed in detail.

6.5 RECEPTOR-MEDIATED UPTAKE OF LYSOSOMAL ENZYMES BY MACROPHAGES

6.5.1 *In vivo* clearance of lysomal hydrolases

Lysosomal enzymes are glycosylated macromolecules whose oligosaccharide chains are rich in mannose and *N*-acetylglucosamine (Touster, 1978). Intravenous infusion of a family of highly purified lysosomal enzymes into the anesthetized rat results in the prompt clearance of enzymatic activity from the circulation (Stahl *et al.*, 1976a). The presence of intact oligosaccharide chains is required for rapid clearance of these enzymes because removal or alteration of the carbohydrate moiety by prior periodate oxidation results in a drastic increase in their plasma survival time (Stahl *et al.*, 1976b). Similar to the clearance of asialoglycoproteins, the liver is an important organ in clearance of lysosomal glycosidases. However, unlike the asialoglycoprotein-recognition system, clearance of injected mannose-terminated glycoproteins is only partially blunted by hepatectomy whereas the same maneuver abolishes clearance of galactose-terminated glycoproteins. The site to which lysosomal glycosidases are cleared (Kupffer cells) was revealed by liver cell fractionate experiments following enzyme clearance (Schlesinger *et al.*, 1978). Clearance of ^{125}I-β-glucuronidase was followed using a protocol designed to separate liver parenchymal from non-parenchymal cells. It was found that liver non-parenchymal cells were the principal targets of injected lysosomal glycosidases. In the absence of the liver, bone marrow and other sites enriched in macrophages take up the injected ligand (Schlesinger *et al.*, 1980).

6.5.2 The mannose-specific glycoprotein receptor

The presence of a cell surface receptor specific for mannose-terminated glycoproteins and certain lysosomal glycosideases (most of which are mannose-rich) was first demonstrated in studies on pulmonary alveolar macrophages (Stahl *et al.*, 1978). Alveolar macrophages bind and internalize a variety of mannose- and *N*-acetylglucosamine-terminated macromolecules. Ligands which are recognized by alveolar macrophages include rat preputial gland and rat liver lysosomal β-glucuronidase, human placental β-glucuronidase, rat liver hexosaminidase, agalacto-orosomucoid (*N*-acetylglucosamine-terminated), ahexosamino-orosomucoid (mannose-terminated), horseradish peroxidase and ribonuclease B. Uptake into macrophages is saturable with increasing ligand concentration and fully inhibited by the presence of an alternative, non-radioactive, ligand such as yeast mannan. Ca^{2+} is required for optimal binding and uptake. The sugar specificity of the macrophage recognition system has been partially revealed by the use of a series of synthetic glycoconjugates called neoglycoproteins. The neoglycoproteins used in these studies were initially prepared by Y.C. Lee and his associates who have done pioneering work in this area (Lee *et al.*, 1976).

Neoglycoproteins are prepared by coupling 2-imino-2-methoxy 1-thioglycosides to the amino groups of proteins. Mannose–BSA (mannose–bovine serum albumin), for example, can be prepared with 30–40 mol of sugar/mol of BSA. The sugar specificity of ligand uptake by macrophages was investigated using ^{125}I-glucuronidase as ligand and the neoglycoproteins as inhibitors (Shepherd *et al.*, 1981). The results in Fig. 6.1 show that man-

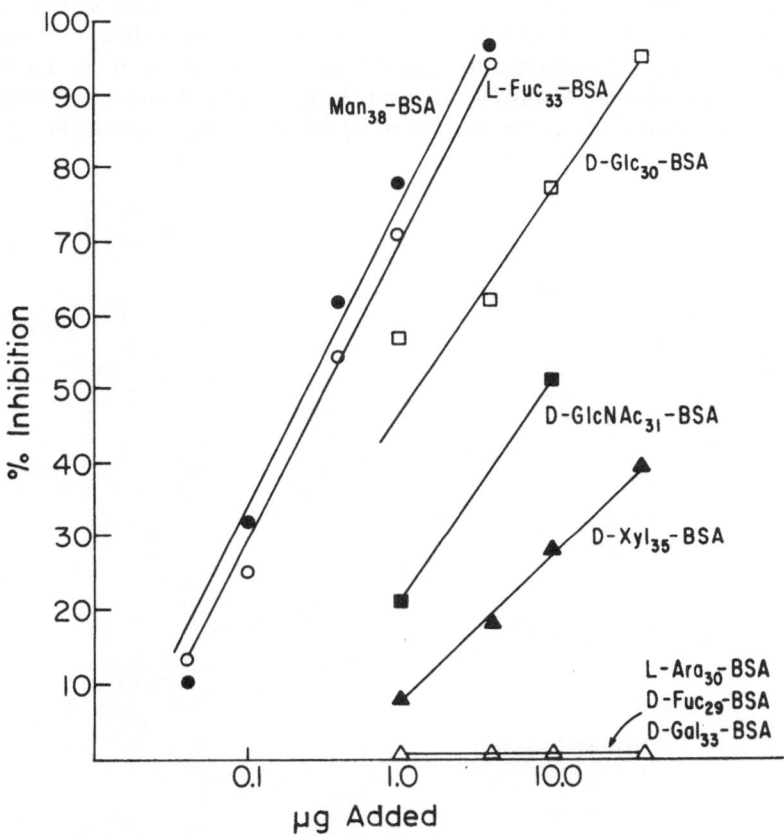

Fig. 6.1 Inhibition of ^{125}I-β-glucuronidase uptake into macrophages by various neoglycoproteins. Cells (5×10^6/ml) in buffered culture media were incubated with 2.5 μg of ^{125}I-β-glucuronidase and increasing concentrations of neoglycoproteins for 10 min at 37°C. The uptake of ligand was terminated by centrifugation through oil. The extent of substitution of the neoglycoprotein is indicated by the subscript. Reproduced with permission from Shepherd *et al.* (1981).

nose–BSA and L-fucose–BSA are the most potent inhibitors of ^{125}I-β-glucuronidase uptake into macrophages. Intermediate in activity was glucose–BSA and N-acetylglucosamine–BSA. Active, but much less so, was D-xylose–BSA. L-Arabinose–BSA, D-fucose–BSA and D-galactose–BSA were inactive. These results suggest that the orientation of hydroxyl groups around C-2, C-3 and C-4, as shown in Fig. 6.2, are critical to the generation of a high-affinity binding site. L-Fucose is interesting because at first glance it would appear to have a structure very different from D-mannose. However, simply by inverting the structure and flipping it 180°, it can be shown that the C-2, C-3 and C-4 sequence of D-mannose is identical to the C-4, C-3 and C-2 sequence of L-fucose. The high-affinity binding site of D-mannose–BSA or L-fucose–BSA undoubtedly results from the presence of a cluster of sugars attached to nearby lysine residues which mimic the structure of an oligosaccharide chain. Maynard and Baenziger (1981) and Holt and Stahl (unpublished observations) have studied the recognition of oligosaccharides by

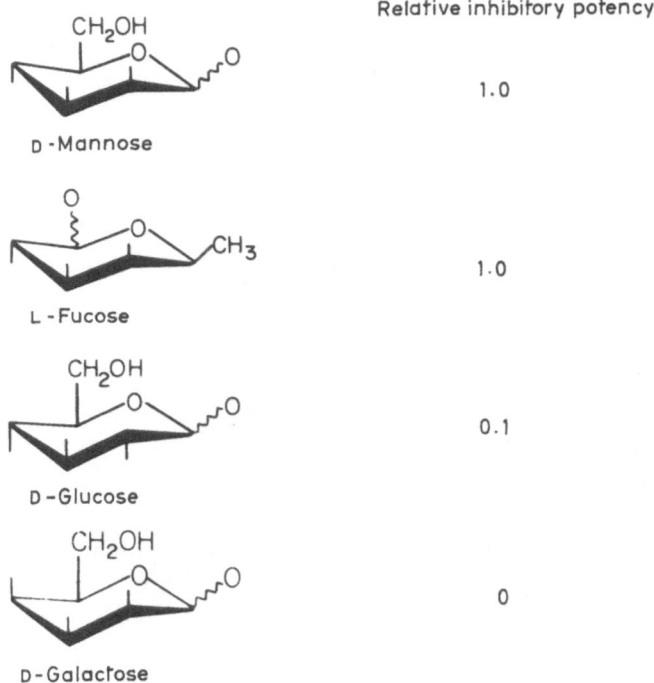

Fig. 6.2 Structural comparison of the neoglycoprotein sugar units in order of inhibitory potency. Neoglycoproteins were added to uptake assays containing ^{125}I-β-glucuronidase as described in Fig. 6.1. Inhibitory potency of mannose–BSA was set at 1.0.

macrophages and the results indicate that all the information necessary for binding and uptake is contained within the oligosaccharide chain.

Macrophages bind and internalize mannose–BSA as shown in Fig. 6.3(a–d). Binding and uptake of iodinated ligands can be operationally separated by cooling cells to 4°C. At this temperature, pinocytosis is essentially

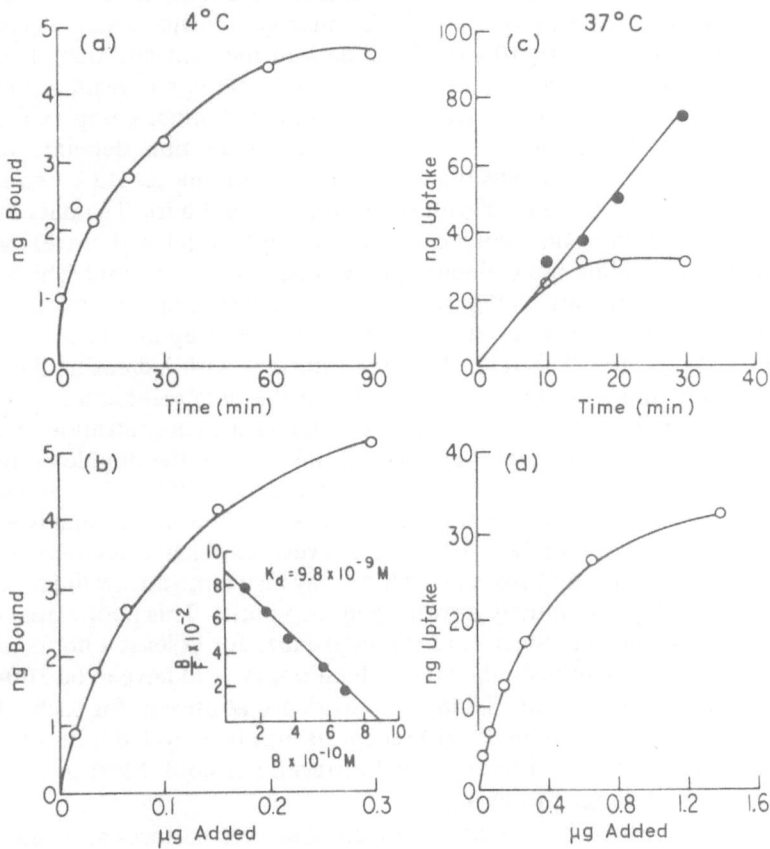

Fig. 6.3 Effect of time and temperature on binding and uptake of ^{125}I-mannose–BSA by alveolar macrophages. Cells (5 × 10[6]/ml) were incubated at 37°C or 4°C in culture media containing ligand. Cells were rapidly separated from the ligand by sedimentation through oil. Non-specific binding and uptake were estimated by adding yeast mannan (1.25 mg/ml) to the assay. (a) Time-dependence of 4°C binding (^{125}I-mannose–BSA, 2.5 μg/ml). (b) Concentration-dependence of 4°C binding at 60 min. (c) Time course of ^{125}I-mannose–BSA uptake (10 μg/ml); O, specific uptake; ●, specific uptake corrected for degradation. (d) Concentration-dependence of ^{125}I-Man–BSA uptake at 37°C and 10 min. Reproduced with permission from Stahl *et al.* (1980).

blocked and binding can be observed in the absence of internalization. Fig. 6.3(a) shows the time course of binding of ^{125}I-mannose–BSA. Using a 60 minute incubation, the concentration-dependence of binding was studied. Saturation was achieved at about 3 μg of mannose–BSA/ml. (All the ligand remains on the surface when cells are incubated with ligand at 4°C and can be removed by incubation with EDTA.) Rearrangement of the binding data in the form of a Scatchard plot (Fig. 6.3(b), insert) indicates a single class of binding sites with a K_d of 10 nM. These data further indicate that alveolar macrophages have about 75 000 sites/cell at 4°C. In comparison, uptake of mannose–BSA into alveolar macrophages is a much more complex process which proceeds briskly at 37°C. Fig. 6.3(c) shows the time-dependence of mannose–BSA uptake by alveolar macrophages. Mannose–BSA uptake by macrophages at 37°C is linear with time over several hours. The data in Fig. 6.3(c) show a 30 min time course where total uptake (closed circles) was determined by adding the cellular activity (open circles) to that which appeared in the mediate as digested bits. The digested ligand appears to be [^{125}I] iodotyrosine. The concentration-dependence of uptake at 37°C is shown in Fig. 6.3(d). The rate of uptake saturates with increasing ligand concentration and a double-reciprocal plot of the data yielded a K_{uptake} of 4×10^{-8}M. The K_{uptake} in this instance is defined as the concentration of ligand which gives half-maximal uptake. The $V_{max.}$ taken from the double-reciprocal plot indicates that each cell can internalize on the order of 2×10^6 molecules of ligand per hour. The number of sites/cell (~75 000) is unreasonably low in view of the rate of uptake (2×10^6/hour). Several explanations could satisfactorily account for this difference. (i) There may be a large intracellular pool of receptors which is continually drawn upon for uptake. This pool would have to be very large indeed. Since cells can internalize for at least 4 hours in the absence of protein synthesis, the intracellular pool would have to be 100-fold larger than the surface pool. There is no evidence at present for such a large intracellular pool. Alternatively, (ii) receptors may be recycled following their internalization, with or without a small intracellular pool. Most of the evidence favors the latter proposal.

^{125}I-Mannose–BSA binds avidly to macrophage cell surfaces and, unless the cells are warmed to 37°C, remains there for extended periods even if the cells are removed from the binding media (the off-rate is slow, $t_{1/2} > 60$ min). Warming the cells to 37°C results in the very rapid internalization of the bound ligand ($t_{1/2} < 5$ min). If it is assumed that receptor–ligand complexes enter the cell and that unoccupied receptor on the surface must be replaced from some intracellular pool, the rate of recovery ought to be measurable via binding experiments on cells which are in the process of internalizing pre-bound ligand. Such an experiment is described in Fig. 6.4. Here, cells were incubated on ice with unlabeled mannose–BSA for 60 min. They were then washed free of excess ligand and warmed to 37°C for 0, 2, 5 and 10 min, after

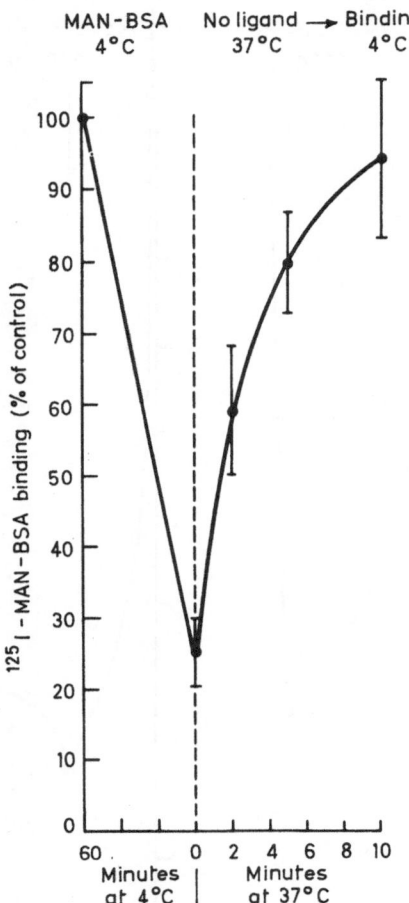

Fig. 6.4 Effect of mannose–BSA preincubation on ^{125}I-mannose–BSA binding following 37°C warm-up. Alveolar macrophages were incubated with mannose–BSA (6 μg/ml) on ice for 60 min as described in Fig. 6.3. The cells were washed free of excess ligand, suspended in fresh media and warmed to 37°C for 0, 2, 5 and 10 min after which they were cooled on ice. Binding was then determined as described in Fig. 6.3. The data are expressed as % of a companion assay where unlabeled mannose–BSA was omitted. Reproduced with permission from Stahl *et al.* (1980).

which they were cooled on ice. Binding was then determined with ^{125}I-mannose–BSA at 4°C. The results show that binding activity is lost following incubation of cells with unlabeled ligand but that brief warming of the cells to 37°C results in rapid recovery of cell surface binding activity. These results suggest that receptors are rapidly replaced from an intracellular pool. To test

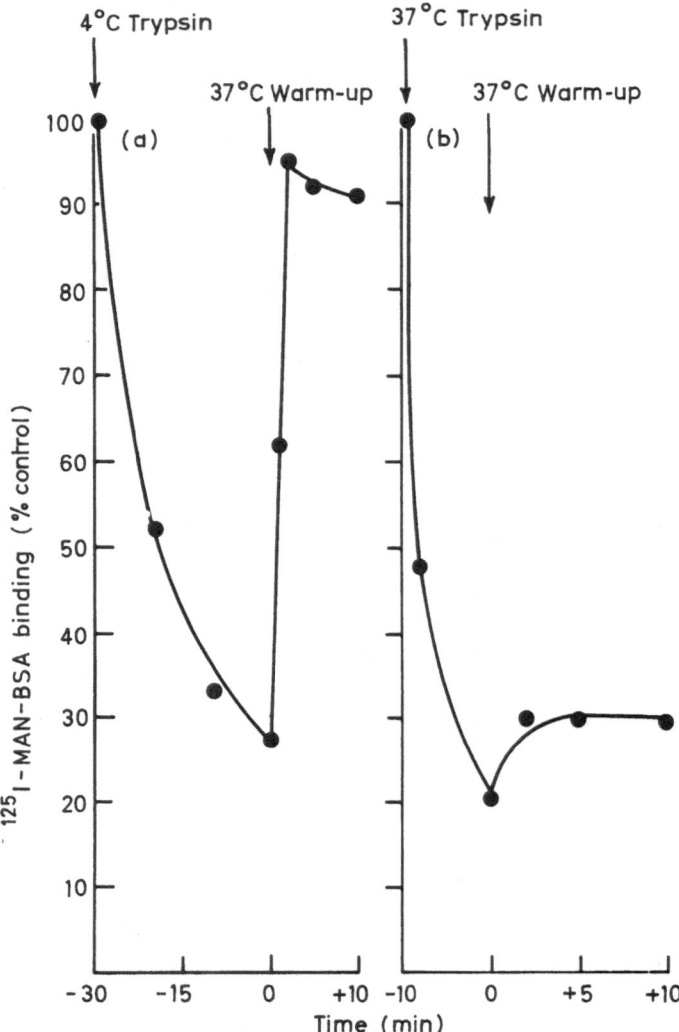

Fig. 6.5 Inactivation of binding of ^{125}I-mannose–BSA by trypsin at 4°C and at 37°C and recovery of binding at 37°C warm-up. Cells were incubated with trypsin (0.01%) in serum-free (pH 7.0) minimal essential medium (MEM) at 4°C (a) and at 37°C (b). At appropriate times the trypsin-containing medium was diluted 1:1 with cold MEM containing 20% fetal calf serum and the cells were washed once by sedimentation. The cells were resuspended in MEM containing 20% fetal calf serum and the cells were washed once by sedimentation. The cells were resuspended in MEM containing 20% fetal calf serum and warmed to 37°C for the times indicated after which they were immediately cooled on ice. Binding activity was determined with ^{125}I-mannose–BSA as described in Fig. 6.3. Data are expressed as a % of a companion set of cells which were carried through the protocol in the absence of trypsin.

for the presence of an intracellular pool, cells were trypsinized at 4°C under conditions where the cell surface mannose–receptor was inactivated. The mannose–receptor is exceptionally trypsin sensitive and the results in Fig. 6.5 show trypsin inactivation of cell surface mannose–receptors at 4°C. Upon warming trypsinized cells to 37°C for a few minutes, however, cell surface binding activity is fully recovered. On the other hand, incubation of cells at 37°C with trypsin results in a complete loss of mannose–receptor activity. Considering that 75% of the cell surface binding activity can be inactivated by trypsin at 4°C, what effect might this have on uptake of mannose–BSA at 37°C? Experimentally this was done by preincubating cells with or without trypsin on ice until ~80% of the cell surface binding activity was annihilated. An uptake experiment was then performed at 37°C. While trypsin treatment had no effect on the K_{uptake}, the maximal velocity of uptake was reduced by only 10–20%. These findings indicate that roughly 85% of the receptor pool is inside the cell protected from trypsin and about 15% is outside. If there are 75 000 receptors on the outside, the total pool of receptors must be ~4 × 10^5/cell. Assuming that each cell can internalize 2 × 10^6 molecules-/hour using 4 × 10^5 receptor sites, each receptor must recycle every 12 minutes.

6.6 AMINES AND RECEPTOR RECYCLING IN MACROPHAGES

6.6.1 Inhibition of uptake by amines

Chloroquine and other weak bases (e.g. NH_4^+, methylamine) have been used by a number of investigators to study the endocytic/lysosomal system (de Duve *et al.*, 1974). Chloroquine has long been known to block lysosomal proteolysis by some unknown mechanism. The novel experiments of Ohkuma and Poole (1978) first showed by direct measurement that lysosomal pH is rapidly increased by the presence of chloroquine and other weak bases. The mechanism by which chloroquine raises lysosomal pH undoubtedly is due to the differential movement of the protonated form of the base as opposed to the unprotonated form across the lysosomal membrane. The base thus accumulates inside lysosomes and raises intralysosomal pH by binding up protons via mass action.

Chloroquine has been shown to inhibit the endocytosis of several ligands which normally enter cells via receptor-mediated endocytosis including α_2-macroglobulin–trypsin complexes (Kaplan and Nielsen, 1979) mannose-phosphorylated enzymes [α-iduronidase (Sando *et al.*, 1979), β-glucuronidase (Gonzalez-Noriega *et al.*, 1980)] and mannose-terminated glycoproteins (Tietze *et al.*, 1980). Given the previous work showing that chloroquine raises lysosomal pH, a query arose about the possible relationship between receptor-mediated uptake and lysosomal pH (or the pH of intracellular vesicles).

Uptake of mannose–BSA by macrophages is sharply inhibited by chloroquine and ammonium ion. The results in Fig. 6.6 show that whereas inhibition of uptake is dependent on the concentration of chloroquine, the binding of mannose–BSA to macrophages at 4°C is unimpaired by the presence of the drug. The blockade of uptake of ligand by chloroquine was found to be unusual with respect to time course of inhibition. Fig. 6.6 shows the linear uptake with time in the absence of the base but that the development of inhibition by chloroquine was concentration- and time-dependent. This result suggested that the inhibition was driven to completion by movement of the system (i.e. use-dependent) and that the initial rate of uptake was unaffected or only slightly reduced. If one imagines that the sequence of events in the uptake of a ligand via receptor-mediated endocytosis is (1) binding, (2) internalization and (3) recovery of fresh binding sites from an intracellular pool simultaneous with internalization of receptor–ligand complex, then the preceding experiment would suggest that amine inhibition is beyond the internalization step. A direct test of this was achieved by binding ligand at 4°C followed by warming to 37°C for various periods of time. The disappearance of cell surface ligand (into the cell) could then be monitored in the absence or presence of chloroquine. This experiment (Tietze *et al.*, 1980) demonstrated

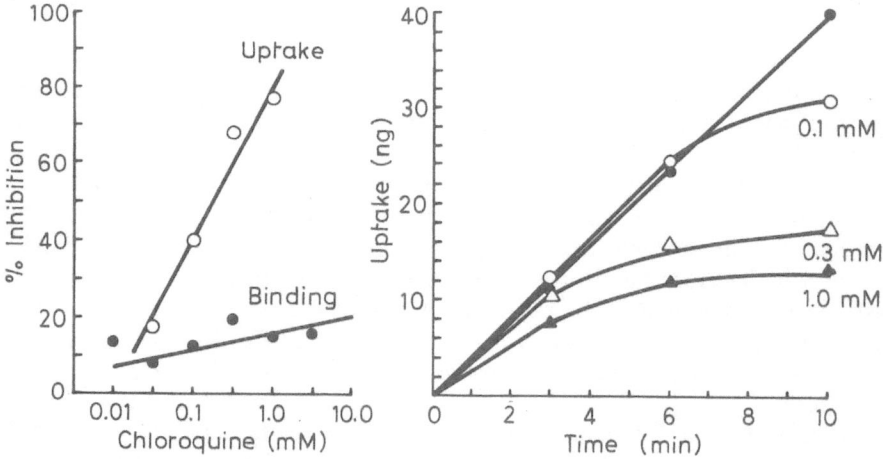

Fig. 6.6 Effect of chloroquine on binding and uptake of ^{125}I-mannose–BSA by alveolar macrophages. Uptake and binding (left) of ^{125}I-mannose–BSA was followed as described in Fig. 6.3. Chloroquine was added to cold media containing cells at least 10 minutes prior to warm-up. Uptake was followed for 10 min at 37°C, binding for 90 min at 4°C. Results are expressed as % inhibition of uptake and binding. Time course (right) of chloroquine inhibition of uptake was determined with 10 μg of ^{125}I-mannose–BSA/ml at the indicated chloroquine concentrations.

that prebound ligand is rapidly internalized whether chloroquine is present or absent. In the control, ligand was transported to lysosomes where it was rapidly degraded. If chloroquine were present, the ligand was not degraded following internalization but simply remained intact within cells. Whether this is due to ligand being trapped inside intracellular vesicles which are not lysosomes or whether it is due to decreased proteolysis secondary to a rise in intralysosomal pH is unclear. Interestingly, incubation of cells with chloroquine or NH_4^+ for a few minutes at 37°C leads to a rapid reduction in cell surface binding (measured at 4°C) (Fig. 6.7). This finding indicates that receptors are internalized in the presence of chloroquine but in the absence of ligand. Moreover, when experiments are conducted with NH_4^+, the drug can be washed out. Incubation of washed cells at 37°C for a few minutes

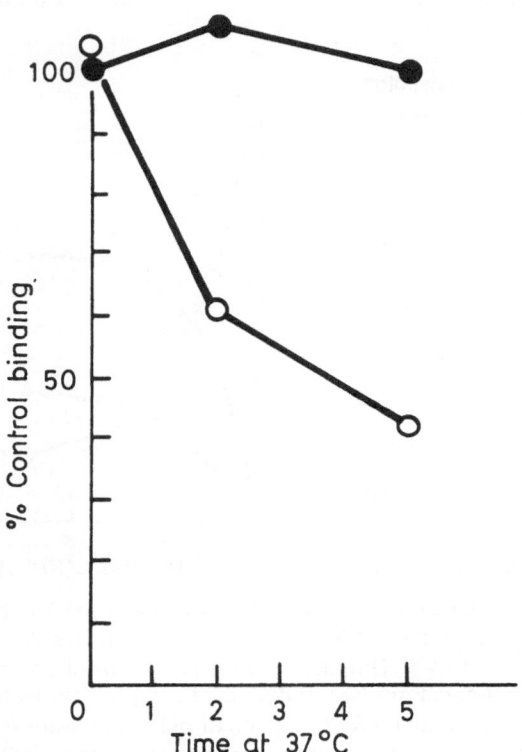

Fig. 6.7 Effect of chloroquine on cell surface binding activity. Cells (1.4×10^7/ml) in MEM containing 20% fetal calf serum with (open circles) or without (closed circles) 1 mM-chloroquine were warmed to 37°C for the indicated times. The cells were then cooled and binding activity was measured at 4°C with ^{125}I-mannose–BSA as described in Fig. 6.3.

results in almost complete recovery of the cell surface binding activity. Other results indicate that the loss of binding activity brought about by incubation of cells with amines is due to loss of binding sites rather than a change in affinity.

6.6.2 Effect of pH on receptor–ligand movement in macrophages

The amine inhibition of mannose–BSA endocytosis suggests that the receptor-bound ligand travels through an acid intracellular environment or compartment where the receptor and ligand dissociate. It is from this compartment then that the receptor may recycle. If a low-pH environment is required for receptor–ligand dissociation then it may be reasonable to ask whether receptor-bound ligand can be internalized from the cell surface at

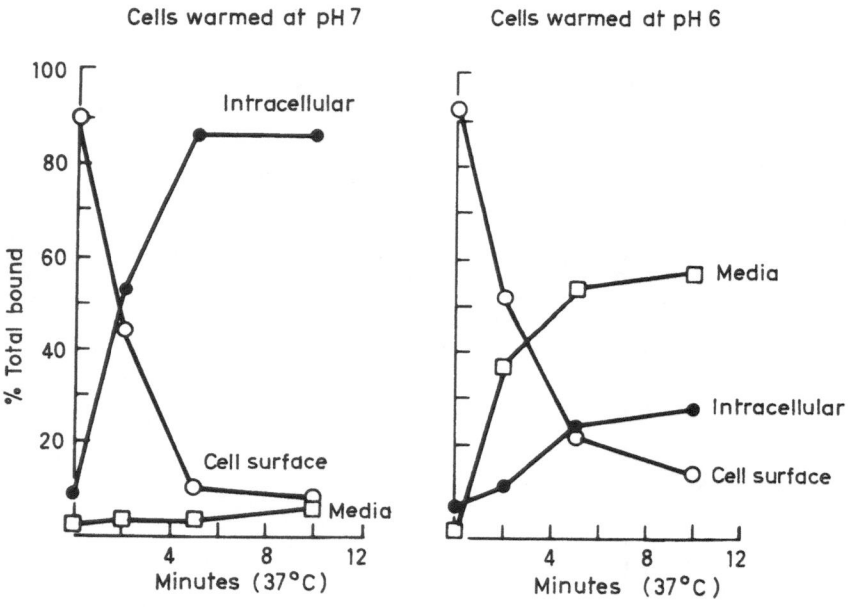

Fig. 6.8 Internalization of prebound ligands at pH 6 and pH 7.0. Cells were prewarmed to 37°C for 30 min in MEM, 10 mg of BSA/ml, 30 mM-sodium phosphate, pH 6.0 and then cooled to 4°C. (Preincubation at pH 6 maximizes binding.) Cells were then bound up with ^{125}I-mannose–BSA as described in Fig. 6.3, washed by sedimentation taken up into the appropriate buffer (i.e. pH 6 or pH 7) and warmed to 37°C for the times indicated. They were then cooled and separated from the 'media'. The cells were then taken up in Ca^{2+}/Mg^{3+}-free balanced-salt solution containing 10 mM-EGTA and 1 mg of trypsin/ml. They were incubated for 15 min at 4°C and then sedimented through oil. Radioactivity released by the trypsin is referred to as 'cell surface'. The trypsin-resistant radioactivity is referred to as intracellular. Reproduced with permission from Tietze *et al.* (1982).

low pH (Tietze *et al.*, 1982). An experiment to test this possibility is shown in Fig. 6.8. Here mannose–BSA was bound to cells at 4°C and the cells were washed free of unadsorbed ligand. The cells were warmed to 37°C in standard media adjusted to pH 6.0 or pH 7.0 and at various times samples were cooled for analysis. Three parameters were followed: (i) the cell surface component (defined as the trypsin-releasable material at 4°C), (ii) the intracellular component (defined as the trypsin-insensitive material), and (iii) the media component (radioactivity which appeared in the warm-up media). Cells warmed up at pH 7.0 rapidly internalized the prebound ligand with a half-time of <3 min. Cells warmed up at pH 6.0 only poorly internalized the prebound ligand. Moreover, the bulk of the prebound mannose–BSA at pH 6.0 was eluted into the media intact. This result is likely to be due to an increased dissociation rate of receptor–ligand complex at pH 6.0. Other experiments with paraformaldehyde-fixed cells (which prevents internalization but not binding) indicate that at 37°C the dissociation rate of prebound ligand is enhanced severalfold by reducing the pH of the media from 7 to 6.

6.6.3 Mobile proton-pump model for receptor recycling

If the pumping of protons were directly linked to the mechanism of receptor recycling it is possible that it may well be the driving force for receptor-mediated endocytosis in macrophages and other cell types. When the accumulation of protons within intracellular vesicles is inhibited by the presence of amines or other agents such as monensin, the receptor–ligand complex and/or the receptor may become trapped within the intracellular compartment through which it normally travels. This compartment may be similar to the receptosome which has been described by Willingham and Pasten (1980). A hypothetical scheme for receptor-mediated endocytosis is presented in Fig. 6.9. Since endocytic vesicles are known to rapidly become acidic (Tycko and Maxfield, 1982), it would seem reasonable to propose that proton transport sites are either a regular feature of endosomes or that they accompany the receptor as it travels through the recycling pathway. The latter has the advantage of having the proton pump in the right place at the right time. This pump site would continually mediate proton transport (mechanism unknown) regardless of its location. When the pump site is present in the plasma membrane with the receptor, a proton gradient would not be generated because the extracellular volume is so large. Since the receptor recycles every 12 minutes, this may represent only a small amount of time. The receptor–ligand complex (and/or the free receptor) along with the transporter (or pump site) moves into a coated pit by lateral diffusion in the plane of the membrane. Coated pits are apparently regularly internalized (Anderson *et al.*, 1977) resulting in both the receptor and the transporter becoming part of the newly formed endocytic vesicle. These newly formed vesicles fuse with each other

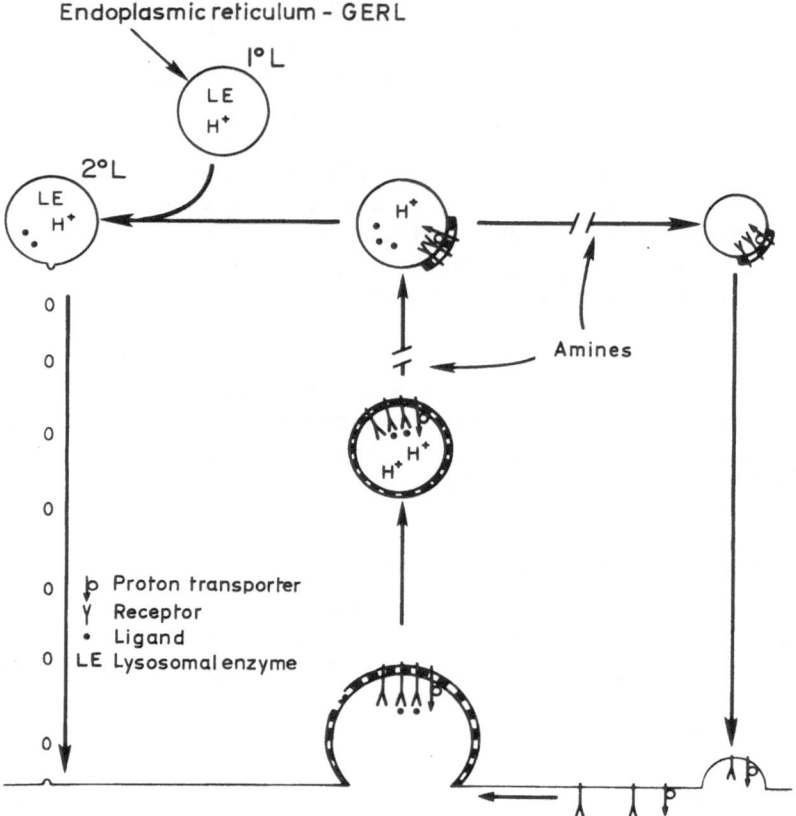

Fig. 6.9 Hypothetical model for receptor recycling with a mobile proton pump. The receptor and the pump travel together into the coated pit. After internalization and through the action of the pump, the newly formed vesicle becomes acid very rapidly. This enhances the fusibility of the vesicle and brings about dissociation of the ligand. The receptor/transporter are subsequently sequestered into small, receptor-rich, transporter-rich, ligand-poor vesicles perhaps under the action of a bit of coat which may remain intact. These small vesicles cycle back to the plasma membrane. The remaining vesicle fuses with lysosomes bringing together ligand, protons and lysosomal enzymes.

and begin to lose their coats perhaps as a consequence of the developing transmembrane pH gradient which forms due to the presence of the proton transporter. Other ions (e.g. Ca^{2+}) may also be co-transported out of this vesicle. The pH of the newly formed vesicle becomes acid leading to an enhanced dissociation of the receptor–ligand complex. Within this larger intracellular vesicle (formed by fusion of numerous smaller vesicles) receptors and transport sites might then cluster and thereby stabilize the adsorbed clathrin and other structural macromolecules. The clustered receptors could then pinch off to form a small (high surface/volume ratio) vesicle. This vesicle would be receptor-rich, transporter-rich and ligand-poor. Vesicles of this sort could then shuttle to the surface and deliver both receptors and transporters to the cell surface for reutilization. The larger vesicles which remain behind would be ligand-rich and acid-filled but receptor-poor. These vesicles might then fuse with primary lysosomes bringing together substrate with enzyme. It could also be that protons are physically transported to the lysosomal system and that pump sites never come into contact with lysosomal proteases. Perhaps the reason pump sites have been difficult to demonstrate in lysosomes is that lysosomes are not their principal location. Rather, pumps may be physically removed from the lysosomal system. Finally, it may be required that membrane is constantly removed from the lysosomal system in a way that there is no net movement of fluid and solute (e.g. enzymes) to the exterior. This may be achieved by very small high surface/volume ratio vesicles which continually shuttle membrane, but very little volume, to the Golgi apparatus or to the cell surface (Thyberg and Stenseth, 1981).

6.7 MODULATION OF ENDOCYTOSIS RECEPTORS IN MACROPHAGES

Receptor-mediated endocytosis, because of its dependence on the presence of a specific protein (i.e. the receptor), might be a cellular process which would be exquisitely sensitive to regulation. This indeed appears to be the case. The receptor for low-density lipoprotein is a good example. The LDL receptor is fully expressed by fibroblasts when the cells are grown in lipoprotein-deficient media. Addition of LDL to up-regulated fibroblasts leads to a decay in receptor activity. The receptor expression is thus modulated to accommodate the cellular need for cholesterol. In the case of the LDL receptor, the loss of receptor activity is undoubtedly due to the turnover of the receptor protein in the absence of synthesis (Goldstein *et al.*, 1979). A second potential mechanism by which cells may control the activity of their pinocytosis receptors would be to shift receptors to a cellular site where the receptors are immobilized. This may be on the surface or in some intracellular depot where they are unable to internalize ligand. [A possible analogy to this concept is found in the recent work on the glucose transporter in fat cells and muscle. Glucose transporters are held within the cell, presumably in an inactive state, until they are called to the surface via the action of insulin (Cushman and Wardzala, 1980).] Alternatively, receptors may be internalized but uncoupled from intracellular transport. These receptors may shuttle into and out of the cell, unable to shed bound ligand perhaps for lack of coming into contact with an acid environment. These receptors would cycle futilely into the cell and back to the surface with ligand still attached. If receptors could be shifted into and out of futile cycles, the advantage to the cell would be substantial since receptors could be mobilized into activity on a moment's notice.

6.7.1 Modulation via futile-cycling pools of receptor

There is now considerable evidence which indicates that the fate of a newly internalized ligand is determined by the cell type and by the nature of the material being transported. For example, endothelial cells are characterized by the presence of numerous vesicles which are involved in transport between blood and the endothelial space (Davies and Kuczera, 1981). Transcellular transport in those cells appears not to involve the lysosomal system. In hepatocytes, the receptor for galactose-terminated glycoproteins mediates transfer of ligands from the extracellular compartment to lysosomes. However, certain ligands, such as asialotransferrin type 3, are not transported to lysosomes and are clearly associated with intracellular vesicles which are not lysosomes. These may be vesicles involved in diacytosis and one can only conclude that the nature of the ligand is a determining factor in the intracellular routing (Debanne and Regoeczi, 1981). Similar observations have been made with synthetic cluster glycosides and their uptake by hepatocytes (Connolly *et al.*, 1982). Macrophages have recently been shown to contain a pool of mannose receptors which are unable to mediate transfer of ligand from cell surface to lysosomes. Instead, ligands appear to cycle into the cell via receptors and return to the surface as receptor-ligand complexes. This pool of receptors accounts for as much as 10–20% of the mannose receptors in alveolar macrophages. The discovery of a cycling pool of receptors in macrophages was a serendipitous finding from studies designed to explore the nature of amine-induced inhibition of receptor-mediated endocytosis. At the outset the possibility was considered that in the presence of amines, the pH of acid intracellular vesicles would be increased and receptor–ligand complex dissociation would be correspondingly reduced. If internalized receptor–ligand complexes remained intact and if receptors were continually recycling, it may have been possible to observe ligand being transported from an intracellular site to an extracellular site, especially if the extracellular pH were reduced, in essence, turning endocytosis around. To examine the possibility that receptor–ligand complexes were returning to the cell surface intact from some intracellular pool, cells were loaded with ^{125}I-β-glucuronidase at 37°C in the presence of NH_4Cl. Initially, ammonium chloride was added to raise the pH of acid intracellular vesicles through which the receptor–ligand complex may be travelling. The results indicated that as much as 40% of the previously internalized ligand could be released into the medium intact and that release was dependent upon conditions which enhanced receptor–ligand dissociation. This was interpreted to be an indication that receptor–ligand complexes were cycling back to the cell surface intact where ligand was dissociated from receptor. Surprisingly, when the same experiment was performed in the absence of NH_4Cl, similar observations were made. This indicated that some receptor–ligand complexes were cycling back to the surface under conditions where receptor-mediated transfer to lysosomes was not blocked. It is not clear

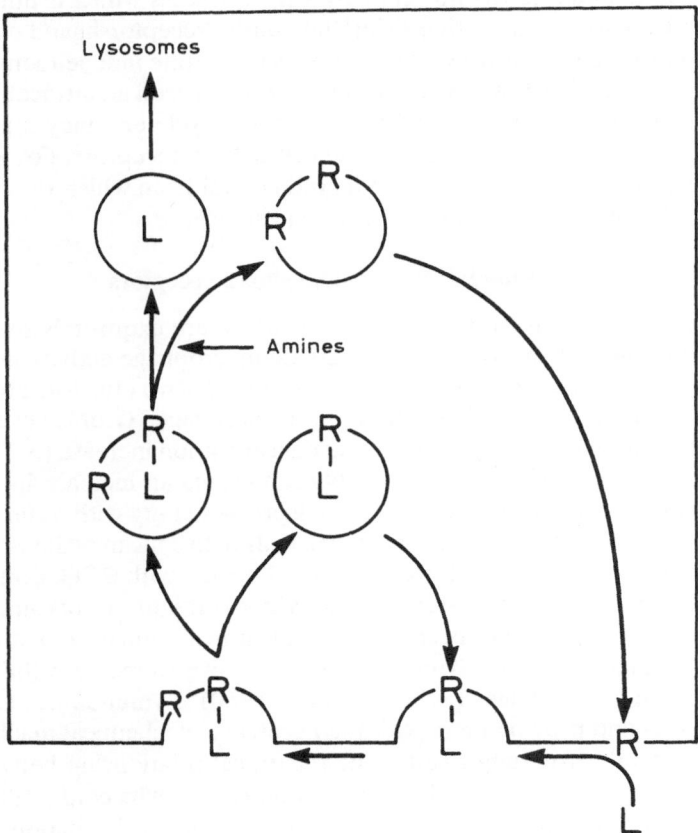

Fig. 6.10 Cycling and recycling pools of pinocytosis receptors. In this model receptor–ligand complexes are internalized into one or two intracellular pathways. The endocytosis or recycling pathway brings receptor–ligand complexes into an acid environment which promotes receptor–ligand dissociation. This pathway is amine-sensitive and is described in Fig. 6.9. A second pathway involves rapid internalization into a non-acid pool through which receptor–ligand complexes move. These receptor–ligand complexes can move back out to the cell surface; however, they may go undetected unless extracellular conditions are such that receptor–ligand dissociation is enhanced (Tietze *et al.*, 1982).

whether this is due to a separate pool of receptors or the nature of the ligand used in these studies. Another possibility, however remote, is that a small population of cells is present which have not yet coupled their receptors to the endocytic process. The latter may be analogous to complement receptors on some mononuclear phagocytes which can mediate binding but are unable to mediate phagocytosis (Griffin and Mullimax, 1981). A diagram which accommodates these observations is found in Fig. 6.10. Here receptor–ligand complexes are entering the normal pathway, a process which is inhibited by amines. Alternatively, or perhaps simultaneously, receptor-ligand complexes are entering a second pathway which appears to operate independently of any acid-intracellular vesicle and which cyles rapidly between an intracellular site and the cell surface. This is undoubtedly a futile cycle and may represent a mechanism for cells to rapidly modulate cell surface receptors. For example, receptors could be held in reserve in a cycling pool from which they could be mobilized by the action of some effector substance.

6.7.2 Effector-induced modulation of pinocytosis receptors

Macrophages are highly differentiated cells which are exquisitely sensitive to their environment. Effectors which modulate macrophage activity may modulate receptor-mediated pinocytosis activity as well. Two effectors are known to modulate mannose-receptor activity in macrophages (Konish *et al.*, 1982). Dexamethasone elicits a time and concentration increase (3–5-fold) in mannose-receptor activity. This probably represents an increase in receptor number since cell surface binding activity increases along with uptake. Lymphokines, produced by treatment of spleen cells with concanavalin A or by *in vitro* stimulation of spleen cells from animals infected with BCG, bring about a 50% reduction in mannose-receptor activity overnight in bone-marrow-derived macrophages. The mechanism of this down regulation is not well understood especially since lymphokines are known to increase fluid-phase pinocytosis in macrophages (Schubert *et al.*, 1980). Lymphokines activate macrophages and provide them with the necessary biochemical machinery to recognize and destroy tumor cells. The reciprocal relationship between the mannose-receptor and macrophage activation (Ezekowitz *et al.*, 1981) is interesting and may provide a clue as to the physiological function of the mannose receptor. However, apart from the immunological implications, the down regulation of the mannose receptor by lymphokine may provide a valuable tool for probing the nature of regulation of receptor-mediated pinocytosis by modulators.

6.7.3 Macrophage hybridomas

It is widely accepted that the mononuclear phagocyte is a highly differentiated cell type having a large repertoire of secretory and endocytic markers (Nathan *et al.*, 1980). The expression of many of these markers is transient and under close regulatory control. Expression of certain traits is influenced by the 'differentiation state' of the macrophage and by environmental and humoral factors. Just as the expression of specific secretory products by B-lymphocytes has been captured by fusion with myeloma tumor lines, it has now been demonstrated that the expression of certain macrophage traits can be captured in macrophage hybrids. Here, primary macrophages were fused by Sendai virus with a drug-sensitive macrophage-like tumor line (HAT-sensitive J774) (Stahl and Gordon, 1982). The primary macrophage parent was highly positive for mannose-receptor activity whereas the J774 line was negative. The mixed hybrids were cloned in agar and colonies were selected which continued to express the mannose receptor over extended periods of time in culture. Fusion of a fibroblast line with a primary macrophage resulted in extinction of the macrophage trait under study (i.e. the mannose receptor), suggesting that macrophage–macrophage fusion is required for continued expression of certain traits. These initial studies with the mannose receptor suggest that, depending upon the state of the primary macrophage employed in the fusion, it may be possible to generate macrophage–macrophage hybrids which express the gamut of phenotypes found in the primary macrophage parent. Moreover, assuming that macrophages are able to shift pinocytosis receptors into different cycling or non-cycling pathways as postulated in Section 6.7.1, it should be possible to capture expression of any one of these pathways in a macrophage–macrophage hybrid. Such hybrids may then be useful experimentally in isolating the types of control through which a cell regulates pinocytosis receptor activity.

REFERENCES

Anderson, R.G.W., Brown, M.S. and Goldstein, J.C. (1977), *Cell*, **10**, 351–364.

Bainton, D. (1981), *J. Cell Biol.*, **91**, 66–76.

Campbell, E.J., White, R.R., Senior, R.M., Rodriguez, R.J. and Kuhn, C. (1979), *J. Clin. Invest.*, **64**, 824–833.

Connolly, D.T., Townsend, R., Kawaguchi, K., Bell, W.R. and Lee, Y.C. (1982), *J. Biol. Chem.*, **257**, 939–945.

Cushman, S.W. and Wardzala, L.J. (1980), *J. Biol. Chem.*, **255**, 4758–4762.

Davies, P. and Bonney, R.J. (1979), *J. Reticulo. Soc.*, **26**, 37–47.

Davies, P.F. and Kuczera, L. (1981), *J. Histochem. Cytochem.*, **29**, 1437–1441.

Dean, R.T., Hylton, W. and Allison, A.C. (1979), *Biochim. Biophys. Acta*, **584**, 57–65.

Debanne, M.T. and Regoeczi, E. (1981), *J. Biol. Chem.*, **256**, 11266–11272.

de Duve, C., de Barsy, T., Poole, B., Trouet, A., Tulkens, P. and van Hoof, F. (1974), *Biochem. Pharmacol.*, **23**, 2495–2531.

Ezekowitz, R.A.B., Austyn, J., Stahl, P. and Gordon, S. (1981), *J. Exp. Med.*, **154**, 60–76.

Fischer, H.D., Gonzalez-Noriega, A., Sly, W.S. and Morre, D.J. (1980), *J. Biol. Chem.*, **255**, 9608–9615.

Griffin, F.M. and Mullimax, P.J. (1981), *J. Exp. Med.*, **154**, 291–305.

Goldberg, D.E. and Kornfeld, S. (1981), *J. Biol. Chem.*, **256**, 13060–13067.

Goldstein, J., Anderson, R.J.W. and Brown, M.S. (1979), *Nature (London)*, **279**, 682–684.

Gonzalez-Noriega, A., Grubb, J.H., Talkod, V. and Sly, W.S. (1980), *J. Cell Biol.*, **85**, 839–852.

Hasilik, A. and Neufeld, E. (1980), *J. Biol. Chem.*, **255**, 4937–4945; 4946–4950.

Hickman, S. and Neufeld, E.F. (1972), *Biochem. Biophys. Res. Commun.*, **49**, 992–999.

Jessup, W. and Dean, R.T. (1980), *Biochem. J.*, **190**, 847–850.

Kaplan, A., Achord, D.T. and Sly, W.S. (1977), *Proc. Natl. Acad. Sci. U.S.A.*, **69**, 2026–2030.

Kaplan, J. and Nielsen, M.N. (1979), *J. Biol. Chem.*, **254**, 7323–7328.

Konish, M., Thomasson, D. and Stahl, P. (1982), *Fed. Proc.* (in press).

Lee, Y.C., Stowell, C. and Krantz, M.J. (1976), *Biochemistry*, **15**, 3956–3962.

LeRoy, J.G., Ho, N.M., MacBrinn, M.C., Zielke, K., Jacob, J. and O'Brien, J.S. (1972), *Pediat. Res.*, **6**, 752–759.

Maynard, Y. and Baenziger, J.U. (1981), *J. Biol. Chem.*, **256**, 8063–8068.

McCarthy, K. and Henson, P. (1979), *J. Immunol.*, **123**, 2511–2517.

Mellman, I., Steinman, R.M., Unkeless, J.C. and Cohn, Z.A. (1980), *J. Cell Biol.*, **86**, 712–722.

Miller, A.L., Kress, R.C., Stein, R., Kinnon, C., Kern, H., Schneider, J.A. and Harms, E. (1981), *J. Biol. Chem.*, **256**, 9352–9362.

Montesano, R., Vassalli, P. and Orci, L. (1981), *J. Cell. Sci.*, **51**, 95–107.

Nathan, C.F., Murray, H.W. and Cohn, Z.A. (1980), *N. Engl. J. Med.*, **303**, 622–626.

Ohkuma, S. and Poole, B. (1978), *Proc. Natl. Acad. Sci. U.S.A.*, **75**, 3327–3331.

Pratten, M.K., Duncan, R. and Lloyd, J.B. (1980) in *Coated Vesicles* (C.S. Ockleford and A. Whyte, eds), Cambridge University Press, pp. 179–187.

Reitman, M. and Kornfeld, S. (1981), *J. Biol. Chem.*, **256**, 4275–4281.

Reitman, M., Varki, A. and Kornfeld, S. (1981), *J. Clin. Invest.*, **67**, 1574–1579.

Riches, D.W.H. and Stanworth, D.R. (1980), *Biochem. J.*, **188**, 933–936.

Sando, G. and Neufeld, E.F. (1977), *Cell*, **12**, 619–627.

Sando, G., Titus-Dillon, P., Hall, C.W. and Neufeld, E.F. (1979), *Exp. Cell Res.*, **119**, 359–364.

Schlesinger, P., Doebber, T., Mandell, B., White, R., DeSchryver, C., Miller, J., Rodman, J. and Stahl, P. (1978), *Biochem. J.*, **176**, 103–111.

Schlesinger, P., Rodman, J., Doebber, T., Stahl, P., Lee, Y.C., Stowell, C.P. and Kuhlenschmidt, T.B. (1980), *Biochem. J.*, **192**, 597–606.

Schneider, D.L. (1981), *J. Biol. Chem.*, **256**, 3858–3864.

Schnyder, J. and Baggiolini, M. (1978), *J. Exp. Med.*, **148**, 435–450.

Schubert, R.D., Wong, J. and David, J.R. (1980), *Cell Immunol.*, **55**, 145–154.

Shepherd, V., Lee, Y.C., Schlesinger, P.H. and Stahl, P. (1981), *Proc. Natl. Acad. Sci. U.S.A.*, **78**, 1019–1022.

Skudlarek, M.D. and Swank, R.T. (1979), *J. Biol. Chem.*, **254**, 9939–9942.

Skudlarek, M.D. and Swank, R.T. (1981), *J. Biol. Chem.*, **256**, 10137–10144.

Smolen, J.E., Korchak, H.M. and Weissman, G. (1981), *Methods Cell Biol.*, **23**, 461–480.

Stahl, P. and Gordon, S. (1982), *J. Cell Biol.* (in press).

Stahl, P., Rodman, J.S., Miller, J. and Schlesinger, P. (1978), *Proc. Natl. Acad. Sci. U.S.A.*, **75**, 1399–1403.

Stahl, P., Six, H., Rodman, J.S., Schlesinger, P., Tulsiani, D. and Touster, O. (1976a), *Proc. Natl. Acad. Sci. U.S.A.*, **73**, 4045–4049.

Stahl, P., Schlesinger, R., Rodman, J.S. and Doebber, T. (1976b), *Nature (London)*, **264**, 86–88.

Stahl, P., Schlesinger, P., Sigardson, E., Rodman, J.S. and Lee, Y.C. (1980), *Cell*, **19**, 207–215.

Stanley, E.R. (1979), *Proc. Natl. Acad. Sci. U.S.A.*, **76**, 2969–2973.

Steinman, R.M., Brodie, S.E. and Cohn, Z.A. (1976), *J. Cell Biol.*, **68**, 665–687.

Tabas, I. and Kornfeld, S. (1980), *J. Biol. Chem.*, **255**, 6633–6639.

Thyberg, J. and Stenseth, K. (1981), *Eur. J. Cell Biol.*, **25**, 308–318.

Tietze, C., Schlesinger, P. and Stahl, P. (1980), *Biochem. Biophys. Res. Commun.*, **93**, 1–8.

Tietze, C., Schlesinger, P. and Stahl, P. (1982), *J. Cell Biol.*, **92**, 417–424.

Touster, O. (1978), in *Protein Turnover and Lysosomal Function* (H. Segal and D. Doyle, eds), Academic Press, New York, pp. 231–250.

Traber, M.G., Defendi, V. and Kayden, H.J. (1981), *J. Exp. Med.*, **154**, 1852–1867.

Tycko, B. and Maxfield, F. (1982), *Cell*, **28**, 643–651.

Waheed, A., Pohlmann, R., Hosilik, A. and Von Figura, K. (1981), *J. Biol. Chem.*, **256**, 4150–4152.

Willingham, M. and Pasten, I. (1980), *Cell*, **21**, 67.

Zidovetzki, R., Yarden, Y., Schlesinger, J. and Jovin, T.M. (1981), *Proc. Natl. Acad. Sci. U.S.A.*, **78**, 6981–6985.

7 Asialoglycoproteins: Hepatic Clearance and Degradation of Serum Proteins

RICHARD J. STOCKERT

Acknowledgement

This work was supported in part by a grant AM-17702 from the National Institutes of Health and the Foundation for the Study of Wilson's Disease, Inc.

Receptor-Mediated Endocytosis
(*Receptors and Recognition*, Series B, Volume 15)
Edited by P. Cuatrecasas and T. F. Roth
Published in 1983 by Chapman and Hall, 11 New Fetter Lane, London EC4P 4EE
© 1983 Chapman and Hall

7.1 INTRODUCTION

A relationship between the structure and composition of the distal portion of the carbohydrate chains of glycoproteins to a potential role in specific recognition processes became evident from a series of studies of the circulatory fate of modified serum glycoproteins. Chemical or enzymatic removal of sialic acid exposing the penultimate galactosyl residues was shown to constitute a recognition signal for the rapid clearance from the circulation of the desialylated glycoproteins. These observations led to the hypothesis that the presence of a normal complement of terminal sialic acid residues was critical for the continued viability of serum glycoproteins in the circulation and provided the impetus for the discovery of the prototype of carbohydrate-directed receptor-mediated endocytosis.

7.2 AN OVERVIEW

Insight into the significance of the structure of the carbohydrate chains of glycoproteins to their continued viability in the circulation first became apparent in experiments in which desialylated ceruloplasmin was injected into rabbits. To prepare a protein doubly labeled with both copper-64 and tritium, for metabolic studies of copper, sialic acid was enzymatically removed and the exposed galactosyl residues were sequentially treated with galactose oxidase and tritiated borohydrite (Morell *et al*., 1966). Although this preparation was indistinguishable from the native glycoprotein, except for the absence of sialic acid, the desialylated ceruloplasmin was removed from the circulation within a few minutes after injection. In contrast, the native fully sialylated protein exhibited a normal serum survival time of the order of 55 hours (Morell *et al*., 1968).

The rapidity of circulatory clearance of the desialylated ceruloplasmin was shown to be due to the presence of the newly exposed terminal galactosyl residues and not the absence of sialic acid. Treatment of the desialylated glycoprotein either with β-galactosidase to remove the terminal galactose or by oxidation of the primary hydroxyl groups of these residues with galactose oxidase prolonged its intravascular survival time toward that exhibited by the native ceruloplasmin (Morell *et al*., 1971). Replacement of the sialic acid residues, by incubation of a partially desialylated preparation with sialyltransferase and cytidine monophosphate sialic acid, restored the physiological half to the resialylated glycoprotein (Hickman *et al*., 1970). Further studies established that although native ceruloplasmin contains approximately ten sialic acid–galactose sequences per molecule, exposure of any two galactosyl residues was sufficient to effect an immediate clearance of the partially desialylated glycoprotein from the circulation (Van den Hamer *et al*., 1970).

169

The generality of terminal galactose recognition was demonstrated by the injection into rats of the desialylated derivatives of a number of plasma proteins (Fig. 7.1) and the glycopeptide hormones human chorionic gonadotropin and follicle-stimulating hormone (Morell *et al.*, 1971). This number has grown over the years and a more complete list is available in a recent review

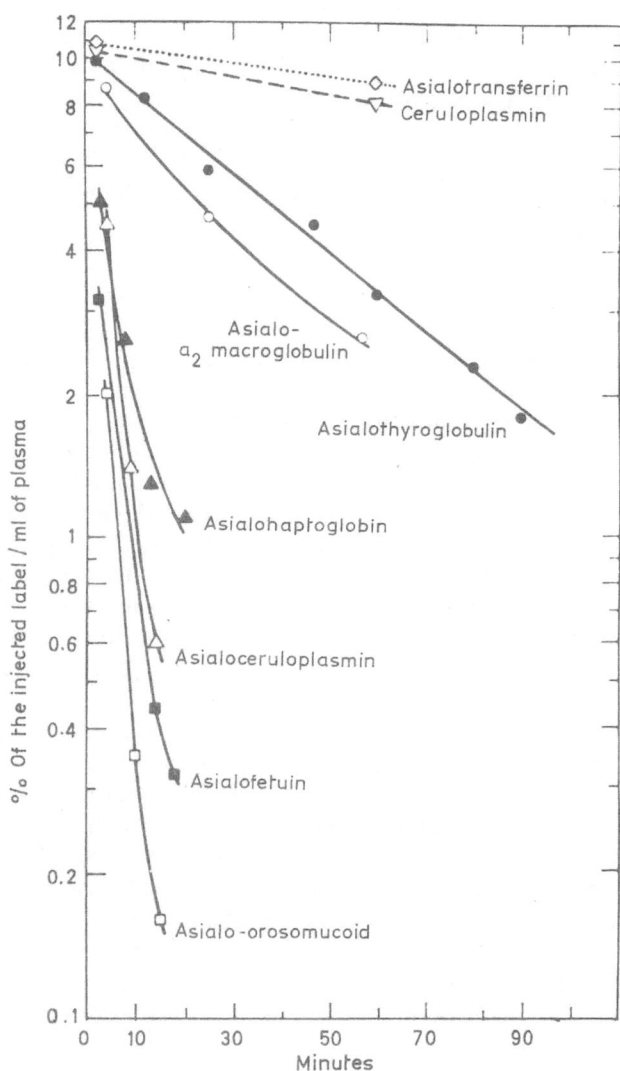

Fig. 7.1 Survival of intravenously injected native ceruloplasmin and desialylated plasma proteins in the rat.

(Neufeld and Ashwell, 1980). The common feature in all these glycoprotein derivatives was the presence of an exposed β-galactosyl residue terminating sequence in their carbohydrate chains. Of notable exception was transferrin, the iron-transporting glycoprotein of plasma, in which exposure of galactose residues did not appear to result in an appreciable reduction of its circulatory survival. It has been subsequently demonstrated that desialylated transferrin was removed from the circulation at a greater rate than the native protein (Regoeczi *et al.*, 1978). However, the rate of clearance of desialylated transferrin is much slower than for the other desialylated glycoprotein derivatives indicating that while necessary, the presence of an exposed terminal galactose is not a sufficient condition for a rapid removal from the circulation.

The rapid clearance of the desialylated glycoprotein from the serum was accompanied by an equally rapid uptake of the modified glycoprotein by the parenchymal cells of the liver. When tritiated desialylated ceruloplasmin is injected, the protein can be demonstrated radioautographically in hepatocytes, with no radioactivity in Kupffer cells (Morell *et al.*, 1968). The results of experiments in which tritium- and copper-64-labeled desialylated ceruloplasmin was injected showed the principal subcellular site of catabolism to be lysosomal (Gregoriadis *et al.*, 1970). At intervals of several minutes individual animals were killed and the livers fractionated by ultracentrifugation. With time, the radioactive label increased in concentration in the lysosome-enriched fraction from which there was a diminishing amount of protein that could be recovered immunochemically.

7.3 HEPATIC RECEPTOR

7.3.1 Isolation and purification

To permit a more detailed investigation of the recognition of desialylated glycoproteins by the liver a sensitive *in vitro* binding assay was developed (Pricer and Ashwell, 1971). Specific binding of desialylated glycoproteins was shown to be uniquely associated with the particulate fraction of rat liver homogenates. The plasma-membrane-enriched fraction was identified as the primary locus of binding activity. This activity was maximal at a neutral pH, and exhibited an absolute requirement for the presence of calcium ions. Although bound desialylated glycoproteins could be readily dissociated by lowering the pH or chelating calcium, no evidence could be obtained for substantial dissociation by the addition of a large excess of desialylated glycoprotein. A spectrum of binding affinities that was greatest for desialylated orosomucoid and, consistent with the *in vivo* observations, lowest, by several orders of magnitude, for desialylated transferrin was demonstrated by this *in vitro* assay system (Van Lenten and Ashwell, 1972).

Binding activity was solubilized with the anionic detergent, Triton X-100, from a lyophilized preparation of a particulate fraction of rabbit liver (Morell and Scheinberg, 1972). This crude preparation exhibited the same charac-teristic binding activity for desialylated glycoproteins as described for rat liver plasma membranes. The hepatic receptor was subsequently isolated and purified by affinity on columns of Sepharose to which desialylated glycopro-tein had been covalently coupled (Hudgin *et al.*, 1974). The purified rabbit liver receptor was characterized as a water-soluble glycoprotein containing 10% carbohydrate by dry weight. As isolated, the molecular weight of the active protein was approximately 250 000. Sodium dodecyl sulfate/polyacrylamide-gel electrophoresis revealed two distinct subunits with apparent molecular weights of 48 000 and 40 000 (Kawasaki and Ashwell, 1976a). Pronase digestion of the glycoprotein resulted in recovery of two different glycopeptides (Kawasaki and Ashwell, 1976b). The major asparagine-linked glycopeptide was shown to consist of a triantennary struc-ture with the terminal oligosaccharides of sialic acid, galactose and *N*-acetylglucosamine linked to a mannose-*N*-acetylglucosamine similar to cir-culatory glycoproteins. The minor glycopeptide recovered from only the 48 000-dalton subunit contained only mannose and *N*-acetylglucosamine.

Analogous receptor proteins have been isolated and purified from rat (Tanabe *et al.*, 1979; Sawamura *et al.*, 1980) and human liver (Baenziger and Maynard, 1980). Unlike the rabbit protein both the rat and human receptors are monomeric with estimated molecular weights of about 47 000 and 40 000 respectively. Recently, the assignment of a single molecular species of the rat protein has been re-examined using *in vivo* or *in vitro* labeling in conjunction with monoclonal antibody techniques (Warren and Doyle, 1981; Schwartz *et al.*, 1981b). These studies revealed at least three molecular weight receptor proteins on sodium dodecyl sulfate/polyacrylamide-gel electrophoresis with common peptide maps demonstrated on two-dimensional gels. Whether all of these species exist *in vivo* or are degradation artifacts of isolation is still not known.

In the case of the human protein, reversibility of ligand binding has been demonstrated, with a dissociation rate for desialylated orosomucoid of 1.7×10^{-3} s^{-1}, while binding to both the rabbit and rat receptor is essentially irreversible. In the other parameters examined, such as specificity for terminal-sugar ligands and assay conditions, the binding properties of the three mammalian binding proteins were indistinguishable.

As was characteristic of the rat plasma membrane preparations (Pricer and Ashwell, 1971), treatment with neuraminidase of the purified protein (Hud-gin *et al.*, 1974) resulted in the complete loss of their binding capacity for galactose-terminated desialylated glycoproteins. This unique property was subsequently clarified by the demonstration that the apparent loss of binding activity was due to the competition for the receptor's active sites by its own newly exposed galactosyl residues. When the receptor's terminal galactosyl

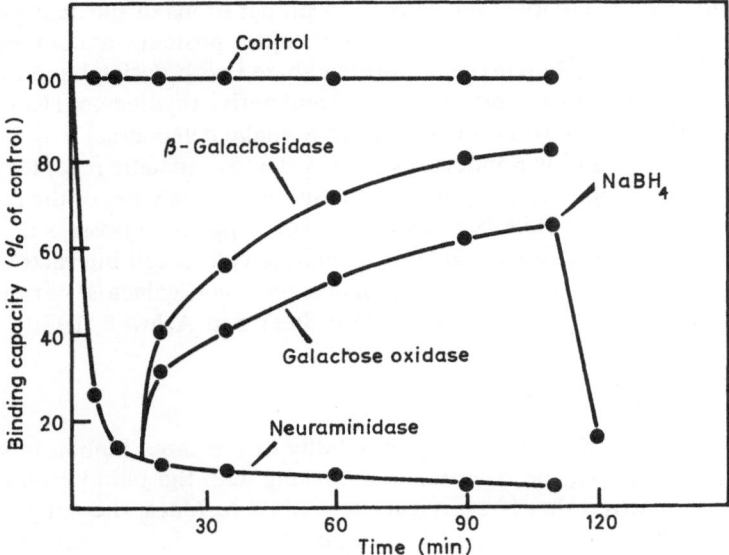

Fig. 7.2 Effects of enzymatic treatment on the binding capacity of the hepatic receptor.

residues were either enzymatically oxidized or hydrolyzed (Stockert *et al.*, 1977), as shown in Fig. 7.2, or masked by resialylation (Paulson *et al.*, 1977), binding activity was restored.

7.3.2 Lectin activity

The ability of the purified rabbit receptor to mediate erythroagglutination (Stockert *et al.*, 1974) and the later demonstration that it induced blastogenesis in desialylated lymphocytes (Novogrodsky and Ashwell, 1977) identified the receptor as a lectin, the first such carbohydrate-binding protein described of mammalian origin. As was the case for the binding of desialylated glycoproteins, the presence of calcium ions was necessary for erythrocyte agglutination. A lower concentration of the rabbit liver receptor was required to agglutinate human type A erythrocytes than type B with type O requiring the highest concentration of receptor for effective agglutination. This recognition by the hepatic receptor of cell surface glycoproteins suggests that, in addition to its possible role in glycoprotein metabolism, such binding may be a factor in certain cellular interactions as well. The ability of the purified lectin to induce desialylated human peripheral blood lymphocytes to mediate mitogen-stimulated cellular cytotoxicity (Vierling *et al.*, 1978) is an example of such possible cell-to-cell interactions.

Receptor lectin activity is not limited to preparations of purified protein. Agglutination activity of the analogous rat receptor protein was demonstrated when freshly isolated hepatocytes were incubated with desialylated erythrocytes, spleen lymphocytes and desialylated and native thymocytes (Kolb *et al.*, 1979). Kupffer cells have also been shown to make cell contact with desialylated erythrocytes and lymphocytes via a D-galactose-specific receptor (Kolb *et al.*, 1980). This observation points to the subtle complexities of the recognition phenomena, for, while both liver cell types appear to possess a galactose-specific receptor capable of mediating cell-to-cell interaction, only the parenchymal cell receptor can recognize and bind galactose-terminated soluble glycoproteins (Morell *et al.*, 1968; Steer and Ashwell, 1980).

7.3.3 Carbohydrate specificity

The quantitative differences in agglutinability of the three human blood groups suggested that the receptor was reacting with the blood group antigens. It was inferred that N-acetylgalactosaminyl residues, the antigenic determinant of blood group A, as well as galactosyl residues, determinants of group B, were both recognized by the receptor. Indeed, agglutination could be reversed by the addition of either of these two monosaccharides and to a lesser extent by L-fucose, the terminal sugar of group O erythrocytes (Stockert *et al.*, 1974). N-Acetylgalactosamine proved to be the most potent monosaccharide inhibitor (Baenziger and Maynard, 1980; Steer and Ashwell, 1980) and when present as the terminal sugar of a macromolecule, such as desialylated bovine submaxillary mucin, is capable of dissociating galactose-terminal glycoproteins from the receptor's binding site (Stockert *et al.*, 1977). Using agarose-immobilized rabbit receptor, an order of glycoside inhibitory capacity was established which confirmed N-acetylgalactosamine as having the highest affinity of any sugar (Sarkar *et al.*, 1979). On the basis of the relative affinity of these and other defined carbohydrate structures, it appears that the receptor's binding site is relatively small, involving the terminal, and extending to at least part of the penultimate, sugar residue.

The presence of N-acetylgalactosaminyl residues on glycoproteins are limited to few examples. Aside from the blood group substances, the most significant example is in the hinge region of immunoglobulin A (IgA). Isolated IgA hinge-region glycopeptides have been shown to inhibit the binding of desialylated glycoproteins (Baenziger and Maynard, 1980) and to be taken up by isolated hepatocytes via this receptor protein (Baenziger and Fiete, 1980). N-Acetylgalactosaminyl residues are, however, common constituents of membrane glycoproteins and therefore may play a role in the cell-to-cell interactions described earlier.

Although glucose was ineffective in reversing agglutination of erythrocytes by purified rabbit receptor or in inhibiting rosette formation between isolated hepatocytes and desialylated lymphocytes, when attached to bovine serum albumin, the glucosyl-albumin derivative proved to be as good, or better, an inhibitor than the galactosyl derivative (Stowell and Lee, 1978). These studies indicated that the receptor protein could not discriminate between D-galacto and D-gluco configurations when coupled to the albumin carrier. Since there are no known examples of glucose-containing circulatory glycoproteins the functional significance of glucose recognition by the rabbit receptor is unclear.

The composition and structure of the glycoprotein's oligosaccharide chains appear to be the sole determinants of receptor recognition (Gan, 1979). Desialylated glycopeptides and the protein-free desialylated oligosaccharides of α_1-antitrypsin were shown to be equipotent in prolonging the plasma survival of the intact desialylated α_1-antitrypsin. Therefore, the failure to recognize the galactosyl residues of desialylated transferrin must reflect a unique structure of its carbohydrate moiety.

The role played by the architecture of the oligosaccharide moiety became evident from studies of binding affinities of bi- and tri-antennary glycopeptides prepared from transferrin (Hatton *et al.*, 1979). Although transferrin contains predominantly two biantennary-type oligosaccharide chains, there is a small quantity of a triantennary glycan which binds with high affinity to the receptor protein. It was suggested that this heterogeneity could be reflected in the several molecular forms of transferrin which, after desialylation, differ significantly in their affinities for the receptor and was the basis for hepatic uptake when measured over an extended period of time (Regoeczi *et al.*, 1979). The appropriate spatial arrangement of the sugar residues, as well as number, has been suggested to explain the differences of affinities of glycoproteins, glycopeptides and oligosaccharides (Baenziger and Maynard, 1980). From these studies it was proposed that recognition at two receptor sites 2.5–3.0 nm apart is the basis for high-affinity binding and is responsible for the observed differences between ligand species. These findings are consistent with the earlier observations that there was a threshold for recognition of synthetic substrata by isolated hepatocytes (Weigel *et al.*, 1979) and that hepatic uptake of partially desialylated ceruloplasmin required exposure of at least two galactosyl residues (Van den Hamer *et al.*, 1970).

7.3.4 Subcellular distribution and reutilization

The original identification of the liver plasma membrane as the major locus of the receptor protein was subsequently expanded to include membranes of the Golgi, a smooth microsomal fraction, and the lysosomes (Riordan *et al.*, 1974; Pricer and Ashwell, 1976). Receptor protein associated with subcellular fractions enriched with these organelles, while not purified to homogeneity, gave similar banding patterns on polyacrylamide-gel electrophoresis. Antibody prepared against purified receptor gave rise to a single coincidence on double immunodiffusion for each fraction. On the basis of these results and their common binding characteristics it was concluded that receptors isolated from these liver subfractions were of common origin (Pricer and Ashwell, 1976). While recent immunocytochemical techniques were able to localize the receptor protein to the plasma membrane no immunodetectable receptor was present on any other subcellular organelle (Geuze *et al.*, 1982).

Regardless of its subcellular distribution there appears to be little question that the bulk of receptor activity is cryptic [i.e. only exposed after solubilization with Triton X-100 (Tanabe *et al.*, 1979)]. Since amounts of desialylated glycoproteins far in excess of the binding activity detectable at the cell surface are transported intracellularly under conditions in which neither synthesis nor degradation of the receptor is enhanced (Tanabe *et al.*, 1979; Warren and Doyle, 1981), it was proposed that cell-surface receptors must either be reutilized, or replenished from the larger pool of intracellular receptor.

In order to differentiate between these two alternatives, the cell surfaces of isolated hepatocytes were uniquely labeled by neuraminidase treatment (Stockert *et al.*, 1980b). This enzyme abolishes binding of galactose-terminated desialylated orosomucoid, but the cells continued to endocytose the higher-affinity ligand, N-acetylgalactosamine-terminated desialylated bovine submaxillary mucin. Only cell-surface-associated receptors were selectively restricted to desialylated bovine submaxillary mucin and the altered specificity of the cell was maintained during continuous endocytosis. From this study it was inferred that any reutilization mechanism in the endocytotic process involves only cell-surface receptors which remain effectively segregated from the bulk of the intracellular (cryptic) pool.

That surface receptor is sufficient to mediate continuing endocytosis was supported by another line of evidence. When mouse L-cells, which were devoid of receptor protein, were fused with rat hepatocyte membrane vesicles (Doyle *et al.*, 1979), or reconstituted vesicles containing purified receptor protein (Baumann *et al.*, 1980), these cells became capable of endocytosis and degradation of desialylated glycoprotein. Consistent with the concept of a functional segregation of receptor pools was the finding that once the initial complement of cell-surface receptors was blocked by a single passage of specific antibody, in the isolated perfused liver, there was no restoration of functional receptor for at least 90 min (Stockert *et al.*, 1981). Since the antibody reaction was limited to the cell-surface receptors, it was inferred that there was no insertion of intracellular receptors in the absence of ligand. It is difficult to assess what effect antibody might have had on the normal recycling of receptor. In fact, experiments using chloroquine (Tolleshaug and Berg, 1979), which appears to reduce surface receptor number in the absence of ligand, suggest that receptors are internalized (cycled) at some constant rate.

Until recently, it remained a distinct possibility that functioning receptors (i.e. receptor–ligand complex) never leave the cell surface and thereby avoid lysosomal degradation. However, on the basis of experiments in which only one cycle of endocytosis was allowed, it was shown that there was a dramatic reduction of available receptor for subsequent continuing endocytosis (Weigel, 1981). Similarly, based on analysis of the rates of internalization, dissociation of the receptor–ligand complex and degradation of the labeled ligand, a model was proposed in which receptor ligand complex remains associated during internalization (Bridges *et al.*, 1982). Furthermore, it was concluded that surface replacement would have had to come from a previously cryptic pool of unoccupied receptors (Bridges *et al.*, 1982).

The mechanism which underlies receptor reutilization is clearly far more complex than the simple two-pool model would allow. The biosynthetic pool of receptor protein could not be available for such a rapid replacement at the surface (Nakada *et al.*, 1981). Receptor localized to subcellular organelles other than the plasma membrane may be performing a yet unknown function such as organelle fusion, as has been suggested (Ashwell, 1977). This leaves the pool localized to the cell periphery of which a variable amount may remain cryptic. It is not unreasonable that only this pool is involved in endocytosis of ligand and that the receptor–ligand complex becomes dissociated just below the plasma membrane for rapid recycling, as recently suggested by immunocytochemical localization (Geuze *et al.*, 1982).

7.4 INTRACELLULAR PATHWAYS

The clearance of desialylated glycoproteins from the circulation is initiated by receptor binding at hepatocyte plasma membrane. Following internalization the ligand is transported to lysosomes where it is degraded. The rate-limiting step in this catabolic pathway was shown to be the translocation between the plasma membrane and lysosomes (Dunn *et al.*, 1979), a process with a mean transit time of approximately 7 minutes. In the isolated perfused liver, this process was shown to be temperature-dependent (Dunn *et al.*, 1980). Uptake and degradation of desialylated fetuin continued but was progressively slowed as the temperature decreased from 35°C to 20°C. Subcellular fractionation and *in situ* electron microscopic radioautography indicated that at 20°C internalized ligand was localized to pinocytic vesicles which did not fuse with lysosomes. When such vesicles were isolated immediately after injection of ligand they were devoid of lysosomal enzyme markers (Pertoft *et al.*, 1978). With time the ligand was found associated with low-density particles identifiable as lysosomal. These lysosomes appear to progressively acquire a higher density during the catabolic process.

The first evidence for lysosomal catabolism of internalized desialylated glycoprotein was provided by studies of the metabolic fate of [^{3}H] galactose-and protein-bound ^{64}Cu-labeled desialylated ceruloplasmin (Gregoriadis *et al.*, 1970). Both labels were shown to migrate with lysosomal enzyme markers in sucrose density gradients. Prior treatment of rats with Triton WR1339 or dextran, varying the specific gravity of the lysosomal fraction, resulted in density shifts of both the ligand and lysosomal enzyme markers. From the ratio of ^{3}H to ^{64}Cu recovered in immunoprecipitable protein it was inferred that enzymatic hydrolysis of galactose preceded the hydrolysis resulting in loss of copper. The peptide component of desialylated glycoproteins are ultimately hydrolyzed to free amino acids (LaBadie *et al.*, 1975). The role of lysosomes as the major site of catabolism was further established by studies of desialylated fetuin degradation in both isolated hepatocytes and the perfused liver (Dunn *et al.*, 1979; Berg and Tolleshaug, 1980). In both cases prior treatment with known inhibitors of lysosomal proteinases resulted in a reduction of ligand degradation and a progressive accumulation of intact ligand in a subcellular fraction identified as lysosomal.

An alternative intrahepatic pathway, which does not lead to degradation of the endocytosed ligand (Ma *et al.*, 1974; Renston *et al.*, 1980), has been proposed to account for the appearance of intact desialylated glycoproteins recovered in bile (Burger *et al.*, 1975; Thomas and Summers, 1978). This observation may be related to the recent finding that at low concentrations of desialylated transferrin, the hepatocytes exocytose the preponderance of the intracellular ligand (Tolleshaug *et al.*, 1981). The authors suggested that lysosomal homing of a pinocytic vessel requires an intracellular target signal. Similar evidence for a short-circuit pathway that does not lead to degradation was obtained using synthetic cluster glycosides (Connolly *et al.*, 1982). While *bi*- and *tri*-galactosides were internalized by isolated hepatocytes, unlike desialylated orosomucoid, these ligands were not degraded. Instead, they were quickly shuttled out of the cell. These results suggest that the architecture of the oligosaccharide participates not only in the recognition process at the cell surface but also plays some role in intracellular targeting of the pinocytic vessel.

A more detailed description of the sequential steps of the pathway was provided by a number of studies in which ligand, demonstrable at the electron microscopic level, was followed from the plasma membrane to lysosomes (Hubbard and Stukenbrok, 1979; Wall *et al.*, 1980; Stockert *et al.*, 1980b; Haimes *et al.*, 1981). From these studies a sequence of events emerge starting with the initial binding of ligand with receptor occurring in coated pits at the sinusoidal surface of the hepatocyte. In one study (Stockert *et al.*, 1980a) uptake was also observed along the lateral surfaces of hepatocytes indicating that receptors are not limited to the sinusoidal surface of the cell. This conclusion was recently confirmed by ligand localization (Kolb-Bachofen, 1981; Zeitlin and Hubbard, 1982) and immunolocalization (Geuze *et al.*, 1982) studies of receptor distribution. In all cases receptor was localized to the entire hepatocyte plasma membrane and in the immunolocalization study receptor was demonstrated in the bile capillary membrane as well. When ligand was followed, a temperature-dependent migration of receptor–ligand complex to coated pits was observed.

The electron microscopic findings document the rapidity of the endocytosis of desialylated glycoproteins. Sections of liver fixed as early as 30 s after injection of ligand linked to horseradish peroxidase show the enzyme reaction product in small vesicular structures in the peripheral cytoplasm near the space of Disses (Wall *et al.*, 1980). As early as 5 minutes following injection, marker enzyme was found in pericanalicular residual bodies which possess morphological characteristics of secondary lysosomes. Using lactosaminated ferritin as ligand, components of the lysosomal compartment were demonstrated (Haimes *et al.*, 1981). These include secretory vacuoles containing lipoprotein-like particles, autophagic vacuoles, residual bodies and GERL, a hydrolase-rich region of smooth endoplasmic reticulum.

The transition of ligand in the coated pit on the cell surface to an intracellular coated vesicle appears to provide the major system for the selective uptake of many macromolecules and peptide hormones which bind to cells (Pearse, 1980). The coat, composed of polymerized protein, is released from the vesicles by partial or total dissociation as a prelude to fusion with new intracellular membrane. Such smooth membraned vesicles, termed 'receptosome' (Willingham and Pastan, 1980), continue to migrate and accumulate in the Golgi region of the cell. Ultimately, ligand is found in lysosomes where degradation takes place. So characteristic is this sequence of events that it suggests that a common biological mechanism underlies all receptor-mediated endocytic processes.

Translocation of pinocytic vesicles across the cytoplasm appears to involve elements of the hepatic cytoskeleton. Direct participation of the hepatocytic microtubules and microfilaments during transport of desialylated fetuin was produced using isolated hepatocytes (Kolset *et al.*, 1979). Treatment of cells with colchicine, an inhibitor of microtubular function, slowed uptake and degradation of ligand which could be accounted for by a reduction of plasma membrane receptor-binding capacity. In contrast, cytochalasin B, a microfilament inhibitor, selectively reduced ligand degradation with no effect on the uptake process suggesting an inhibition of intracellular transport to, or fusion with, the lysosomal compartment.

7.5 ALTERATION OF RECEPTOR FUNCTION

Sera from normal persons and patients with a variety of illnesses were tested for their ability to inhibit desialylated glycoprotein binding to plasma membranes (Marshall *et al.*, 1974). The small amounts of inhibitory substances in the sera of controls or patients without liver disease, which were presumed to represent a low level of circulating desialylated glycoprotein, were significantly increased in sera obtained from patients with clinically confirmed cirrhosis or hepatitis. Serum inhibitors were isolated and identified as desialylated glycoproteins by affinity chromatography on a column of purified receptor covalently linked to agarose (Lunney, 1976). Resolution of the inhibitors by gel filtration indicated that they were a heterogeneous population of desialylated glycoproteins (Marshall and Williams, 1978).

An increased serum titer of desialylated glycoproteins in patients with hepatocellular disease could result from either the retention in the circulation due to reduced receptor activity or from a defect in the terminal stages of glycosylation during biosynthesis of serum glycoproteins. If the former were true, it would support the concept that recognition by the hepatic receptor is an obligatory step in regulating plasma protein turnover. A direct link between loss of receptor activity and appearance of circulating inhibitor has been demonstrated in the rat model of acute hepatitis induced by D-galactosamine injection (Sawamura *et al.*, 1981).

Evidence for a reduction in receptor protein as a consequence of hepatocellular pathology was obtained in studies using the chemical carcinogen, *N*-2-acetylaminofluorene (Stockert and Becker, 1980). The level of receptor-binding activity of neoplastic nodules induced by the carcinogen was approximately 37% of that of normal rat liver. The binding capacity of primary hepatocellular carcinomas which resulted from this regimen was reduced by 95%. This loss of binding activity was found to be proportional to the decreased concentration of antigenically detectable receptor present in the altered tissue. In an earlier study, it was shown that six Morris hepatoma cell lines, transplanted into rats, exhibit negligible binding activity (Hickman and Ashwell, 1974). That transformation results in loss of receptor activity is not without exception. Recently, the screening of a number of hepatoma cell lines led to the discovery of a continuous human hepatoma line (HepG2) which contains the receptor (Schwartz *et al.*, 1981a). This valuable line provides a well-defined system to study the molecular details of receptor-mediated endocytosis, topology and biosynthesis.

The effects of cellular replication on expression of receptor function were studied in the isolated perfused liver system (Gartner *et al.*, 1981). While the rat hepatocyte divides approximately once per year, rapid cellular proliferation occurs throughout the liver remnant following a two-thirds hepatectomy. Analysis of influx rate constants revealed a dramatic decrease for desialylated glycoprotein uptake during the first 48 hours of regeneration, a time that corresponds to the period of greatest cellular proliferation. Transport function remained depressed for 4 days with a gradual return to normal by the sixth day when regeneration was substantially completed. These results sug-

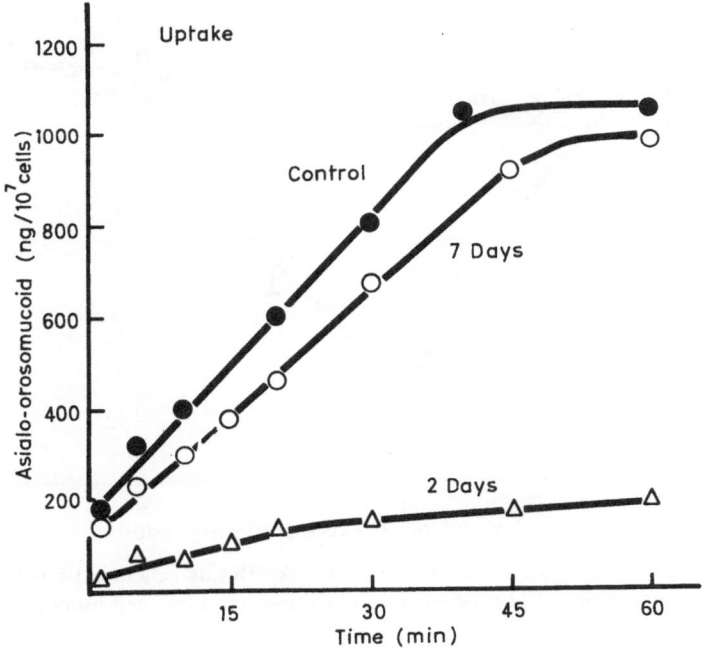

Fig. 7.3 Uptake of desialylated orosomucoid (AsOR) by hepatocytes isolated 2 days (△) or 7 days (○) following a partial hepatectomy and from sham-operated controls (●).

gest that the liver cell undergoes retrodifferentiation to an immature stage, fetal-like liver devoid of receptor (Hickman and Ashwell, 1974) and that a maturation process follows cellular division for restoration of specific hepatocytic function.

To determine whether the reduction of binding activity after two-thirds hepatectomy was restricted to the plasma membrane receptor, or affected all receptor pools as in hepatocellular carcinoma, surface binding was selectively measured in isolated hepatocytes (Howard *et al.*, 1982). Two days after a two-thirds hepatectomy endocytosis of desialylated orosomucoid was reduced by 80% when compared to cells isolated from sham-operated controls (Fig. 7.3). As was seen in the isolated perfused liver system, normal uptake was shown to be the result of a selective decrease in plasma-membrane-associated receptor (Fig. 7.4). There was an 80% reduction of surface receptor activity at 2 days. At this time point total activity assayed in cell homogenates, following solubilization with Triton X-100 was reduced by only 30%. Thus, during the peak time of cell division induced by two-thirds hepatectomy, binding activity of the plasma-membrane-associated receptor was disproportionately reduced.

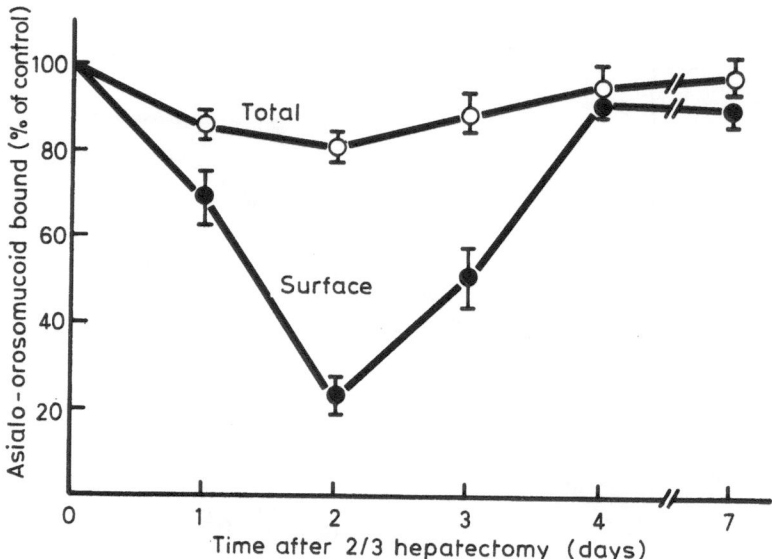

Fig. 7.4 Binding of desialylated orosomucoid (AsOR) by (○) intact hepatocytes (at 4°C when binding does not proceed to endocytosis) and (●) cell homogenates in Triton X-100.

These results indirectly support the concept of separate receptor pools which may have different cellular functions.

7.6 CONCLUSION

The original findings describing a galactose-specific binding protein in the liver have opened up to reveal a remarkable collection of cellular receptors for glycoproteins; a detailed review of these is available (Neufeld and Ashwell, 1980). Four recognition systems have been described for receptor-mediated endocytosis of glycoproteins in the liver alone. In mammals, both galactose and fucose (Prieels *et al.*, 1978) receptors are localized to the parenchymal cells, while receptors for *N*-acetylglucosamine or mannose are present in the macrophages (Kupffer cell) (Stahl *et al.*, 1978). An analogous receptor protein, specific for only *N*-acetylglucosamine, has been isolated from avian liver (Lunney and Ashwell, 1976; Kawasaki and Ashwell, 1977). This list almost certainly will grow further, pointing to the significant role of carbohydrate structure as a means of information storage and retrieval.

REFERENCES

Ashwell, G. (1977), in *The Glycoconjugates*, Butterworths, London, Vol. 4, pp. 57–71.

Baenziger, J.V. and Fiete, D. (1980), *Cell*, **22**, 611–620.

Baenziger, J.V. and Maynard, Y. (1980), *J. Biol. Chem.*, **255**, 4607–4613.

Baumann, H., Hou, E. and Doyle, D. (1980), *J. Biol. Chem.*, **255**, 10001–10002.

Berg, T. and Tolleshaug, H. (1980), *Biochem. Pharmacol.*, **29**, 917–925.

Bridges, K., Harford, J., Klausner, R. and Ashwell, G. (1982), *Proc. Natl. Acad. Sci. U.S.A.*, **79**, 350–354.

Burger, R.L., Schneider, R.J., Mehlman, C.S. and Allen, R.H. (1975), *J. Biol. Chem.*, **250**, 7707–7713.

Connolly, D.T., Townsend, R.R., Kawaguchi, K., Bell, W.R. and Chuan, Y. (1982), *J. Biol. Chem.*, **257**, 939–945.

Doyle, D., Hou, E. and Aarren, R. (1979), *J. Biol. Chem.*, **254**, 6853–6856.

Dunn, W.A., Hubbard, A.L. and Aronson, N.N., Jr. (1980), *J. Biol. Chem.*, **255**, 5791–5978.

Dunn, W.A., LaBadie, J.H. and Aronson, N.N., Jr. (1979), *J. Biol. Chem.*, **254**, 4191–4196.

Gan, J.C. (1979), *Int. J. Biochem.*, **11**, 481–486.

Gartner, U., Stockert, R.J., Morell, A.G. and Wolkoff, A.W. (1981), *Hepatology*, **1**, 99–106.

Geuze, H.J., Slot, J.W., Strous, G.J.A.M., Lodish, H.F. and Schwartz, A.L. (1982), *J. Cell Biol.*, **92**, 865–870.

Gregoriadis, G., Morell, A.G., Sternlieb, I. and Scheinberg, I.H. (1970), *J. Biol. Chem.*, **245**, 5833–5837.

Haimes, H., Stockert, R.J., Morell, A.G. and Novikoff, A.B. (1981), *Proc. Natl. Acad. Sci. U.S.A.*, **78**, 6936–6939.

Hatton, M.W.C., Marz, L. and Berry, L.R. (1979), *Biochem. J.*, **181**, 633–638.

Hickman, J. and Ashwell, G. (1974), in *Enzyme Therapy in Lysosomal Storage Disease* (Tager, J.M., Hooghwinkel, H.J.M. and Daems, W.T.H., eds.), North-Holland Publishing Co., Amsterdam, pp. 169–192.

Hickman, J., Ashwell, G., Morell, A.G., Van den Hammer, C.J.A. and Scheinberg, I.H. (1970), *J. Biol. Chem.*, **245**, 759–766.

Howard, D.J., Stockert, R.J. and Morell, A.G. (1982), *J. Biol. Chem.* (in press).

Hubbard, A.L. and Stukenbrok, H. (1979), *J. Cell Biol.*, **83**, 65–81.

Hudgin, R.L., Pricer, W.E., Ashwell, G., Stockert, R.J. and Morell, A.G. (1974), *J. Biol. Chem.*, **246**, 1461–1467.

Kawasaki, T. and Ashwell, G. (1976a), *J. Biol. Chem.*, **251**, 1296–1302.

Kawasaki, T. and Ashwell, G. (1976b), *J. Biol. Chem.*, **251**, 5292–5298.

Kawasaki, T. and Ashwell, G. (1977), *J. Biol. Chem.*, **252**, 6536–6543.

Kolb, H., Kolb-Bachofen, V. and Schlepper-Schafer, J. (1979), *Biol. Cell.*, **36**, 301–308.

Kolb, H., Vogt, D. and Herbertz, L. (1980), *Hoppe Seylers Z. Physiol. Chem.*, **361**, 1747–1750.

Kolb-Bachofen, V. (1981), *Biochim. Biophys. Acta*, **645**, 293–299.

Kolset, S.O., Tolleshaug, H. and Berg, T. (1979), *Exp. Cell Res.*, **122**, 159–167.

184   *Receptor-Mediated Endocytosis*

LaBadie, J.H., Chapman, K.P. and Aronson, N.N., Jr. (1975), *Biochem. J.*, **152**, 271–279.

Lunney, J. (1976), *Studies on the Regulation of Serum Glycoprotein Homeostasis*, Thesis, The Johns Hopkins University, Baltimore.

Lunney, J.K. and Ashwell, G. (1976), *Proc. Natl. Acad. Sci. U.S.A.*, **73**, 341–343.

Ma, M.H., Laird, W.A. and Scott, H. (1974), *J. Histochem. Cytochem.*, **22**, 160–169.

Marshall, J.S., Green, A.M., Pensky, J., Williams, S., Zinn, A. and Carlson, D.M. (1974), *J. Clin. Invest.*, **54**, 555–562.

Marshall, J.S. and Williams, S. (1978), *Biochim. Biophys. Acta*, **543**, 41–52.

Morell, A.G., Gregoriadis, G., Scheinberg, I.H., Hickman, J. and Ashwell, G. (1971), *J. Biol. Chem.*, **246**, 1461–1467.

Morell, A.G., Irvine, R.A., Sternlieb, I., Scheinberg, I.H. and Ashwell, G. (1968), *J. Biol. Chem.*, **243**, 155–159.

Morell, A.G. and Scheinberg, I.H. (1972), *Biochem. Biophys. Res. Commun.*, **48**, 808–815.

Morell, A.G., Van den Hamer, C.J.A. and Scheinberg, I.H. (1966), *J. Biol. Chem.*, **241**, 3745–3749.

Nakada, H., Sawamura, T. and Tashiro, Y. (1981), *J. Biochem. (Tokyo)*, **89**, 135–141.

Neufeld, E.F. and Ashwell, G. (1980), in *The Biochemistry of Glycoproteins and Proteoglycans* (Lennarz, W.J., ed.), Plenum Publishing Corp., New York, pp. 241–266.

Novogrodsky, A. and Ashwell, G. (1977), *Proc. Natl. Acad. Sci. U.S.A.*, **77**, 676–678.

Paulson, J.C., Hill, R.L., Tanabe, T. and Ashwell, G. (1977), *J. Biol. Chem.*, **252**, 8624–8628.

Pearse, B. (1980), *Trends Biol. Sci.*, **5**, 131–134.

Pertoft, H., Warmegard, B. and Hook, M. (1978), *Biochem. J.*, **174**, 309–317.

Pricer, W.E., Jr. and Ashwell, G. (1971), *J. Biol. Chem.*, **246**, 4825–4833.

Pricer, W.E., Jr. and Ashwell, G. (1976), *J. Biol. Chem.*, **251**, 7539–7544.

Prieels, J.P., Pizzo, S.V., Galsgow, L.R., Paulson, J.C. and Hill, R.L. (1978), *Proc. Natl. Acad. Sci. U.S.A.*, **75**, 2215–2219.

Regoeczi, E., Taylor, P., Hatton, M.W.C., Wong, K.L. and Koj, A. (1978), *Biochem. J.*, **174**, 171–178.

Regoeczi, E., Taylor, P., Debanne, M.T., Marz, L. and Hetton, M.W.C. (1979), *Biochem. J.*, **184**, 399–407.

Renston, R.H., Maloney, D.G., Jones, A.L., Hradek, G.T., Wong, K.Y. and Goldfine, I.D. (1980), *Gastroenterol.*, **78**, 1373–1388.

Riordan, J.R., Mitchell, L. and Slavik, M. (1974), *Biochem. Biophys. Res. Commun.*, **59**, 1373–1379.

Sarkar, M., Liao, J., Kabat, E.A., Tanabe, T. and Ashwell, G. (1979), *J. Biol. Chem.*, **254**, 3170–3174.

Sawamura, T., Kawasto, S., Shiozaki, Y., Sameshima, Y., Nakada, H. and Tashio, Y. (1981), *Gastroenterol.*, **81**, 527–533.

Sawamura, T., Nakada, H., Fujii-Kuriyama, Y. and Tashiro, Y. (1980), *Cell Struct. Funct.*, **5**, 133–146.

Schwartz, A.L., Fridovich, S.E., Knowles, B.B. and Lodish, H.F. (1981a), *J. Biol. Chem.*, **256**, 8878–8881.

Schwartz, A.L., Marshak-Rothstein, A., Rup, D. and Lodish, H.F. (1981b), *Proc. Natl. Acad. Sci. U.S.A.*, **78**, 3348–3352.

Stahl, P.D., Rodman, J.S., Miller, M.J. and Schlesinger, P.H. (1978), *Proc. Natl. Acad. Sci. U.S.A.*, **75**, 1399–1403.

Steer, C.J. and Ashwell, G. (1980), *J. Biol. Chem.*, **255**, 3008–3013.

Stockert, R.J. and Becker, F.F. (1980), *Cancer Res.*, **40**, 3632–3634.

Stockert, R.J., Gartner, U., Morell, A.G. and Wolkoff, A.W. (1981), *J. Biol. Chem.*, **255**, 3830–3831.

Stockert, R.J., Haimes, H.B., Morell, A.G., Novikoff, P.M., Novikoff, A.B., Quintana, N. and Sternlieb, I. (1980a), *Lab. Invest.*, **43**, 556–563.

Stockert, R.J., Howard, D.J., Morell, A.G. and Scheinberg, I.H. (1980b), *J. Biol. Chem.*, **255**, 9028–9029.

Stockert, R.J., Morell, A.G. and Scheinberg, I.H. (1974), *Science*, **186**, 365–366.

Stockert, R.J., Morell, A.G. and Scheinberg, I.H. (1977), *Science*, **197**, 667–668.

Stowell, C.P. and Lee, Y.C. (1978), *J. Biol. Chem.*, **256**, 2230–2234.

Tanabe, T., Pricer, W.E., Jr. and Ashwell, G. (1979), *J. Biol. Chem.*, **254**, 1038–1043.

Thomas, P. and Summers, J.W. (1978), *Biochem. Biophys. Res. Commun.*, **80**, 335–339.

Tolleshaug, H. and Berg, T. (1979), *Biochem. Pharmacol.*, **28**, 2919–2922.

Tolleshaug, H., Chindemi, P.A. and Regoeczi, E. (1981), *J. Biol. Chem.*, **256**, 6526–6528.

Van den Hamer, C.J.A., Morell, A.G., Scheinberg, I.H., Hickman, J. and Ashwell, G. (1970), *J. Biol. Chem.*, **245**, 4397–4402.

Van Lenten, L. and Ashwell, G. (1972), *J. Biol. Chem.*, **247**, 4633–4640.

Vierling, J.M., Steer, C.J. and Hickman, J.W. (1978), *Gastroenterology*, **75**, 456–461.

Wall, D.A., Wilson, G. and Hubbard, A.L. (1980), *Cell*, **21**, 79–93.

Warren, R. and Doyle, D. (1981), *J. Biol. Chem.*, **256**, 1346–1355.

Weigel, P. (1981), *Biochem. Biophys. Res. Commun.*, **101**, 1419–1425.

Weigel, P.H., Schnaar, R.L., Kuhlenschmidt, M.S., Schmell, E., Lee, R.T., Lee, Y.C. and Roseman, S. (1979), *J. Biol. Chem.*, **254**, 10830–10838.

Willingham, M.C. and Pastan, I. (1980), *Cell*, **21**, 67–77.

Zeitlin, P.L. and Hubbard, A.L. (1982), *J. Cell Biol.*, **92**, 634–647.

8 Entry of Toxic Proteins into Cells

SJUR OLSNES and KIRSTEN SANDVIG

Acknowledgements

The authors are grateful to Professor A. Pihl for his critical reading of the manuscript.

Receptor-Mediated Endocytosis
(*Receptors and Recognition*, Series B, Volume 15)
Edited by P. Cuatrecasas and T. F. Roth
Published in 1983 by Chapman and Hall, 11 New Fetter Lane, London EC4P 4EE
© 1983 Chapman and Hall

8.1 INTRODUCTION

The cell surface membrane constitutes a very efficient permeability barrier between the extracellular and intracellular compartments. Most hydrophilic molecules do not penetrate this barrier unless specific uptake mechanisms are available. Hydrophilic molecules as large as proteins are usually not able to enter into the cytosol. However, since the late 1960s compelling evidence has accumulated that several protein toxins do penetrate the cell membrane and damage components in the cytosol. The toxins all act enzymatically by modifying and inactivating intracellular targets. In some cases this inactivation causes cell death. In other cases the toxins induce in the affected cells physiological changes which may be deleterious to the organism as a whole.

Some of the toxins discussed here are pathogenic factors in major epidemic diseases. Thus, in diphtheria, the toxin causes the most serious complications such as heart failure. Cholera toxin and *Escherichia coli* heat-labile toxin cause excessive water release from the ileum, resulting in serious and often lethal diarrhoea. The exotoxin A appears to play a role in infections with *Pseudomonas aeruginosa*, e.g. after serious burns. Although not proved, it is likely that *Shigella* toxin plays a role in dysenteria. Serious and even lethal intoxications by the plant toxins abrin, modeccin, ricin and viscumin have often been reported.

The toxic proteins described in this chapter all consist of two functionally different parts which are termed A and B and which are connected by a single disulfide bridge. In each case the A-part is an enzyme which enters the cytosol and modifies intracellular targets, whereas the B-part is necessary for binding of the toxin to cell surface receptors. Such binding is a requirement for toxic effect in living cells. In broken cell systems, however, the A-part alone is sufficient for toxic effect. Clearly, the binding of the toxin to cell surface receptors is important for the penetration of the A-part into the cytosol.

Several years ago it was suggested that endocytosis plays a role in toxin entry into the cytoplasm (Nicholson, 1974; Olsnes *et al.*, 1974; Refsnes *et al.*, 1974). This suggestion was based partly on electron microscopic examinations which showed extensive endocytosis of ferritin-labeled ricin in toxin-sensitive cells, and partly on the finding that abrin and ricin were not toxic to reticulocytes, although the toxins were readily bound to these cells which have very low endocytic activity. For several years, attempts to elucidate whether endocytosis is indeed involved in toxin entry failed to give unambiguous answers (see Olsnes 1978). However, recently evidence has been accumulating that the toxins or their A-parts indeed enter the cytosol from endocytic vesicles. It must be kept in mind, however, that even if endocytosis is involved in the uptake, the main problem, *viz.* how the A-part of the toxin is translocated across the lipid bilayer, remains.

189

In studying the entry of molecules with a biological activity as high as that of the toxins here described (a single toxin molecule may kill a cell), it must be born in mind that the physical and chemical methods commonly used to follow molecules in cells (electron microscopy, immunological and isotope techniques) are not readily applicable. The high number of toxin molecules that must be added to the cells to allow detection in such studies, may enter the cells by several different routes. Therefore, the molecules which are later detected may not have entered in the same way as those that cause the intoxication. Studies with physical and chemical methods should therefore be correlated with other studies where the biological effect, i.e. toxicity, is followed. Toxic effect is so far the only reliable parameter for measuring toxin entry into the cytosol.

8.2 STRUCTURE AND MECHANISM OF ACTION OF TOXIC PROTEINS

8.2.1 Abrin, ricin, modeccin and viscumin

These four toxins which are isolated from plant material have a similar structure and mechanism of action. In all cases the binding activity and the enzymatic activity are carried on two separate polypeptide chains (Fig. 8.1). The length of the chains varies slightly between the four toxins (see Table 8.1). In each case the A-chain migrates in sodium dodecyl sulfate (SDS)/polyacrylamide gels slightly more rapidly than the B-chain (for review, see Olsnes *et al.*, 1974; Olsnes and Pihl, 1976, 1982). Ricin has been the most extensively studied and its primary structure is known. The amino acid sequence of ricin A-chain contains stretches of hydrophobic amino acids as well as stretches consisting of hydrophilic amino acids. Low-resolution (0.4 nm) X-ray crystallographic studies of ricin revealed that the B-chain has a bilobal structure and that each domain is able to bind galactose. One binding site was more highly occupied than the other one (Villafranca and Robertus, 1981).

Treatment of the plant toxins with sodium dodecyl sulfate did not reduce the enzymatic activity of the A-chain, whereas the binding property of the

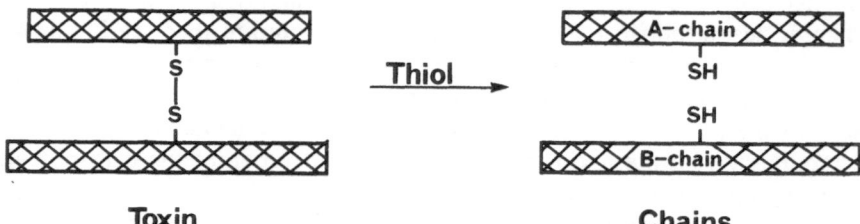

Fig. 8.1 Schematic structure of abrin, modeccin, ricin and viscumin and their chains.

B-chain was lost after such treatment. All four plant toxins are glycoproteins. The majority of carbohydrate is bound to the B-chain.

The A-chains of all four toxins inactivate the 60S ribosomal subunits enzymatically in an as yet still unknown way. As a result, elongation factor 2 (EF2) is not able to bind to the subunit and protein synthesis stops. Abrin and ricin A-chains inactivate pure ribosomes in simple buffer solution. There are certain differences between the toxins. Thus, the presence of EF2, particularly together with GTP, protects ribosomes against inactivation by abrin and ricin, whereas EF2 (in the absence of GTP) sensitizes the ribosomes to modeccin, possibly by slightly changing their conformation (Olsnes and Abraham, 1979).

Each A-chain molecule of abrin and ricin inactivates pure ribosomes at a rate of 1500 ribosomes per min. The K_m with respect to ribosomes is about 2×10^{-7}M, which ensures that abrin and ricin A-chains act at close to their maximal rates (V_{max}) in the cytosol. Even when EF2 is present, modeccin acts more slowly in cell-free systems than abrin and ricin A-chains. Since modeccin is as toxic to cells as abrin and ricin, it is possible that modeccin must be somehow activated, e.g. by proteolytic cleavage, to achieve its maximal enzymatic activity. Such cleavage could occur in the cell, e.g. in lysosomes, before the A-chain enters the cytosol. Abrin, ricin and modeccin were found to kill cells even if only a single molecule enters the cytosol (Eiklid *et al.*, 1980).

The B-chains of abrin, ricin, modeccin and viscumin are lectins which bind to carbohydrates containing terminal galactose residues (Olsnes *et al.*, 1978). Binding is a requirement for toxin action and cells can be partially protected by the addition of galactose or lactose to the medium.

8.2.2 Diphtheria toxin and *Pseudomonas aeruginosa* exotoxin A

Among the toxins discussed here, diphtheria toxin has been studied in most detail. It is synthesized by *Corynebacterium diphtheriae* as a single polypeptide chain. The toxin contains an arginine-rich region which is easily split by proteolytic enzymes present in the bacterial culture medium as well as in serum. The two polypeptide fragments thus formed are linked by a disulfide bridge (Fig. 8.2). The shorter fragment (fragment A) carries the enzymatic activity, whereas fragment B binds the toxin to cell surface receptors.

Diphtheria toxin fragment A is highly hydrophilic and resistant to boiling, extreme pH and sodium dodecyl sulfate. Fragment B is much more sensitive to such treatment. It contains a region which is highly hydrophobic, but not exposed under neutral, non-denaturing conditions (Boquet *et al.*, 1976). After denaturation, or when pH drops below 4.5, the hydrophobic region in the B-fragment is exposed (Sandvig and Olsnes, 1981a).

Lambotte *et al.* (1980) showed that diphtheria toxin fragment B contains

Table 8.1 Physical and biological properties of toxins and their chains and fragments

Toxin	Molecular weight	pI	Intracellular target	Receptor	Receptor number ($n \times 10^5$)	Binding constant ($K_a \times 10^7 M^{-1}$)
Diphtheria toxin	62 000	6.0	EF2	Glycoprotein	0.05–2	50
A-fragment	21 500	5.2	EF2			
B-fragment	40 000	6.8	EF2			
Pseudomonas aeruginosa toxin	70 000		EF2	?		
A-fragment	26 000		EF2			
Abrin	65 000	6.1	60S ribosomes	Galactose	300	12
A	30 000	4.6	60S ribosomes			
B	35 000	7.2	60S ribosomes			
Ricin	62 057	7.1	60S ribosomes	Galactose	300	2.6
A	30 625	7.5	60S ribosomes			
B	31 432	4.8	60S ribosomes			

	Molecular weight	pI	Target	Receptor		
Modeccin	63000	6.2–7.1	60S ribosomes	Galactose		
A	28000	5.8–6.1	60S ribosomes			
B	38000	7.3–8.2	60S ribosomes			
Viscumin	~57000		60S ribosomes	Galactose		
A	29000		60S ribosomes			
B	33000		60S ribosomes			
Shigella toxin	65000		60S ribosomes	?	~10	~1000
A_1	27500		60S ribosomes			
A_2	3000		60S ribosomes			
B	6–7 × 5000		60S ribosomes			
Cholera toxin	83000		Adenylate cyclase	GM_1	0.0006–100	100
A_1	23500		Adenylate cyclase			
A_2	5500		Adenylate cyclase			
B	5 × 11500		Adenylate cyclase			
E. coli heat-labile toxin	~85000		Adenylate cyclase	Galactose		
A_1	21000		Adenylate cyclase			
A_2	7000		Adenylate cyclase			
B	5 × 11500		Adenylate cyclase			

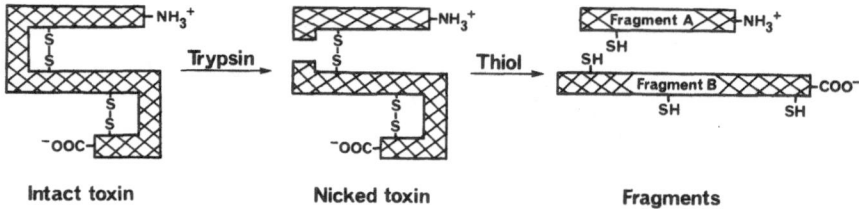

Intact toxin　　　　**Nicked toxin**　　　　**Fragments**

Fig. 8.2 Schematic structure of diphtheria toxin and its fragment.

two lipid-associating domains. One of these domains is located in the highly hydrophilic 9000 mol.wt. *N*-terminal region, and has a structure similar to that of the phospholipid headgroup-binding domain of human apolipoprotein 1. It can be considered as a surface lipid-associating domain. The other lipid-associating domain, which is highly hydrophobic, is located in the middle of fragment B. Its structure resembles that of the membranous domain of intrinsic membrane proteins and it can be considered as a transverse lipid-associating domain. This domain, which has an α-helix length (3.5 nm) approximately that of the thickness of the hydrocarbon region of the lipid bilayer (Kayser *et al.*, 1981), has the ability to insert itself into membranes and to increase the conductance across the bilayer. The *C*-terminal 8000-dalton region, which is assumed to bind to the receptor, does not show similarities with lipid-associating domains.

Pseudomonas aeruginosa exotoxin A is also synthesized by the bacterium as a single polypeptide chain (mol.wt. 70 000). Its structure is much less well known than that of diphtheria toxin and it is not clear if the enzymatic function and the binding function are located at different domains of the protein. Like diphtheria toxin, the enzymatic activity of *Pseudomonas aeruginosa* toxin is strongly increased by treatment with thiols in the presence of denaturing agents like urea (Leppla *et al.*, 1978). This suggests that the enzymatic region is not exposed in the intact native toxin. Enzymatically active fragments of 26 000 mol.wt. (Vasil *et al.*, 1977; Chung and Collier, 1977a) and 48 000 mol.wt. (Sanai *et al.*, 1980) have been identified. A receptor binding domain has so far not been detected.

The intracellular mechanisms of action of diphtheria toxin and *Pseudomonas aeruginosa* exotoxin A appear to be identical and consist of enzymatic ADP-ribosylation of a certain amino acid, diphthamide, present in elongation factor 2 (Van Ness *et al.*, 1980). Elongation factor 2 (EF2) is required to translocate the growing polypeptide chain from the A-site back to the P-site on the ribosome after the peptide bond has been formed. Therefore, when EF2 is inactivated, protein synthesis stops.

The amino acid which is ADP-ribosylated, diphthamide, has only been found in EF2. It is probably produced by post-translational modification of a histadine residue and has the structure shown in Fig. 8.3. At least three

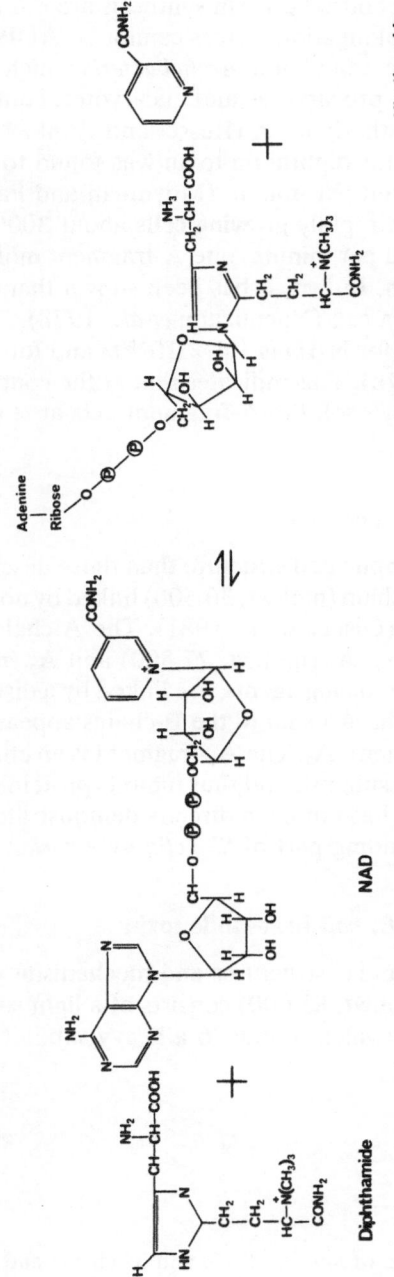

Fig. 8.3 ADP-ribosylation of diphthamide.

enzymes appear to be involved in the post-translational synthesis of diph-thamide (Van Ness *et al.*, 1980).

Procaryotic and mitochondrial protein synthesis are not inhibited by diph-theria toxin because the elongation factors cannot be ADP-ribosylated (Pappenheimer, 1977). On the other hand, *Archebacteria* which are considered to be intermediary between procaryotes and eucaryotes, contain EF2 which is ADP-ribosylated by diphtheria toxin (Kessel and Klink, 1980).

The turnover number for diphtheria toxin was found to be 2000 EF2 molecules ADP-ribosylated per minute (Moynihan and Pappenheimer, 1981). Estimating that in rapidly growing cells about 3000 new EF2 molecules are synthesized per minute, one A-fragment molecule may therefore prevent proliferation. In fact, it has been shown that one molecule of diphtheria toxin can kill a cell (Yamaizumi *et al.*, 1978).

The K_m of fragment A for NAD is 1.4×10^{-6} M and for EF2 1.5×10^{-7} M (Chung and Collier, 1977b). This indicates that at the concentrations of EF2 and NAD found in the cytosol, the A-fragment acts at close to its maximal rate (V_{max}).

8.2.3 *Shigella dysenteriae* cytotoxin

This toxin has a more complicated structure than those described above (Fig. 8.4). It consists of one A-chain (mol.wt. 30 500) linked by non-covalent bonds to six or seven B-chains (Olsnes *et al.*, 1981). The A-chain is easily split by trypsin into two fragments, A_1 (mol.wt. 27 500) and A_2 (mol.wt. 3000), which, in the absence of reducing agents, are linked by a disulfide bridge. The non-covalent linkage of the A-chain to the B-chains appears primarily to be through the smaller fragment, A_2. The A_1-fragment is an enzyme which inactivates the 60S ribosomal subunits and thus inhibits protein synthesis (Reisbig *et al.*, 1981). Although it has not been directly demonstrated, it is likely that the B-chains form the binding part of *Shigella dysenteriae* cytotoxin.

8.2.4 Cholera toxin and *E. coli* heat-labile toxin

These two toxins have similar structures and mechanisms of action (Gill, 1978). Cholera toxin (mol.wt. 83 000) consists of a light subunit, A (mol.wt. 29 000), linked by non-covalent bonds to a heavy subunit, B (mol.wt.

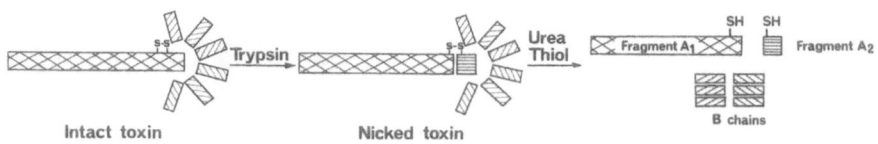

Fig. 8.4 Schematic structure of *Shigella* toxin and its chains and fragments.

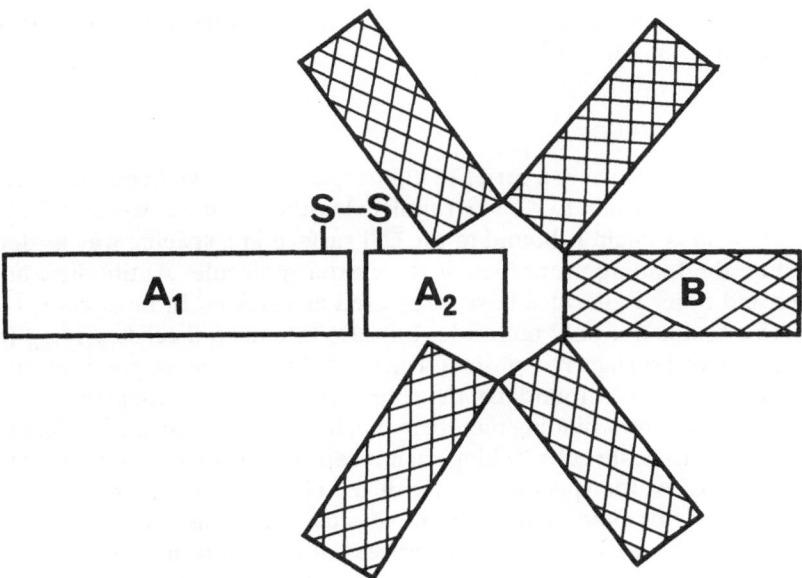

Fig. 8.5 Schematic structure of cholera toxin and *E. coli*, heat-labile toxin and their chains or fragments.

54 000), consisting of five B-protomers of mol.wt. 11 500 each (Fig. 8.5). The A-chain is easily split by trypsin into two fragments, A_1 and A_2, which are linked by a disulfide bridge. The A-subunit is linked to the B-subunit by the A_2-fragment.

The A_1-fragment is an enzyme which ADP-ribosylates the guanine nucleotide regulatory component of adenylate cyclase. When GTP is bound to this component, the catalytic subunit of adenylate cyclase is active. When GTP is hydrolyzed as a result of GTPase activity of the guanine nucleotide regulatory component, the adenylate cyclase is inactivated. The cholera toxin-induced ADP ribosylation blocks the GTPase reaction and thus stabilizes the adenylate cyclase in the active state. It is not clear which amino acid in the guanine nucleotide regulatory component is ADP-ribosylated by the A_1-fragment, but it could be arginine, since the toxin is able to ADP-ribosylate pure arginine. Also, a 12 500-mol.wt. fragment of A_1 was claimed to be able to ADP-ribosylate polyarginine (Lai *et al.*, 1981).

The B-subunit of cholera toxin binds to a particular ganglioside (GM_1). There is good evidence that GM_1 represents the cholera toxin receptor on the cell surface.

The lengths of the different polypeptide chains in *E. coli* heat-labile toxin differ slightly from those of cholera toxin (Table 8.1), but in general the structure and function of the two toxins are the same. *E. coli* heat-labile toxin binds to structures containing terminal galactose residues, but the nature of the receptor is so far not known.

8.2.5 Chimaeric toxins

During the last few years a series of chimaeric toxins have been prepared by linking a toxin A-chain to a carrier molecule other than the B-chain. In this way the toxin A-chain is bound to the cell surface in a specific way as determined by the binding properties of the carrier molecule. Antibodies, hormones and other molecules have been used as carriers. In most cases the chimaeric toxins not only bound to, but also intoxicated cells carrying the appropriate cell surface receptors (Olsnes, 1981). In general, good selectivity for the target cells was found, i.e. the chimaeric toxins were toxic to cells with receptors for the new binding moiety in much lower concentrations than those required to intoxicate cells lacking such receptors. However, compared with the native toxins, the specific toxicity of the chimaeric toxins was in most cases, low. In this chapter we will only discuss a few cases where chimaeric toxins have been used to study the process of toxin internalization.

8.2.6 Structural and functional similarities between the toxins

Between the toxins described above there are a few striking similarities which deserve consideration. As first pointed out by Gill (1978), SS bonds have not been found in any of the enzymatically active polypeptides (A-chains or A_1-fragments), whereas disulfide bonds appear to be common in the B-part of the toxins. Thus, in diphtheria toxin there is one internal SS bond in the B-fragment, in abrin B-chain there is one, in ricin B-chain four and in cholera toxin there is one internal disulfide per B-protomer.

Another characteristic of the A-chains (and the A_1-fragments) is their resistance to denaturation by sodium dodecyl sulfate (Gill, 1978). Most proteins, including the binding moiety of the toxins, lose their biological activity when they are treated with 1% sodium dodecyl sulfate, whereas the enzymatically active moieties of the toxins are fully active after such treatment. Diphtheria toxin A-fragment is an exceptionally stable molecule which also tolerates boiling and extreme pHs. Also, cholera toxin A_1-fragment is temperature-stable.

As will be discussed below, it is possible that the A-chains pass the cell membrane as unfolded polypeptide chains. The absence of internal disulfide bridges clearly facilitates the unfolding necessary in such a mechanism. It is clear that if such unfolding occurs, the A-chain must be able to refold to its active configuration once it is in the cytosol. The ability of the A-chains to recover their activity after denaturation is in accordance with this model.

In ricin A-chain an oligosaccharide is bound to amino acid no. 10, and this clearly would interfere with the transport of the extended protein through the membrane. It has been shown, however, that the first 40 amino acids of ricin A-chain can be removed without loss of enzymatic activity and it is possible that such cleavage occurs before the A-chain enters the cytoplasm (see Olsnes and Pihl, 1982).

In all toxins so far studied, hydrophobic stretches occur either in the A-chain or in the B-chain. In diphtheria toxin a hydrophobic region is present in the B-fragment, whereas in ricin two hydrophobic regions are present in the A-chain. Also in cholera toxin the A subunit is more hydrophobic than the B subunit. Such hydrophobic stretches may be involved in toxin transport across the membrane.

As will be discussed below, the toxin receptors differ and in most cases their structure is not known in detail. A terminal galactose residue seems, however, to be involved in the receptors for most of the toxins. Thus, abrin, modeccin, ricin and viscumin bind to oligosaccharides (possibly glycoproteins) carrying terminal non-reducing galactose residues. This also seems to be the case for *E. coli* heat-labile toxin, and the terminal galactose residue in ganglioside GM_1 is required for binding of cholera toxin. There is no evidence that terminal galactose residues are involved in binding of diphtheria toxin, *Pseudomonas aeruginosa* exotoxin A and *Shigella* toxin.

In the case of diphtheria toxin, abrin and ricin the enzymatically active part of the toxins acts at a high rate in cell-free systems. Thus, diphtheria toxin fragment A inactivates 2000 molecules of EF2 per minute. As mentioned above, this is just enough to prevent a net accumulation of EF2 in rapidly growing cells and therefore to stop cell division. Abrin and ricin A-chains inactivate 1500 pure ribosomes per minute and approximately 10 times less in a total cell lysate. This is, however, still enough to stop proliferation as the inactivation of one ribosome per polysome is sufficient to stop protein synthesis. The K_m of the enzymes is in all the cases sufficiently high to allow the toxin A-chains to operate close to $V_{max.}$ under the concentrations of ribosomes, EF2 and NAD present in the cytosol. Apparently, nature has made the toxins just as efficient as necessary for one A-chain molecule to kill the cell. It seems logical that there has not been any evolutionary pressure to make the toxin A-chains even more efficient.

8.3 TOXIN BINDING TO CELL SURFACE RECEPTORS

8.3.1 Nature of the receptors

(a) *The cholera toxin receptor is ganglioside GM$_1$*

Only in the case of cholera toxin is the nature of the receptor known in detail (for review see Holmgren, 1981). This toxin binds to ganglioside GM_1 (Fig. 8.6). The following observations support the contention that GM_1 is the

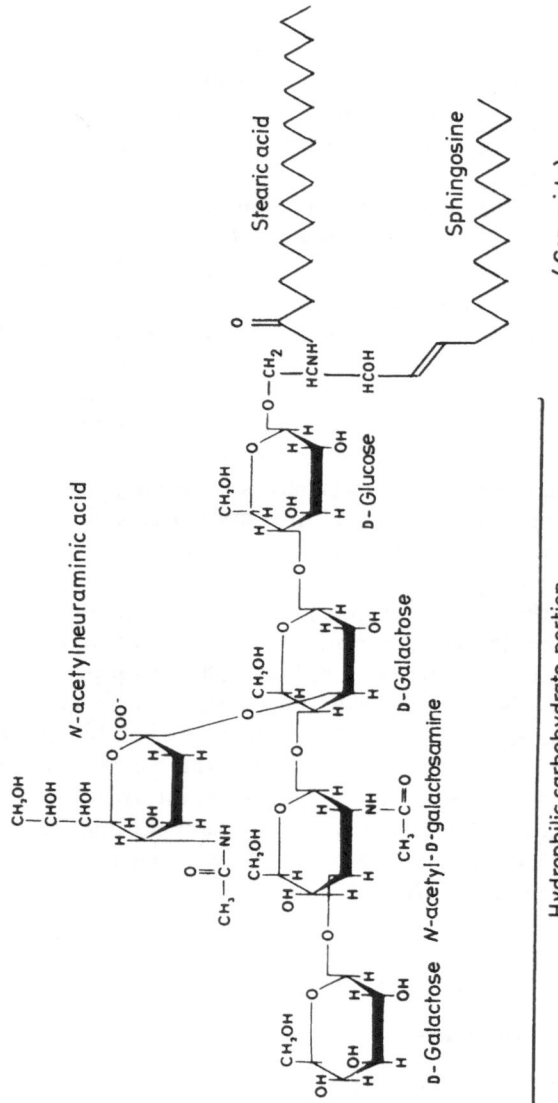

Fig. 8.6 Structure of ganglioside GM_1.

functional receptor for the toxin. There is in different cell lines a direct relationship between the content of GM$_1$ and the number of cholera toxin molecules that can be bound. Exogenous GM$_1$ molecules can be incorporated into the cell membrane and serve as functional binding sites. Furthermore, the GM$_1$ oligosaccharide can compete for binding to cells. The terminal galactose residue in GM$_1$ is decisive for binding, and removal of this sugar abolishes the binding. Finally, when cholera toxin is bound, GM$_1$ cannot be labeled by the galactose oxidase–NaB^{3}H$_4$ method.

Although there is ample evidence that cholera toxin binds to GM$_1$ on the cell surface, the receptor complex may also comprise other components in the membrane, like proteins. It is in accordance with this, that cholera toxin induces patching and capping on the surface of lymphocytes, a finding which is hard to reconcile with the view that the toxin binds only to monovalent, freely movable receptors in the membrane (Gill, 1977).

(b) *Binding sites for toxic lectins*

The receptors for the toxic lectins abrin, ricin, modeccin and viscumin are not known in detail. The toxins bind to oligosaccharides which may be bound to a variety of glycoprotein and glycolipid species at the cell surface. The main requirement for binding is that the carbohydrate carries a terminal galactose residue (Fig. 8.7).

Lactose and glycoproteins with terminal galactose residues inhibit binding of the toxic lectins as well as their toxic effect on cells (Olsnes *et al.*, 1978). Among those glycoproteins examined, desialylated fetuin was the best inhibitor for abrin and ricin. Binding of abrin to this glycoprotein was found to be as strong as the binding of the toxin to cell surface receptors (Sandvig *et al.*, 1978b). Modeccin receptors are different from the receptors for abrin and

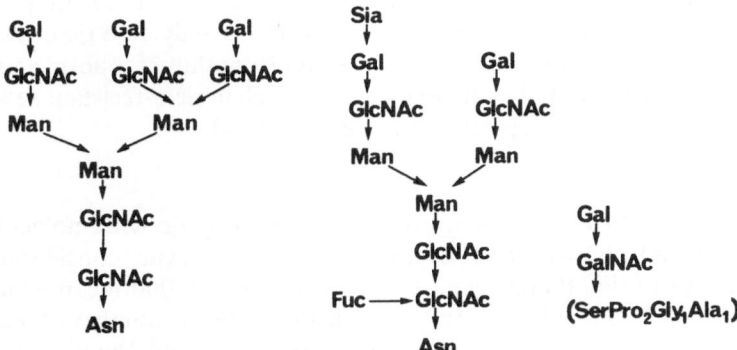

Fig. 8.7 Structure of different glycopeptides binding ricin. Reproduced with permission from Baenziger and Fiete (1979).

ricin and they are present on the cell surface in a much lower number (Olsnes *et al.*, 1978).

The oligosaccharide chains of many cell surface glycoproteins and glycolipids are terminated by sialic acid and have a penultimate galactose residue. Therefore, after neuraminidase treatment the number of terminal galactose residues on the cell surface is strongly increased and so is the number of binding sites for abrin, ricin, modeccin and viscumin. The new receptors are equally efficient as, and in some cases even more efficient (Olsnes *et al.*, 1978; Gottlieb and Kornfeld, 1976; Sandvig *et al.*, 1978a) than, the receptors available before the neuraminidase treatment.

The finding that several ricin-resistant cells have a reduced number of terminal galactose residues and a reduced ability to bind ricin constitutes further evidence that oligosaccharides containing terminal galactose residues represent functional toxin receptors. The reduced number of receptors is in some cases due to abnormal addition of sialic acid to terminal galactose residues resulting in masking of the receptors (Gottlieb and Kornfeld, 1976; Olsnes and Refsnes, 1978). In other cases it is due to incomplete synthesis of the corresponding oligosaccharide chains. In three cases (Gottlieb and Kornfeld, 1976; Stanley *et al.*, 1975; Meager *et al.*, 1976) this was due to lack of the same enzyme, *viz.* an *N*-acetylglucosaminyl transferase which links *N*-acetylglucosamine to mannose. Such residues are required for the subsequent addition of galactose (Fig. 8.7).

In some ricin-resistant cell lines, neuraminidase treatment increased the sensitivity to toxins to a much greater extent than it increased the number of toxin binding sites. Apparently, some of the binding sites previously masked by sialic acid are particularly efficient in facilitating toxin uptake. It is possible that sialylation of different receptors is under separate genetic control, and that in the ricin-resistant cells, sialic acid is added preferentially to those sites which are most active in toxin internalization.

Abrin-resistant cells are resistant to ricin and vice versa, but they are not resistant to modeccin. This is also the case with oversialylated ricin-resistant cells. Clearly, the receptors for modeccin differ from those for abrin and ricin and they are not sialylated by those enzymes which in ricin-resistant cells add sialic acid on to ricin receptors (Olsnes *et al.*, 1978).

(c) *Diphtheria toxin receptor*

The diphtheria toxin receptor is probably a glycoprotein with molecular weight 153 000 (Proia *et al.*, 1981). Different cell lines vary considerably in their number of diphtheria toxin receptors (4000–200 000 receptors/cell) (Middlebrook *et al.*, 1978) and in toxin-sensitive cells the number of receptors is roughly related to the sensitivity. However, Chang and Neville (1978) detected diphtheria toxin receptors also in purified surface membranes of liver and mammary glands from mice and rats which are resistant to this toxin.

Boquet and Pappenheimer (1976) were unable to demonstrate receptors on intact diphtheria-toxin-resistant mouse L-cells.

The binding of diphtheria toxin to its receptor, which occurs by the B-fragment, is inhibited by adenosine 5'-tetraphosphate (Middlebrook *et al.*, 1978) and also by other polyphosphate compounds. It is well established that the toxin has high affinity for NAD ($K_d \sim 9 \times 10^6$ M) because of a site on the A-fragment (Lory *et al.*, 1980a,b; Proia *et al.*, 1980). Recent data have shown that the toxin has another site with affinity for phosphate-containing compounds, the P-site (Lory and Collier, 1980), which is apparently located in the C-terminal, 8000-dalton cationic cyanogen bromide fragment of the B-fragment (Lory *et al.*, 1980b; Proia *et al.*, 1980). This fragment also carries the second SS bridge.

Although binding of nucleotides and other phosphate-containing compounds to the P-site inhibits binding to toxin receptors (Proia *et al.*, 1979), it is possible that this site is not directly involved in binding to the receptor (Lory *et al.*, 1980b). In binding of nucleotides like ATP, both the NAD site on the A-fragment and the P-site on the B-fragment appear to be involved (Lory *et al.*, 1980b; Proia *et al.*, 1980). Also, polycations interfere with the binding of diphtheria toxin to its receptor, apparently by competing for the receptor with the cationic P-site on the B-fragment (Proia *et al.*, 1981). The P-site could also bind to membrane phospholipids and thus stabilize the association of diphtheria toxin with the cells. In fact, diphtheria toxin was found to bind to the phosphate portion of some, but not all, kinds of phospholipids in liposomes (Alving *et al.*, 1980). It is also in accordance with this that the binding was reduced when the cells were treated with phospholipase C (Mochring and Crispell, 1974), an enzyme which removes phosphate-containing polar groups from phospholipids.

After neuraminidase treatment cells become ~3-fold more sensitive to diphtheria toxin (Sandvig *et al.*, 1978a; Mekada *et al.*, 1979). This does not necessarily mean that diphtheria toxin receptors are masked by sialic acid. Possibly, the increased sensitivity is due to a reduced number of negatively charged groups at the cell surface. This may allow a more ready attachment of the cationic site on the B-fragment to a polyphosphate structure on the receptor.

(d) *Binding sites for other toxins*
The receptor for *Pseudomonas aeruginosa* exotoxin A is not known, but the observation that after neuraminidase treatment cells became more sensitive to this toxin (Mekada *et al.*, 1979) and that the binding was inhibited by concanavalin A (FitzGerald *et al.*, 1980) suggests that the receptor contains carbohydrates.

Also the nature of the receptor for *Shigella dysenteriae* cytotoxin is not known in detail. Different cell lines exhibit widely different sensitivities to this toxin. Most cell lines are highly resistant, and tolerate 10^6 times more Shigella toxin than sensitive cells. So far, only primate epithelial cells were found to be sensitive. *Shigella* toxin receptors appear to be present both on sensitive and resistant cells (Eiklid and Olsnes, 1980).

8.3.2 Binding kinetics

The strength of the binding of the toxins to cell surface receptors varies, but it falls approximately within the same range ($K_a = 10^7$–10^{10} M^{-1}) as that of the binding of protein hormones to their receptors. The number of toxin receptors varies greatly (60–3×10^7/cell) in different cell types.

Different cell lines vary considerably in their sensitivity to diphtheria toxin. Cell lines derived from African green monkey kidney are the most sensitive ones so far found and they appear to contain the highest number of receptors. Thus, diphtheria toxin binds to 1.5×10^5 receptors on Vero cells, whereas on HeLa cells, which are 100 times less sensitive, there are only 5×10^3 receptors/cell. With Vero cells a K_a of 9×10^8 M^{-1} was found (Middlebrook *et al.*, 1978), in agreement with Moynihan and Pappenheimer (1981), who found a K_a of 5×10^8 M^{-1} for the Vero-related CV cells and 9×10^7 M^{-1} for BHK cells.

Binding of diphtheria toxin can be competed for by the mutant toxin, crm 197, which contains an intact B-fragment, and a defect A-fragment. After incubation with excess unlabeled toxin, bound toxin is released with a half-time of 7 h at 4°C. At pH 9 very little diphtheria toxin is bound and the cells are almost resistant. With decreasing pH more toxin is bound and the cells become increasingly sensitive (Middlebrook *et al.*, 1978).

When uptake was measured over a prolonged period of time (hours), considerably more diphtheria toxin was associated with cells at 4°C than at 37°C. At 37°C the binding was found to be biphasic and reached a maximum after 1–2 h. After that it was reduced to less than one-half (Middlebrook *et al.*, 1978). This is probably due to a combined effect of toxin degradation and toxin-induced consumption of receptors.

A strong reduction in the number of diphtheria toxin receptors was reported after treatment of cells with several different metabolic inhibitors. Particularly efficient was fluoride (Middlebrook, 1981). The affinity of the remaining receptors was not changed. The receptors remaining in the presence of the drugs internalized toxin at a normal rate (half-life of bound toxin ~30 min) (Middlebrook, 1981). Even in the presence of protein synthesis inhibitors the receptors rapidly reappeared when the drugs were removed and the cells were incubated at 37°C.

Abrin and ricin bind rapidly to cells (Sandvig *et al.*, 1976). The K_a values for binding to HeLa cells in suspension were found to be 1.2×10^8 M^{-1} (37°C) and 2.1×10^8 M^{-1} (0°C) for abrin and 2.6×10^7 M^{-1} (37°C) and 3×10^8 M^{-1} (0°C) for ricin. When HeLa cells were grown in monolayers, the binding constant decreased with increasing cell density (Sandvig, 1978). The binding of abrin and ricin to HeLa cells at 37°C is maximal at pH around 7 (Sandvig *et al.*, 1976). Dissociation kinetics indicated that the toxins bind to extended binding sites (Olsnes *et al.*, 1978; Sandvig *et al.*, 1976, 1978b).

Cells contain many fewer binding sites for modeccin than for abrin and ricin. Thus, HeLa cells contained only 2×10^5 binding sites per cell for modeccin (Olsnes *et al.*, 1978). After neuraminidase treatment the number of binding sites for abrin, modeccin and ricin was strongly increased (Sandvig *et al.*, 1978a; Rosen and Hughes, 1977; Nicolson *et al.*, 1975a). The K_a for binding to the new receptors did not differ from that of binding to the receptors available before the neuraminidase treatment (Sandvig *et al.*, 1978a).

Cholera toxin binds to cells with $K_a \sim 10^9$ M^{-1}. Each B-chain binds one GM$_1$ molecule (five per toxin). Although there is little variation between cells in the strength of the binding (Holmgren, 1981), the number of binding sites varies strongly (between 60 and 10^7 per cell) (Gill, 1977). As little as 10 molecules bound per cell may be sufficient to activate the adenylate cyclase.

The binding of *Shigella* toxin to cell surface receptors is particularly strong ($K_a \sim 10^{10}$ M^{-1}) and sensitive cells were found to have a high number of binding sites ($\sim 10^6$/cell) (Eiklid and Olsnes, 1980).

8.3.3 Role of the receptor in toxin entry

The presence on the cell surface of receptors with high affinity for toxins ensures that toxin added to the medium is concentrated at the cell surface. To secure a high local concentration of toxin could be the only function of the receptor. However, it is also possible that the receptor plays a more direct role in the entry, like directing the toxin to a particular vesicular compartment or to participate in the formation of a channel in the membrane through which the toxin can be transported.

Experiments with ricin containing covalently bound mannose 6-phosphate groups suggest that the natural, galactose-containing receptor plays an important role in the entry (Youle *et al.*, 1981). Thus, when the galactose-binding site of ricin was selectively inactivated by *O*-acetylation of a tyrosine residue, most of the toxic activity was lost in spite of the fact that the toxin was bound to mannose 6-phosphate receptors on the fibroblasts. In the presence of lactose, ricin–mannose 6-phosphate which had not been acetylated, was also bound exclusively by the mannose 6-phosphate receptor. In spite of this, a strong cytotoxic effect was obtained. Possibly, the lactose content of the vesicle is reduced after endocytosis. The mannose 6-phosphate–ricin could then become attached to the galactose-containing receptor.

8.4 ENDOCYTOSIS OF TOXINS

8.4.1 Ultrastructural studies

The most detailed ultrastructural studies of cellular uptake of toxins have been carried out with ricin and cholera toxin. Gonatas *et al*. (1975, 1977) studied the uptake and distribution in cells of complexes of ricin and horse-radish peroxidase which were visualized as a precipitate of oxidized diaminobenzidine–osmium black. Upon incubation at 4°C only a continuous peripheral rim, the plasma membrane, was stained (Fig. 8.8a). However, if the cells were washed and subsequently incubated at 37°C, the plasma membrane staining first acquired a patchy pattern and then diminished in intensity. After 30 min at 37°C, various degrees of cytoplasmic staining appeared in round, oval or elongated vesicles of 0.1–1 μm (Fig. 8.8b). Clusters of stained vesicles were found adjacent to the elongated cisternae of the Golgi apparatus. After 0.5–1 h, vesicles were usually found near the concave (*trans*) aspect of the Golgi cisternae and at the edges of the cisternae.

Even after 3 h at 37°C some small patches of staining at the cell surface remained, but most label was now found in the cytoplasm, particularly in one or two of the parallel cisternae of the Golgi apparatus. Peroxidase activity was not present in the large, dense bodies believed to be neuronal lysosomes. Only after incubation for more than 3 h a few dense bodies (lysosomes) were labeled. Essentially the same pattern was observed with horseradish peroxidase linked to cholera toxin (Joseph *et al*., 1979). Also in this case very little material was found in the lysosomes and endocytosed toxin remained essentially intact for 24 h. A potent uncoupler of oxidative phosphorylation strongly inhibited the internalization of ricin–horseradish peroxidase complexes.

Comparison with the staining pattern obtained with acid phosphatase indicated that ricin and cholera toxin accumulate in those vesicles which belong to the GERL apparatus (Gonatas *et al*., 1977; Joseph *et al*., 1979). The pattern of endocytosis of ricin and cholera toxin linked to horseradish peroxidase was definitely different from that obtained with free horseradish peroxidase which was found to be endocytosed in lysosomes and small vesicles adjacent to the lysosomes.

The rate of uptake of ricin from the cell surface is much slower than the uptake of epidermal growth factor (EGF) and low-density lipoprotein (LDL) which, within minutes, undergo endocytosis via coated pits into multivesicular bodies and lysosomes. Clearly, the diffuse adsorptive endocytosis of ricin is quantitatively and qualitatively different from that of EGF and LDL.

Although endocytosed abrin and ricin were found to be released into the medium to some extent (Sandvig *et al*., 1978a), direct morphometric evidence for recycling of toxin to the cell surface was not obtained (Gonatas *et al*., 1980). Probably, when the internalized toxin is recycled back to the surface membrane it is rapidly released into the medium (Sandvig *et al*., 1978a).

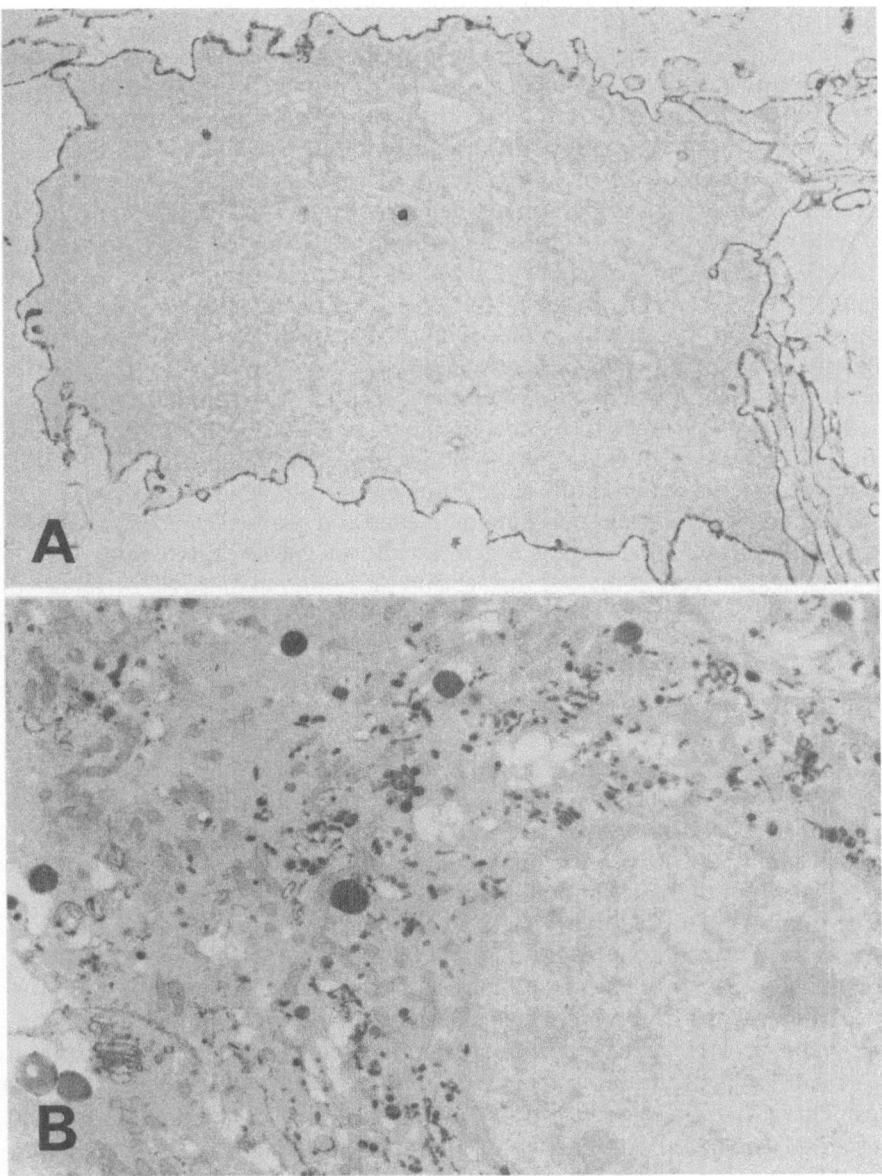

Fig. 8.8 Electron micrograph of ricin–horseradish peroxidase complexes internalized by dorsal root ganglion cells. (A) The cells were incubated with ricin–horseradish peroxidase for 1 h at 4°C, washed, fixed and stained for peroxidase (× 6390). (B) The cells were incubated with ricin–horseradish peroxidase for 1 h at 4°C, washed and incubated in medium for 3 h at 36°C. Then the cells were fixed and stained for peroxidase (× 8520). The electron micrographs are published with the kind permission of Dr N.K. Gonatas, Philadelphia.

Also studies in living animals showed that ricin is endocytosed. ^{125}I-ricin injected into the submandibular gland was transported retrograde to the nerve cell bodies present in the cervical sympathetic ganglion where it induced degenerative changes of these cells (Harper *et al.*, 1980). Such retrograde transport, which occurs in cytoplasmic vesicles, has been demonstrated for rabies virus and for many macromolecules including tetanus toxin. In spite of heavy damage of the affected cells by ricin, only a certain fraction (15–25%) of the total cells in the ganglion developed degenerative changes (only 40–50% of the neurons in the superior cervical ganglion have post-synaptic fibers in the submandibular gland). The most likely reason for this is that when the toxin reached the ganglion it was not excreted into the extracellular compartment where it presumably would have been taken up equally well by all cells. Rather it selectively intoxicated those cells from which the neurons originated. If this explanation is correct, it implies that the toxin is able to enter the cytosol from intracellular vesicles.

Experiments with ricin–ferritin complexes and mouse 3T3 cells showed essentially the same pattern as described above. At 4°C the binding was disperse, exclusively at the cell surface and the bound toxin could be released by galactose. In contrast, with increasing time at 37°C, an increasing fraction of the bound toxin was internalized and could no longer be released by galactose (Nicolson, 1974; Nicolson *et al.*, 1975b). The toxin first appeared clustered at the cell surface and then it was taken into endocytic vesicles. After 60 min most of the toxin was inside vesicles. Also in these cells no fusion with lysosomes was observed.

In contrast to the findings with ricin and cholera toxin, ferritin-labeled *Pseudomonas aeruginosa* exotoxin A was accumulated in coated pits (Fitz-Gerald *et al.*, 1980). Also ferritin-labeled *Pseudomonas aeruginosa* exotoxin A and ^{125}I-diphtheria toxin remained at the cell surface when the incubation temperature was 4°C, but after a 2 h chase at 37°C, most of the surface-bound toxin was internalized (FitzGerald *et al.*, 1980; Dorland *et al.*, 1981). A similar degree of internalization of diphtheria toxin occurred in the presence of NH$_4$Cl which protects the cells against the toxin. On the other hand methylamine, NH$_4$Cl and chloroquine inhibited clustering and internalization of *Pseudomonas aeruginosa* toxin. In the presence of concanavalin A which protects against diphtheria toxin, the majority of the labelled diphtheria toxin remained associated with the plasma membrane even at 37°C, indicating that the lectin prevents toxin internalization (Dorland *et al.*, 1981).

8.4.2 Biochemical studies on endocytosis and its importance for toxin entry

With abrin, ricin, modeccin and viscumin, uptake of toxin from the cell surface can be easily measured by biochemical methods using labeled toxins. Thus, when the cells were washed with lactose all the toxin bound to the cell surface was released, whereas toxin in endocytic vesicles or other compartments in the cell interior remained associated with the cells. Lactose-resistant toxin, i.e. toxin which could not be washed off the cells with lactose, was accessible neither to anti-toxin nor to extracellular ^{125}I labeling (Sandvig *et al.*, 1978a; Sandvig and Olsnes, 1979) and was probably internalized. Part of the internalized toxin was later released from the cells in an apparently intact form, supporting the view that only a few toxin molecules are actually transported to the cytoplasm (Sandvig *et al.*, 1978a; Sandvig and Olsnes, 1979). Ricin B-chain was taken up in the same way as intact ricin.

A quite different approach to measure the uptake of ricin B-chain from the cell surface was taken by Houston (1981). He pretreated mouse leukemia cells with ricin B-chain and then, after washing them to remove unbound B-chain, he added A-chain under conditions allowing reconstitution of ricin at the cell surface. The development of cytotoxic effect was taken as evidence that the A-chain had been internalized. It was found that B-chain bound to the cells was initially very efficient in internalizing the A-chain, but after 90–120 min preincubation at 37°C before the A-chain was added, there was no toxic effect, indicating that by then the bound B-chain had been internalized.

Also in the case of diphtheria toxin is it possible with biochemical methods to distinguish between toxin at the cell surface and toxin which is internalized. Thus, the combined treatment with proteolytic enzymes and inositol hexaphosphate appears to release selectively diphtheria toxin present at the cell surface. After incubation of cells with ^{125}I-diphtheria toxin at 4°C, essentially all radioactivity can be released by such treatment, whereas after incubation at 37°C an increasing fraction cannot be released (Middlebrook *et al.*, 1979). The results were interpreted to mean that the bound diphtheria toxin is internalized with a half time estimated to be 25 min.

Concanavalin A was found to strongly inhibit internalization of ^{125}I-diphtheria toxin, although it did not inhibit binding to cell surface receptors (Middlebrook *et al.*, 1979). This is in accordance with ultrastructural studies (Dorland *et al.*, 1981). Concanavalin A also protected cells against intoxication. This indicates that there is a causal relationship between the internalization inhibited by concanavalin A and toxin entry into the cytosol. Wheat germ agglutinin and lentil lectin had an effect similar to concanavalin A.

8.4.3 Endocytosis in toxin-resistant cells

Toxin-resistant cell mutants have been isolated in several laboratories. Some of these cells have a modified intracellular target for the toxins, whereas others appear to be deficient in the uptake mechanism for the toxins. Best studied are diphtheria-toxin-resistant cells first isolated by Moehring *et al.* (1980). These authors described two classes of resistant cells. One type, Dip^{rI}, which appears to be deficient in diphtheria toxin uptake, is only resistant to moderate toxin concentrations and not to *Pseudonomas aeruginosa* toxin (Moehring *et al.*, 1980; Gupta and Siminovitch, 1980). Another type, Dip^{rII}, contains toxin-resistant elongation factor II (EF2). Mutants of this type are resistant even to very high concentrations of diphtheria toxin and they are also resistant to *Pseudomonas aeruginosa* toxin.

The mutants containing toxin-resistant EF2 are again divided into two subgroups. One subgroup appears to have a mutation in the codon of the EF2 gene specifying the amino acid, presumably histidine, which is later post-translationally modified to diphthamide. The second group appears to bear defects in enzymes involved in the post-translational modification of this residue. The latter group is subdivided into three complementation groups, suggesting that three or more enzymes are involved in the post-translational modification. When EF2 from such cells is treated *in vitro* with active enzymes from toxin-sensitive cells, it is modified to contain diphthamide which can then be ADP-ribosylated.

So far, extensive studies have not been carried out with mutant cells which are resistant to diphtheria toxin due to inefficient toxin entry. Out of 24 Dip^{rI} strains tested, none was cross-resistant to *Pseudomonas aeruginosa* toxin. This indicates that the mutation affects some steps which are specific to the uptake of diphtheria toxin.

Also in several ricin-resistant cell lines the resistance appears to consist of an inability of the toxin to enter the cytosol (Gottlieb and Kornfeld, 1976; Sandvig *et al.*, 1978a; Meager *et al.*, 1976). Nicolson *et al.* (1976, 1978), Hyman *et al.* (1974) and Robbins *et al.* (1977) described a ricin-resistant lymphoma cell line with sensitive ribosomes (Nicolson and Poste, 1978) which contained nearly a normal number of binding sites but which endocy-tosed bound toxin at a strongly reduced rate. The clustering of bound toxin on the cell surface was delayed at low toxin concentrations, but it did take place at higher toxin concentrations to which the cells were sensitive. The modification was apparently selective for ricin since clustering and endocytosis of concanavalin A by these cells occurred at the same rate as in the parent cells. Two surface proteins with mol.wts. of 80 000 and 35 000 were missing in the resistant cell line. Only the 80 000-mol.wt. protein was found to be a ricin receptor. The resistant cells have instead a 70 000-mol.wt. protein. Also, in several other toxin-resistant cells, changes in surface glycoproteins have been found (Olsnes and Refsnes, 1978; Meager *et al.*, 1976; Pena *et al.*, 1979). Unfortunately, in the different laboratories several different cell lines have been used to isolate mutants. This makes comparison and search for consistent alterations difficult.

Although in several abrin- and ricin-resistant cell lines the resistance can be accounted for by a reduced number of toxin receptors, this is not always the case. Thus, many cell lines are much more resistant to abrin, ricin and modeccin than would be expected from the reduction in the number of toxin receptors (Sandvig *et al.*, 1978a). Since toxin-resistant ribosomes have not been found in these cells, it appears that the transport of the A-chain into the cytosol is inhibited. In most of these cells endocytosis of toxins is not reduced and the degradation of internalized toxin does not occur at a higher rate than in the parent cells. It therefore appears that in these cells it is the transport of the A-chain across the membrane which is impaired (Sandvig *et al.*, 1978a). Alternatively, the toxin molecules may be directed to a compartment from which they are unable to enter the cytoplasm. In any event, some specific mechanism must be affected, since there is no cross-resistance between cells resistant to abrin and ricin on the one hand, and diphtheria toxin and modeccin on the other (Olsnes *et al.*, 1978).

8.4.4 Fate of endocytosed toxin

Diphtheria toxin which has entered endocytic vesicles is rapidly degraded if the cells are kept at 37°C, but not if they are transferred to 4°C (Middlebrook *et al.*, 1978). After degradation, most of the radioactivity is recovered as [^{125}I]monoiodotyrosine. Concanavalin A and anti-(diphtheria toxin), which both inhibit internalization of diphtheria toxin, also inhibited the degradation. This indicates that only internalized toxin is degraded (Dorland *et al.*, 1979). The degradation is assumed to take place in the lysosomes. This is supported by the finding that chloroquine and methylamine inhibited the degradation of diphtheria toxin, although they did not inhibit binding and endocytic uptake of the toxin (Leppla *et al.*, 1980; Mekada *et al.*, 1981). Also, salicylate inhibited the degradation of diphtheria toxin (Middlebrook, 1981). Surprisingly, NH_4Cl, which does not inhibit binding and uptake of the toxin, was reported *not* to inhibit diphtheria toxin degradation (Dorland *et al.*, 1981).

After endocytosis, diphtheria toxin appears to retain the ability to enter the cytosol only for a short period of time (Sandvig and Olsnes, 1981a; Draper and Simon, 1980; Sandvig and Olsnes, 1980), whereas endocytosed abrin, ricin and modeccin appear to be able to enter the cytosol for hours after endocytosis (Sandvig and Olsnes, 1982b). This will be discussed in detail below.

Abrin and ricin are not easily degraded and only small amounts appear to be accumulated in lysosomes. Two hours after internalization, about 90% of the internalized ricin remained intact (Sandvig *et al.*, 1978a) and abrin was even more resistant to degradation. The limited degradation which did take place, probably occurred in lysosomes since it could be inhibited by NH_4Cl (Sandvig and Olsnes, 1979). Part of the endocytosed toxin was later released into the medium, possibly owing to recycling of the receptor–toxin complex back to the cell surface (Sandvig and Olsnes, 1979). The release of intact ricin from cells follows completely different kinetics from those of uptake. Thus, there was a strong increase in the release rate around 20°C and the rate approached its maximum at about 30°C (Sandvig and Olsnes, 1979).

8.5 PENETRATION THROUGH THE MEMBRANE

8.5.1 Evidence that the penetration does not occur by unspecific rupture of vesicles

It has been suggested (Nicolson 1974, 1975; Nicolson *et al.*, 1975a) that ricin-containing endocytic vesicles rupture and release their content into the cytosol. This mechanism, which was suggested on the basis of electron micrographs (Nicolson, 1974), is unlikely in the light of more recent data. Thus, toxin-resistant cell lines, which are resistant to one toxin, are usually not resistant to the other toxins described in this chapter. Most of these cell lines endocytosed normal amounts of toxin. If unspecific rupture of vesicles occurred, it would be most likely to release all the toxins contained in the same vesicles. Since it has been shown in other cases of adsorptive endocytosis that different ligands are internalized by the same vesicle, selective resistance to one toxin would be very unlikely if the toxin entry involved simple rupture of some of the vesicles.

The contention that the toxins do not enter by unspecific vesicle rupture is also supported by the fact that low pH and absence of Ca^{2+} protect well against abrin, modeccin and ricin, but not against diphtheria toxin, although endocytosis of all four toxins occurs under these conditions (Sandvig and Olsnes, 1982b). In fact, low pH which strongly protects against the plant toxins, is necessary for diphtheria toxin to enter the cells.

Also calculations of the amount of toxin endocytosed by fluid-phase endocytosis indicates that vesicle rupture does not occur frequently (Pappenheimer, 1977). Mouse L-cells are resistant to diphtheria toxin, although they internalize by bulk endocytosis the same amount of toxin as the sensitive HeLa cells. Cell-free systems from both cell lines are equally sensitive to ADP-ribosylation by diphtheria toxin fragment A. From the extracellular fluid uptake it can be calculated that at the lowest concentration of diphtheria toxin which is toxic to L-cells (10^{-5} M), 2×10^6 toxin molecules are internalized in 5 h. In the case of HeLa cells less than 10^{-9} M diphtheria toxin is cytotoxic. At such low concentration, a bulk uptake of only 200 toxin molecules takes place during 5 h (Boquet and Pappenheimer, 1976). If this is the minimum number of toxin molecules required to inhibit protein synthesis after 5 h, rupture of only one out of 10 000 vesicles would be sufficient to induce cytotoxic effect in L-cells. Clearly, unspecific rupture of vesicles cannot be a frequent event. Therefore, if the rate of vesicle rupture is as low in HeLa cells as in L-cells, only one out of 50 cells could be intoxicated within 5 h by this mechanism. However, most of the HeLa cells were intoxicated.

Only under special conditions, such as virus infection, do toxins or their A-chain appear to enter the cells by a mechanism involving unspecific leakage (Fernández-Puentes and Carrasco, 1980; Yamaizumi *et al.*, 1979). During the entry of several viruses, the cells become transiently leaky and proteins resembling toxin A-chains, such as the pokeweed anti-viral protein, gelonin and dianthin may then enter the cells.

8.5.2 Lag time

It is characteristic for all the toxins here described that even in the presence of very high toxin concentrations, a certain lag period (10–30 min) elapses before any effect on the cells can be observed. In contrast, in cell-free systems the toxins appear to act immediately. It was suggested that the lag time could be the time required for the toxin to be taken up by endocytic vesicles and released into the cytoplasm (Refsnes *et al.*, 1974).

The lag time was found to decrease with increasing toxin concentrations. Measurements of inhibition of protein synthesis in cells treated with abrin and ricin (Olsnes *et al.*, 1976) indicated a minimum lag time of about 30 min. Youle and Neville (1979) and Moynihan and Pappenheimer (1981) measured the rate of ADP-ribosylation of EF2 in diphtheria-toxin-treated cells and found a minimum lag time of about 12–15 min. Similarly, a lag period of 15 min to 2 h elapsed from the addition of cholera toxin to cells until stimulation of adenylate cyclase was measurable (Gill, 1977). The lag period was found to decrease not only with increasing toxin concentration but also with increasing concentration of GM_1 on the cells. Furthermore, it decreased with increasing temperature (Fishman, 1980).

8.5.3 Effect of temperature

The uptake of abrin and ricin by cells is strongly temperature-dependent, but there does not appear to be any critical temperature where the rate of uptake changes abruptly. The activation energy for uptake was essentially the same (15–21 kcal/mol, 63–88 kJ/mol) when the uptake was measured in two entirely different ways. In the one case the gross uptake of ^{125}I-labeled abrin was measured, i.e. mainly endocytosis, whereas in the other case the rate of transfer of toxin to a compartment where it could no longer be inactivated with antiserum was measured. Since in the latter case toxic effect on cells was assayed, it reflects a step in the true entry route of toxin to the cytosol. In both cases some uptake was measurable even at 5°C (Sandvig and Olsnes, 1979), and linear Arrhenius plots were obtained. Linear Arrhenius plots have previously been found for fluid-phase pinocytosis but are unusual for uptake through membranes. This, together with a similar activation energy for toxin uptake measured by the two methods, suggests that endocytosis of toxin is a process involved in toxin entry into the cytosol. Completely different kinetics were found for toxin release from the cells (Sandvig and Olsnes, 1979).

With cholera toxin different results were obtained. Thus, at temperatures up to 15°C cholera toxin remained in a position where it could be inactivated with anti-toxin, whereas at higher temperatures the toxin entered the cells and activated adenylate cyclase (Fishman, 1980). If the toxin had been allowed to enter at higher temperatures, cyclic AMP formation went on even at 15°C, indicating that the effect of temperature is primarily on the toxin entry.

8.5.4 Effect of pH

(a) *Penetration of diphtheria toxin at low pH*
It was first observed by Kim and Groman (1965) that NH_4Cl protects cells against diphtheria toxin at neutral pH, although it does not inhibit toxin binding. They assumed that the inhibition was due to free NH_3 since at lower pH (pH 6.5), much higher concentrations of ammonium chloride were required for protection (Kim and Groman, 1965). When cells with preadsorbed diphtheria toxin (adsorbed at 4°C) were incubated at 37°C and then, after increasing periods of time, NH_4Cl or anti-toxin were added, the extent of intoxication was in both cases the same (Kim and Groman, 1965). This was taken as evidence that NH_4Cl inhibits toxin entry. Recently, chloroquine and several amines were also found to protect against diphtheria toxin in a similar way to NH_4Cl (Leppla *et al.*, 1980; Sandvig *et al.*, 1979). Common to NH_3, chloroquine and related amines is the ability to pass through membranes in their uncharged form. When they enter vesicles with low pH such as lysosomes and receptosomes, they become protonated and are then trapped in the vesicles as they are unable to pass through the membrane in their protonated form. As a result, the pH in the vesicles increases.

Cells are most sensitive to diphtheria toxin at low pH and they are almost insensitive at pH 9 (Middlebrook *et al.*, 1978; Duncan and Groman, 1969). We and others recently found that the protective effect of NH_4Cl was abolished if cells with preadsorbed diphtheria toxin were incubated at pH below 4.6 for a few seconds at 37°C (Sandvig and Olsnes, 1981a) or for 30 min at 4°C (Draper and Simon, 1980). Also toxin in intracellular vesicles is able to intoxicate the cells. This follows from experiments where cells containing toxin endocytosed in the presence of NH_4Cl, were chilled to 4°C, treated in the cold with anti-toxin to inactivate extracellular toxin, and then incubated at 37°C in the absence of NH_4Cl. Under these conditions only toxin endocytosed in the presence of NH_4Cl would be able to intoxicate cells. This was indeed found to be the case (Draper and Simon, 1980). In a similar experiment we allowed toxin to be endocytosed in the presence of NH_4Cl, then we treated the cells with anti-toxin at 0°C and subsequently exposed them briefly to pH 4.5 at 37°C (Sandvig and Olsnes, 1980). Also in this case the cells were intoxicated.

When cells with prebound, nicked toxin were exposed to medium at pH 4.5, inhibition of protein synthesis rapidly occurred (Sandvig and Olsnes, 1981a). The proteolytic cleavage ('nicking') between the A- and B-fragment must occur before the toxin enters the cell. Thus, even at low pH, very little toxic effect was obtained with unnicked toxin. The 'nicking' usually occurs by proteases excreted by the bacterium or by such present in the serum of the cell culture medium (Sandvig and Olsnes, 1981a). At low pH, nicked toxin apparently enters directly through the surface membrane. The rate of toxin entry increased with decreasing pH, and it was maximal at pH 4.5 and below. Furthermore, it was strongly temperature-dependent.

Diphtheria toxin fragment B contains a hydrophobic domain which appears to be important for the entry. Exposure of the hydrophobic region can be measured by the ability of the toxin to bind [^{3}H] Triton X-100 (Boquet *et al.*, 1976). In the incomplete toxin, crm 45, the hydrophobic region appears to be exposed even at neutral pH (Boquet *et al.*, 1976), whereas in intact diphtheria toxin no Triton X-100 was bound under these conditions. However, at pH below 4.5 intact diphtheria toxin also bound Triton X-100, indicating that the hydrophobic region in the B-chain is exposed under these conditions (Sandvig and Olsnes, 1981a). When toxin in solution was exposed to pH 4.5, about 90% of the toxic activity was lost. However, if the toxin had first been bound to receptors on the cell surface, the toxic activity was not reduced. This suggests that diphtheria toxin first binds to cell surface receptors, then it is transferred to a compartment with low pH and the hydrophobic region on the B-chain is exposed. This region may then become inserted into the membrane where it possibly forms a hydrophobic channel through which the A-chain can pass.

Experiments with planar lipid bilayer membranes have shown that under appropriate conditions diphtheria toxin can indeed insert itself into the membrane and form channels which are at least permeable to ions. Such channels were formed when a positive potential was applied across the membrane and when the pH was low on the *cis* side, i.e. the same side as the toxin. When the incomplete diphtheria toxin molecule, crm 45, which lacks the *C*-terminal 17 000 daltons, was used, channels were formed when the pH on the *cis* side was 5.5 or lower (Fig. 8.9). Similar results were obtained with B-45, the part of the B-fragment present in crm 45 (Kagan *et al.*, 1981). When the whole diphtheria toxin was used, pH 4.5 was required for channel formation (Donovan *et al.*, 1981). The channels opened only when the potential was *cis*-positive and there was evidence of opening and closure of single channels. A requirement for channel formation was that the phospholipids in the membrane were negatively charged (Kagan *et al.*, 1981). The presence of phosphatidylinositol phosphate in the membrane strongly increased the extent of channel formation (Donovan *et al.*, 1981).

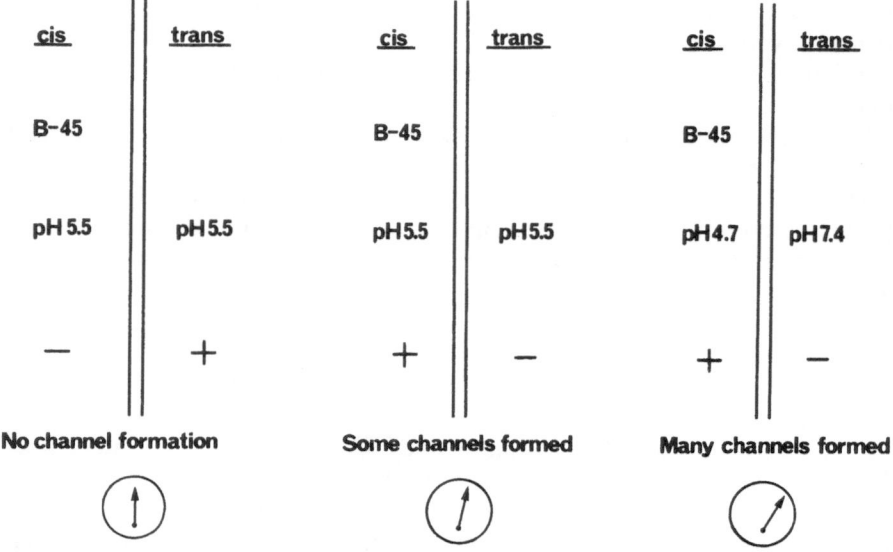

Fig. 8.9 Formation of channels in membranes by the B-fragment of crm 45.

The channels appeared to span the membrane since treatment with pronase from the *trans* side destroyed the channel activity (Kagan *et al*., 1981). It is of interest that the hydrophobic domain in the B-fragment consists of an α-helix with a length of 3.5 nm, which is enough to span the membrane. A cyanogen bromide fragment containing this region was found to insert itself into membranes and form ion-conductive channels (Kayser *et al*., 1981).

Although it was found (Boquet, 1979) that B-45 was not able to release glucose from liposomes at neutral pH, B-45 inserted itself into liposomes at low pH and formed channels permeable to solutes with mol.wt. up to 1500 (Kagan *et al*., 1981). Such permeability indicates a diameter of $\geqslant 1.8$ nm, which is just enough to allow diphtheria toxin fragment A to penetrate in its extended form.

When added to the mucosal side of turtle bladder, *Pseudomonas aeruginosa* exotoxin A, diphtheria toxin as well as cholera toxin induced rapid changes in the short circuiting current, the conductance and the potential across the epithelium. Apparently, toxin binding increased the ion permeability of the cells (Brodsky *et al*., 1979). For this effect to develop, active receptor binding properties of the toxins were required. The changes appear to be too rapid to be caused by the intracellular action of the toxins. It is also difficult to explain the increased ion permeability of the epithelium by formation of ion-permeable channels like those occurring in model membranes. At least in the case of diphtheria toxin such channels may only be formed in the membrane of intracellular acidic vesicles.

An indication that the hydrophobic region of diphtheria toxin fragment B inserts itself into membranes as suggested above is the observation that at neutral pH crm 45 is much more toxic to Schwann cells which have an extended surface membrane, than to several other cells tested in culture (Pappenheimer and Harper, 1982). Furthermore, crm 45, which is almost non-toxic when given intravenously, was as toxic as intact diphtheria toxin when administered intracerebrally. Possibly, the exposed hydrophobic domain inserts itself into the extended surface membrane of the Schwann cells.

(b) *Effect of amines and pH on the entry of other toxins*
NH_4Cl, chloroquine and other amines protect against several toxins other than diphtheria toxin, but higher concentrations are required for full protection. Besides increase in the pH in acidic vesicles, amines affect a variety of cellular processes (see below), and it is possible that the protection against some of the toxins is not due to increase in pH in acidic vesicles. Although low concentrations of methylamine sensitized cells to *Pseudomonas aeruginosa* exotoxin A (Mekada *et al.*, 1981), higher concentrations (20 mM) provided complete protection (FitzGerald *et al.*, 1980). In contrast to the finding with diphtheria toxin, the protective state lasted for some time after methylamine had been removed. Thus, if *Pseudomonas aeruginosa* toxin was added immediately after removal of methylamine, no toxic effect was seen. Full protection was also obtained if methylamine or chloroquine was added as late as 10 min after a 2 min exposure to *Pseudomonas aeruginosa* toxin.

NH_4Cl and chloroquine also protected against modeccin (Sandvig *et al.*, 1979). However, lowering of the pH in the medium did not overcome the protection against modeccin in contrast to the case with diphtheria toxin. A possible explanation for the difference is that modeccin may require processing in the lysosomes before entry into the cytosol (see above) and that this processing is inhibited by NH_4Cl and chloroquine. A modeccin-resistant HeLa cell variant tolerating $\sim 10^4$ times more modeccin than the parent cells was not further protected against modeccin by NH_4Cl. This indicates that modeccin can be internalized by two mechanisms, one efficient and NH_4Cl-sensitive, and another NH_4Cl-resistant and inefficient. If so, only the latter one is retained in the modeccin-resistant variant (Sandvig *et al.*, 1979).

With cholera toxin conflicting results have been obtained with amines. Some authors (Houslay and Elliott, 1981) found protection with several amines, others (Hagmann and Fishman, 1981) only with dansylcadaverine. As shown by Gill *et al.* (1981), the effect is probably exerted on the adenylate cyclase as such rather than on cholera toxin entry.

Surprisingly, NH_4Cl, chloroquine and methylamine sensitized rather than protected cells to abrin, ricin and a hybrid of *Wistaria floribunda* lectin and diphtheria toxin fragment A (Mekeda *et al.*, 1981; Sandvig *et al.*, 1979; Ray and Wu, 1981a). This could be related to the fact that these toxins are most active at pH above neutrality. When the cell culture medium was adjusted to pH 6 or lower, Vero cells were completely protected against abrin, modeccin and ricin, although toxin binding and endocytosis were not much reduced. With increasing pH up to 8.5, the sensitivity to abrin and ricin increased, while modeccin exerted its maximal effect between pH 7 and pH 8 (Fig. 8.10).

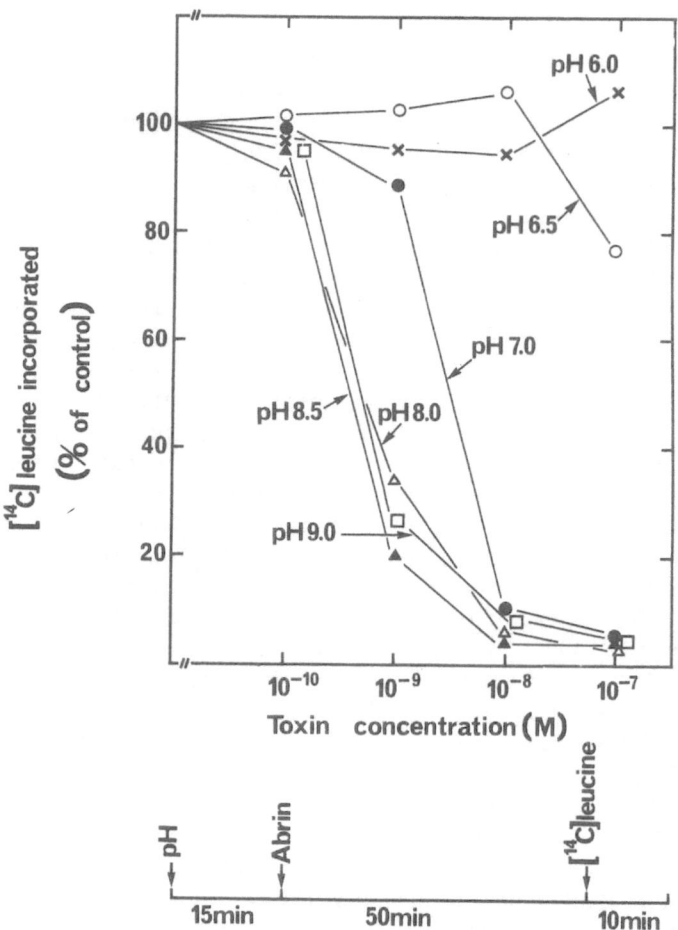

Fig. 8.10 Effect of pH on the sensitivity of cells to abrin. The indicated amounts of abrin were added to Vero cells in 0.14 M-NaCl/20 mM-Hepes/2 mM-CaCl$_2$ and pH as indicated. After incubation for 50 min at 37°C, the ability of the cells to incorporate [^{14}C]leucine into trichloroacetic acid-precipitable material during 10 min was measured.

The increased sensitivity of cells to abrin and ricin at pH 8 could be due to the higher Ca^{2+} flux occurring at this pH compared to that at pH 7. In the presence of Co^{2+}, which inhibits the uptake of Ca^{2+}, there was no difference in sensitivity of cells to ricin at pH 7 and pH 8 (Sandvig and Olsnes, 1982b). As will be discussed below, Ca^{2+} influx into the cytosol appears to be required for the entry of abrin and modeccin and, to a lesser extent, for ricin. The reason for the increase in sensitivity to abrin and ricin in the presence of NH_4Cl and chloroquine could be due to a higher Ca^{2+} efflux from intracellular vesicles into the cytosol when the pH of the vesicles is increased by the amines.

Although transport of Ca^{2+} into cells is reduced at pH 6, the reduction of Ca^{2+} flux cannot alone account for the protection against ricin which also to some extent enters cells in the absence of Ca^{2+}. Apparently, the protection against abrin and modeccin by low pH is not only mediated by an effect on Ca^{2+} transport (Sandvig and Olsnes, 1982b). Supporting this idea is the fact that, at pH 6, 10 mM $CaCl_2$ in the medium did not increase the sensitivity of the cells.

(c) *Effect of amines in other systems*

Primary amines are efficient inhibitors of receptor-mediated uptake of a variety of proteins. Most efficient is dansylcadaverine which inhibits internalization of α_2-macroglobulin, insulin, T_3, epidermal growth factor and low-density lipoprotein (Pastan and Willingham, 1981). At least in the case of α_2-macroglobulin, low-density lipoprotein and *Pseudomonas aeruginosa* toxin, dansylcadaverine and other amines prevent the formation of clusters in coated pits, which appears to be a requirement for entry by the receptosome pathway. It has been suggested, but not proved, that transglutaminase activity is required for the clustering and that the amines act as competitive inhibitors of this enzyme.

NH_4Cl, chloroquine, dansylcadaverine and related compounds are also potent inhibitors of the internalization of certain viruses, such as Semliki Forest virus, Vesicular stomatitis virus and others. These viruses require low pH to fuse with cellular membranes and release their nucleocapsid into the cytosol. Such low pH is obtained in lysosomes and receptosomes. When amines diffuse into the vesicles and increase their pH, the viruses become trapped (Helenius *et al.*, 1980).

Lysosomotropic amines inhibit the uptake of lysosomal enzymes in fibroblasts (Sando *et al.*, 1979) and the uptake of mannose–protein conjugates by macrophages (Tietze *et al.*, 1980). The latter effect appears to be due to inhibition of receptor recycling. Lysosomotropic amines also inhibit the stimulation of DNA synthesis by epidermal growth factor. In this case the internalization is not inhibited, whereas the amines strongly inhibit degradation of the hormone–receptor complexes (Michael *et al.*, 1980; King *et al.*, 1981).

The best-studied effect of NH₄Cl, chloroquine and related compounds is their ability to inhibit degradation of proteins in lysosomes, which is apparently due to their ability to increase pH in the vesicles (Seglen, 1975; Seglen and Gordon, 1980; Poole and Ohkuma, 1981). Altogether, it is clear that amines have a variety of effects on cells and the protective and sensitizing effects observed with some of the toxins described above are therefore at the present time difficult to interpret.

8.5.5 Role of calcium

(a) *Requirement for calcium in the medium*
We recently found that abrin and modeccin do not inhibit protein synthesis when there is no Ca^{2+} present in the cell culture medium (Fig. 8.11(a)). Ca^{2+}

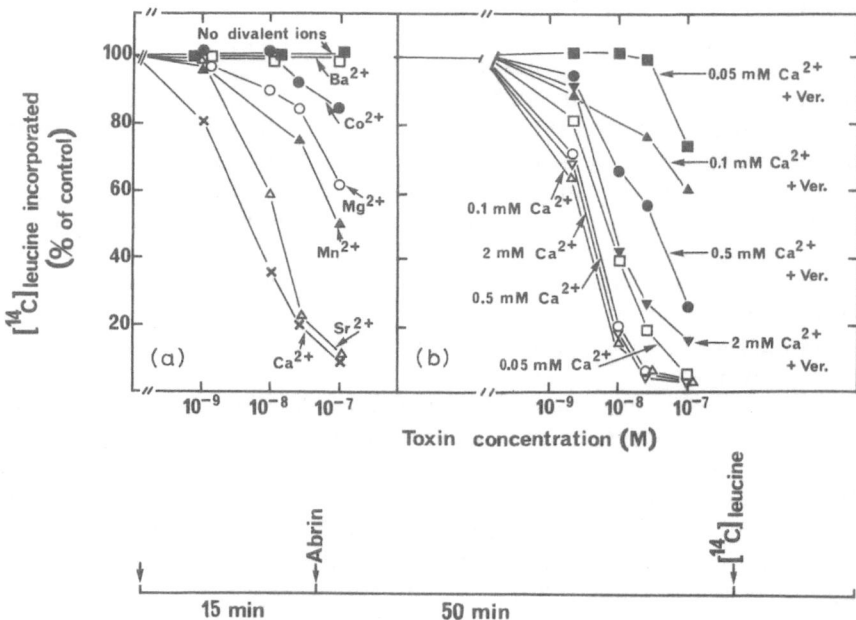

Fig. 8.11 Ability of Ca^{2+} to sensitize cells to abrin in the absence and presence of Verapamil. Vero cells in disposable trays were rinsed with EGTA and incubated with iso-osmotic NaCl/Hepes buffer, pH 7.3. The indicated cations were added to a final concentration of 2 mM in (a) whereas in (b) Ca^{2+} was added to the concentrations given on the figure. When indicated, 0.25 mM-Verapamil was added. After incubation for 15 min at 37°C, increasing amounts of abrin were added. The cells were further incubated for 50 min and then the ability of the cells to incorporate [¹⁴C]leucine during 10 min was measured.

deprivation also reduced the sensitivity of the cells to ricin, but this toxin exhibited some effect even in the absence of Ca^{2+}. The protective effect of Ca^{2+} deprivation was not due to reduced binding and endocytic uptake of the toxins, and the data are compatible with the possibility that Ca^{2+} is somehow required for transport of the A-chain across the membrane (Sandvig and Olsnes, 1982a).

Ca^{2+} may also be involved in entry of *Pseudomonas aeruginosa* exotoxin A and cholera toxin. In the presence of 2 mM EGTA the cells were protected against *Pseudomonas aeruginosa* toxin (FitzGerald *et al.*, 1980). However, since transglutaminase may be involved in the clustering and endocytosis of this toxin, the effect of Ca^{2+} deprivation could be due to inhibition of this Ca^{2+}-dependent enzyme rather than on the entry of the toxin through the membrane.

The rate of cholera-toxin-induced adenylate cyclase stimulation in glial tumor cells was increased when Ca^{2+} was present in the medium (Broström *et al.*, 1981). Once the adenylate cyclase had been maximally stimulated by the toxin, removal of Ca^{2+} did not reduce the rate of cAMP accumulation. This suggests that the requirement of Ca^{2+} is for toxin entry.

Also the effect of diphtheria toxin was somewhat reduced after removal of divalent cations from the cells by EGTA, but this was apparently due to reduced toxin binding to the cells. The binding could be restored by addition of several cations other than Ca^{2+} (Sandvig and Olsnes, 1982a).

(b) *Is Ca^{2+} influx necessary?*

Verapamil and Co^{2+}, which both inhibit uptake of $^{45}Ca^{2+}$ into cells, protected Vero cells against abrin and modeccin when the Ca^{2+} concentration was low (Fig. 8.11(b)). This protection was overcome when the Ca^{2+} concentration in the medium was increased, indicating that a Ca^{2+} flux into the cells may be necessary for transport of A-chains into the cytoplasm. If this is so, the Ca^{2+} influx must occur in some defined manner. Thus, the Ca^{2+} ionophore A23187, which strongly increased the influx of $^{45}Ca^{2+}$, did not sensitize the cells but strongly protected them against abrin and modeccin. A23187 also protected to some extent against diphtheria toxin, but not against ricin (Sandvig and Olsnes, 1982a). The reason for the protection against abrin and modeccin is not clear, but it could be due to inhibition of the natural Ca^{2+} influx routes. Thus, it was reported that high intracellular Ca^{2+} concentrations may lead to closure of physiological Ca^{2+} channels (Brehn and Eckert, 1978; Standen, 1981). Influx of Ca^{2+} through such channels may be required for entry of abrin and modeccin.

The requirement of Ca^{2+} for entry appears to be associated with the A-chain. Thus, a hybrid toxin consisting of abrin A/ricin B was, like abrin, not toxic in Ca^{2+}-free medium, whereas the hybrid ricin A/abrin B was, like ricin, less affected by the Ca^{2+} deprivation. Furthermore, the Ca^{2+} ionophore

A23187 protected well against abrin A/ricin B, but not against ricin A/abrin B (Sandvig and Olsnes, 1982a).

Also lanthanides have been used to block Ca^{2+} transport. As expected, La^{3+} did block the entry of abrin and modeccin into cells. However, this ion protected equally well against ricin and diphtheria toxin (Sandvig and Olsnes, 1982a). This is probably due to the fact that La^{3+} has a more general effect on membrane permeability and fluidity (Gordon *et al.*, 1978). Also another trivalent cation, Fe^{3+}, which has no effect on Ca^{2+} transport, protected against abrin, ricin, modeccin and diphtheria toxin (Sandvig and Olsnes, 1982a).

(c) *Effect of trifluoperazine*

The requirement of Ca^{2+} for toxin entry suggested that calmodulin might somehow be involved. However, even if this is so, it is unlikely to be the only reason why Ca^{2+} is required for toxin entry. Thus, the inhibitor of calmodulin-dependent processes, trifluoperazine, had almost no effect on the sensitivity of cells to abrin and ricin (Sandvig and Olsnes, 1982a) in spite of the fact that Ca^{2+} facilitated their entry. Trifluoperazine protected strongly against modeccin which required Ca^{2+}, but also against diphtheria toxin which has no Ca^{2+} requirement for entry. Possibly, the protective effect of trifluoperazine against modeccin and diphtheria toxin does not consist in a direct effect on toxin penetration, but rather in diverting the intracellular transport of the toxin to vesicular compartments where they are not able to penetrate into the cytosol.

Trifluoperazine also delayed the cholera-toxin-induced accumulation of cAMP in cells (Broström *et al.*, 1981), but it is not clear if this is due to an effect on toxin entry or on the adenylate cyclase as such.

8.5.6 Protective effect of various compounds

(a) *Metabolic inhibitors*

It has been reported from several laboratories that metabolic inhibitors prevent the action of toxins. Thus, the inhibitor of glycolysis, NaF, was reported to protect HeLa cells completely against diphtheria toxin, whereas inhibitors of oxidative phosphorylation did not protect (Duncan and Groman, 1969; Ivins *et al.*, 1975; Middlebrook and Dorland, 1977). In the presence of glucose, glycolysis is an important source of energy in HeLa cells.

Vero cells were completely protected against all toxins studied (abrin, modeccin, ricin, viscumin, *Shigella* toxin and diphtheria toxin) when both an inhibitor of glycolysis (2-deoxyglucose) and an inhibitor of oxidative phosphorylation (NaN_3) were present in the medium (Fig. 8.12), whereas each inhibitor alone provided only partial (2-deoxyglucose) or no (NaN_3) protection (Sandvig and Olsnes, 1982b). It is well established that metabolic

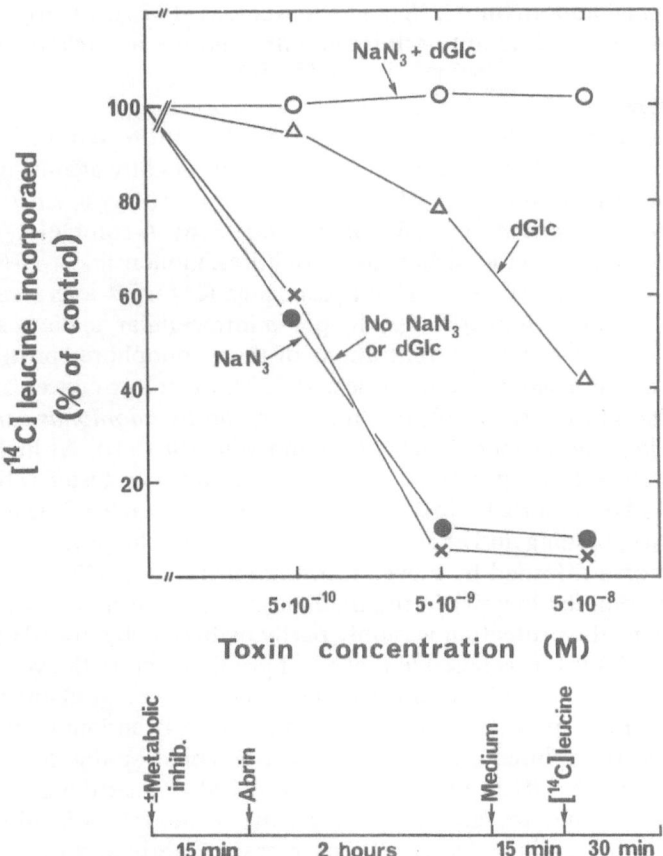

Fig. 8.12 Ability of metabolic inhibitors to protect cells against abrin. Vero cells were incubated for 15 min with and without 50 mM-2-deoxyglucose and 10 mM-NaN$_3$ as indicated. Then increasing amounts of abrin were added and the incubation was continued for 2 h more. Subsequently medium without inhibitors was added and the cells were allowed to recover for 15 min. Finally, the ability of the cells to incorporate [^{14}C]leucine during 30 min was measured.

inhibitors strongly decrease endocytosis, an effect which could account for their protective effect as endocytosis appears to be involved in toxin entry (see below). However, other processes must also be affected. Thus, the entry of diphtheria toxin at low pH, which appears to occur directly through the surface membrane, is blocked in the presence of metabolic inhibitors.

More difficult to interpret is the finding that NaN$_3$, NaF and dinitrophenol (DNP) inhibited the stimulation of adenylate cyclase in rat lymphocytes treated with cholera toxin (Craig and Cuatrecasas, 1975). Clearly, the metabolic inhibitors could inhibit both toxin entry and the adenylate cyclase.

(b) *Ionophores*

As mentioned above, the Ca^{2+} ionophore A23187 protected well against abrin and modeccin, but not against ricin and only slightly against diphtheria toxin (Sandvig and Olsnes, 1982a). It also did not protect against *Pseudonomas aeruginosa* toxin (Ray and Wu, 1981a). A completely different pattern was found with the carboxylic ionophores, monensin, Br-X-537A and nigericin. These ionophores are able to exchange K^+ for H^+ ions across membranes and therefore they increase the pH in intracellular acidic vesicles. We found that even very low concentrations of these ionophores protected against modeccin (Sandvig and Olsnes, 1982b). Such low concentrations increased the sensitivity of cells to abrin, ricin and *Pseudomonas aeruginosa* toxin (Sandvig and Olsnes, 1982b; Ray and Wu, 1981a,b). At higher concentrations Br-X-537A protected against all of four toxins tested (abrin, ricin, modeccin and diphtheria toxin) (Sandvig and Olsnes, 1982b). The ionophore did not inhibit binding and endocytosis of the toxins. The protection against diphtheria toxin afforded by low concentrations of Br-X-537A was completely overcome by low pH in the medium, whereas at higher ionophore concentrations the protection was only partly overcome by low pH (Sandvig and Olsnes, 1982b). It is possible that a pH gradient across the vesicle membrane is required for diphtheria toxin entry and that the gradient cannot be maintained in the presence of high concentrations of the ionophore.

In the case of diphtheria toxin the protection could be due to ionophore-induced increase of pH in the vesicles. This could also be the case with modeccin which may require low pH for entry, as its action is inhibited by NH_4Cl and other amines. However, in the case of abrin and ricin, which do not enter at low pH, such a mechanism can only explain the sensitizing effect seen at low ionophore concentrations. The protection seen at higher ionophore concentrations must be explained in another way. Possibly, it is related to the fact that carboxylic ionophores affect the membrane traffic in cells. Thus, monensin and nigericin disrupt the normal flow of export proteins from the Golgi apparatus to the cell surface, and they induce swelling of the Golgi cisternae which appear as large vacuoles. Furthermore, monensin was found to inhibit the return of certain receptors to the cell surface, apparently because the receptors were trapped in the Golgi apparatus (Basu *et al.*, 1981; Tartakoff and Vassalli, 1978; Uchida *et al.*, 1980). Clearly, if the toxins can only penetrate the membrane in a particular vesicular compartment, carboxylic ionophores may prevent the toxins from entering this compartment.

(c) *Other compounds*

Retinoic acid and some other retinoids protected cells against abrin, modeccin, ricin and diphtheria toxin, whereas the tumor promoter 12-*O*-

tetradecanoyl phorbol 13-acetate increased the sensitivity to abrin, modeccin and ricin (Sandvig and Olsnes, 1981b). Neither of the compounds affected binding and endocytosis of the toxins and presumably the compounds interfered with entry of the toxins into the cytosol.

The polycation poly-L-ornithine as well as local anesthetics and Ruthenium Red protected somewhat against diphtheria toxin, but not against *Pseudomonas aeruginosa* toxin. Lidocaine and chlorpromazine sensitized the cells to *Pseudomonas aeruginosa* toxin (Middlebrook and Dorland, 1977).

Protein synthesis appears to be required for stimulation of adenylate cyclase by cholera toxin. Thus, in cells treated with cycloheximide or puromycin, the ability of the adenylate cyclase to be stimulated by cholera toxin rapidly decreased, whereas it responded well to isoproterenol and prostaglandin E_1. When protein synthesis was inhibited there was also no reduction of the disulfide bridge linking the A_1-fragment to the A_2-fragment (Hagmann and Fishman, 1981). Also the degradation of the toxin was blocked. Possibly, cholera toxin entry is somehow linked to protein synthesis.

8.5.7 Interaction of cholera toxin with model membranes

After binding of cholera toxin to membranes containing GM_1 the toxin somehow interacts with the lipid bilayer. Thus, cholera toxin was found to form ion-permeable channels in lipid bilayers containing GM_1, but not in those lacking the ganglioside (Tosteson and Tosteson, 1978).

Recently penetration of cholera toxin into membranes has been studied with photoreactive glycolipid and phospholipid compounds in two laboratories (Tomasi and Montecucco, 1981; Wisnieski and Bramhall, 1981). The compounds were allowed to enter the membrane in the dark and then the photoreactive group was activated by light. The free radical formed reacted with adjacent molecules including proteins.

Since the photoreactive compounds had been radioactively labeled, binding to defined proteins could be studied by dissolving the membranes and analyzing the proteins by polyacrylamide gel electrophoresis. Photoreactive compounds with the reactive group located at different levels of the fatty acid chains were used to measure the depth in the membrane to which the different parts of the toxin had penetrated.

Wisnieski and Bramhall (1981) used as a model membrane Newcastle disease virus with a glycolipid probe incorporated. This probe carried the reactive group at a position that would only label proteins which had penetrated deeply into the membrane. When cholera toxin was bound to the virus, only the A-chain was labeled. The labeling was most intense when the light exposure occurred about 1 min after addition of toxin to the virus and over the next 5 min it was reduced to about half.

Tomasi and Montecucco (1981) found that compounds carrying the reactive group at a position that would locate it to the surface of the membrane

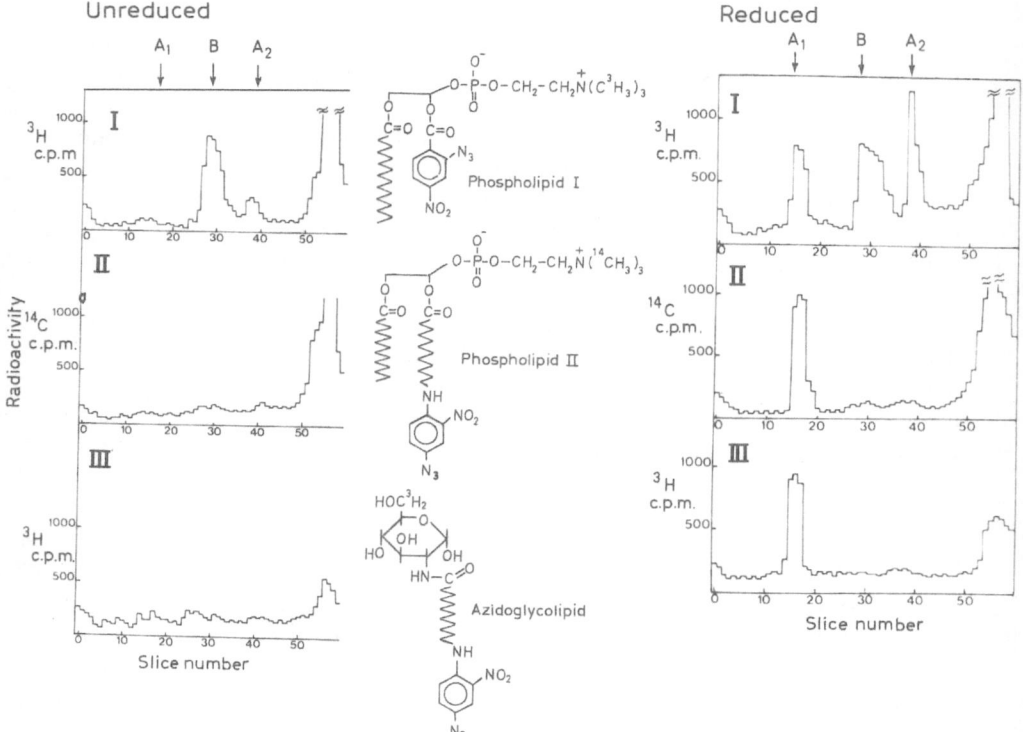

Fig. 8.13 Labeling of cholera toxin bound to liposomes with photoreactive probes before and after reduction of the A subunit. Unreduced and reduced cholera toxin were added to liposomes containing radioactively labeled phospholipids (I and II) or a glycolipid (III) which were carrying a photoreactive group. When incorporated into the lipid bilayer, the photoreactive group in I was located at the surface, whereas in II and III it was located more deeply inside the membrane. After treatment with light to activate the photoreactive group, the membranes were dissolved. The proteins were isolated, treated with sodium dodecyl sulfate and 2-mercaptoethanol and analyzed by polyacrylamide gel electrophoresis. (Redrawn and modified from Tomasi and Montecucco, 1981.)

labeled both the B-chains and the A-chain, whereas compounds with the reactive group penetrating more deeply into the membrane reacted only with the A-chain. In contrast to the results of Wisnieski and Bramhall (1981), Tomasi and Montecucco (1981) found labeling of the A-chain only after reduction of the disulfide bridge linking the A_1- and the A_2-fragments together (Fig. 8.13). Possibly this reduction exposes a hydrophobic domain which enters the membrane in a similar way to diphtheria toxin at low pH (Fig. 8.14). Also the A-chain of ricin undergoes considerable conformational changes when it is split from the B-chain (Olsnes and Saltvedt, 1975).

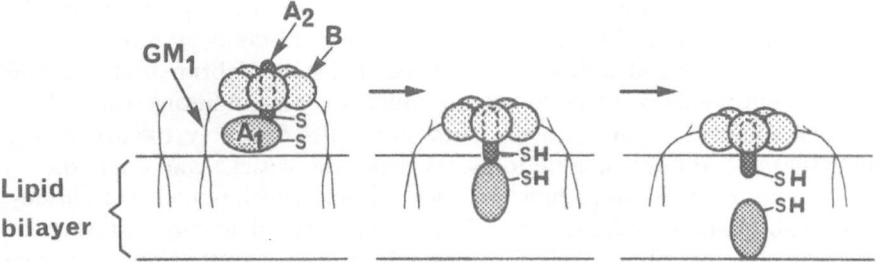

Fig. 8.14 Schematic model of cholera toxin penetration. (Redrawn and modified from Wisnieski and Bramhall, 1981.)

8.6 PRESENT MODEL FOR TOXIN ENTRY

8.6.1 Adsorptive endocytosis

From the circumstantial evidence available it now appears most likely that all toxins discussed in this chapter enter the cytosol from endocytic vesicles. However, it must be stressed that there is still no direct evidence that this is so. The best evidence that endocytosis is involved is the finding that diphtheria toxin requires low pH to enter the cytosol. Under artificial circumstances when the pH of the medium is reduced to pH 4.5, diphtheria toxin apparently enters directly through the surface membrane (Sandvig and Olsnes, 1981a). Under normal conditions, however, such low pH does not occur at the cell surface, but only in intracellular vesicles like lysosomes and receptosomes (Maxfield and Tycko, 1981). Therefore it is likely that the transfer across the membrane occurs from such vesicles.

We believe that the transfer of diphtheria toxin to the cytosol occurs from non-lysosomal acidic vesicles like the receptosomes. The main argument against penetration from lysosomes is that only previously nicked diphtheria toxin, is able to intoxicate cells. If the transfer were to occur from lysosomes where proteolytic enzymes are abundant, it is likely that unnicked toxin would also be active. The fact that the toxin is taken up and degraded in lysosomes indicates that the A-fragments produced during this process are either unable to penetrate the membrane surrounding the lysosome or that they are degraded too rapidly.

It should also be noted that endocytosed diphtheria toxin retains the ability to enter the cytosol and intoxicate cells only for a very short period of time. Thus, after preincubation of cells with toxin for short periods, NH_4Cl which inhibits the exit of toxin from endocytic vesicles, was not more efficient in inhibiting the toxic effect on cells than addition of anti-toxin which only inactivates toxin on the cell surface (Sandvig and Olsnes, 1981a).

Abrin, modeccin and ricin also appear to be able to enter the cytosol from endocytic vesicles, but, in contrast to diphtheria toxin, these toxins retain their ability to enter the cytosol for several hours after being endocytosed (Sandvig and Olsnes, 1982c). To study the ability of endocytosed toxin to enter the cytoplasm, advantage was taken of the fact that abrin, modeccin and ricin are continually being endocytosed under several conditions where these toxins are unable to inhibit protein synthesis in cells. Probably, the toxins are, under these conditions, unable to escape from the vesicles and enter the cytosol. These conditions include Ca^{2+} deprivation, which renders cells insensitive to abrin and modeccin, and pH 6.0 and intermediate concentrations of the carboxylic ionophore Br-X-537A which protect against abrin, modeccin and ricin. In these experiments the toxins were endocytosed under the protective conditions, then anti-toxins were added to inactivate toxin not endocytosed and normal medium was added. The next day the cells were found to be intoxicated, presumably by toxin which had accumulated in endocytic vesicles.

Also *in vivo* data indicate that toxin may enter the cytosol from endocytic vesicles (Harper *et al.*, 1980). Thus, it was found that ricin is endocytosed at the nerve endings and transported in vesicles retrograde back to the ganglion where it appears to intoxicate only those cells from which the neurons originated and not adjacent cells in the ganglion.

The finding that endocytosed toxins appear to be able to intoxicate cells does not necessarily mean that this is the main entry mechanism. However, the available data suggest that this is the case. To test this, cells were exposed for 1 h to abrin or modeccin in the absence and presence of Ca^{2+}. In the absence of Ca^{2+} the toxin accumulates in endocytic vesicles, but it does not enter the cytosol, whereas in the presence of Ca^{2+}, both processes take place. Then the cells were incubated overnight in the presence of anti-toxin. If exit from endocytic vesicles were only an unimportant side stream in toxin entry, the cells exposed to toxin in the presence of Ca^{2+} should be much more intoxicated than those exposed to toxin in Ca^{2+}-free medium. The results showed, however, that the same toxic effect was obtained in the two cases, indicating that approximately equal amounts of toxin had been internalized under the two conditions. It therefore appears that there is not another more efficient entry mechanism which bypasses endocytic uptake. Furthermore, the uptake of toxin into vesicles must occur rapidly compared to the subsequent penetration into the cytosol.

Macromolecules other than toxins are transferred to the cytosol from intracellular vesicles. It has recently been established that certain enveloped viruses enter the cytosol from endocytic vesicles. This is discussed in detail in Chapter 5 of this volume. In these viruses the entry of the nucleocapsid occurs by fusion of the virus membrane with the membrane of the cell. In Sendai virus the fusion is initiated by a viral fusion protein which has some similarities with toxins, particularly with diphtheria toxin. Thus, like diphtheria toxin, the fusion protein must be proteolytically nicked to be active and the two fragments obtained are linked by an SS bond. The fusion protein contains a hydrophobic region which is exposed to a much greater extent after proteolytic cleavage (Hsu, Scheid and Choppin, 1981). With Sendai virus, membrane fusion occurs at neutral pH, but in other cases, like Semliki Forest virus, vesicular stomatitis virus, and influenza virus, the membrane fusion occurs only at acidic pH. As the viruses are readily taken up by endocytosis and delivered to the lysosomes it has been proposed that the membrane fusion and nucleocapsid release into the cytosol take place in the lysosomes (Helenius *et al.*, 1980; White *et al.*, 1981). With the present information that receptosomes also have low internal pH (Maxfield and Tycko, 1981), it is possible that the fusion occurs in these organelles.

Endocytosis may prove to be a common process in internalization of macromolecules destined for the cytosol. If not only toxins and viruses but also physiologically important molecules such as certain protein hormones must enter the cytosol to exert their effect, entry from endocytic vesicles may be advantageous. Thus, the transfer of such large molecules may induce transient leakiness of the membrane, as has indeed been shown to be the case during virus entry (Fernández-Puentes and Carrasco, 1980; Yamaizumi *et al.*, 1979). Therefore, if the entry occurred from the cell surface, concentration gradients across the plasma membrane could be dissipated. A strong increase in intracellular Ca^{2+} concentration and disappearance of the membrane potential could be among the results. If, however, the macromolecules are first taken into endocytic vesicles, the transfer across the membrane may occur only after the vesicle is sealed off from the cell surface, i.e. the vesicle may act as a lock.

8.6.2 Transfer of the A-moiety in its extended form

Whereas data have accumulated that the transfer of the A-moiety across the membrane occurs from the inside of intracellular vesicles, we are left with speculations as to how the transfer occurs at the molecular level. We will here consider some possibilities. As outlined above, it is highly unlikely that the release occurs by unselective leakage or by rupture of vesicles. Probably some elaborate mechanism is involved.

The fact that hydrophobic regions are present in the toxins suggests that they are able to enter into membranes. However, the target for most of the toxins is free in the cytosol and the toxin A-chains must therefore somehow be able to escape from the inner side of the membrane. Proteolytic cleavage of the inserted protein to separate the hydrophobic part from the enzymatic part has been suggested as a possible mechanism for release (Pappenheimer, 1977).

Structural studies suggest that the A-moieties may be transferred across the membrane in their extended form. The lack of internal SS bridges and the ability of the A-moiety to recover activity after exposure to denaturing conditions indicate that unfolding and refolding can take place (Gill, 1978). Measurements of the size of channels formed by the hydrophobic part of diphtheria toxin B-fragment indicated that the channels are wide enough to accommodate diphtheria toxin A-fragment in its extended form (Kagan *et al.*, 1981). Also, normally occurring pores, such as the gap junctions between cells, have a similar pore size.

The best-characterized case of transfer of proteins across membranes is that of proteins synthesized for export. In this case it is now clear that the protein is transferred across the membrane as an extended polypeptide chain (Walter and Blobel, 1981). The messenger RNA for such a protein is first attached to free ribosomes in the cytoplasm. The polysome complex becomes membrane-attached as soon as the first part of the polypeptide is synthesized, because this part of the polypeptide, the signal sequence, has affinity for a signal-recognition protein on the cytosolic side of the endoplasmic reticulum (Walter and Blobel, 1981). Once the polysomes are bound, the peptide chain is transferred in its extended form through 'pores' in the endoplasmic reticulum into the lumen. Here the protein is folded into its three-dimensional structure and various post-translational modifications are carried out. Finally, the protein is transported along intracellular tubular and vesicular structures like the endoplasmic reticulum and the Golgi apparatus, until it reaches the cell surface where it is released. The entry of toxin A-chains into the cytosol may to some extent be considered as a reversal of this mechanism.

A question analogous to that of how toxin A-chains enter through membranes is raised by the fact that most of the proteins found in the matrix of mitochondria (Neupert and Schatz, 1981) and chloroplasts (Dobberstein *et al.*, 1977) are coded for by nuclear genes and synthesized on free polysomes in the cytoplasm. The proteins are usually synthesized as slightly larger precursor molecules which are bound to 'receptors' on the mitochondria and chloroplasts. Like the signal sequence of proteins destined for excretion, the cytosolic precursors for mitochondrial proteins also contain their additional sequence at the *N*-terminal, and it is cleaved off upon entry into the organelle where the proteins acquire the correct conformation (Neupert and Schatz, 1981). It is possible that the proteins are transported into the organelles as extended polypeptide chains which rapidly refold.

8.6.3 What is the driving force for transfer through the membrane?

The entry of the A-moiety into the cytosol appears to be a process requiring metabolic energy. Thus, the uptake of all toxins here discussed is prevented if both glycolysis and oxidative phosphorylation are stopped by treatment with metabolic inhibitors. One reason for this is that such inhibitors stop endocytosis. However, in systems where endocytosis is not required, the transfer of proteins across membranes also appears to be energy-dependent.

The rapid entry of nicked diphtheria toxin, which occurs when the cell culture medium is adjusted to low pH, does not involve endocytosis, as the toxin appears to enter directly through the surface membrane. In spite of this, the entry does not occur when glycolysis and oxidative phosphorylation are inhibited. This indicates that the transfer across the membrane as such is an energy-dependent process.

We have suggested that the membrane potential could be a driving force in the entry (Sandvig and Olsnes, 1981a). At pH below 5 the carboxyl groups of the side chains of glutamic and aspartic acids are protonated, and diphtheria toxin A-fragment will have a net positive charge. The positively charged groups will be attracted by the negatively charged cell interior. Thus, the membrane potential could initiate the transfer of the A-fragment through a hydrophilic channel formed by diphtheria toxin fragment B (Fig. 8.15). As the protonated carboxyl groups reach the cytosolic side of the membrane, where the pH is close to neutrality, they lose their protons and can then form bonds with the amino groups of the protein. This could initiate refolding of the A-fragment. This tendency of the A-fragment to refold into its correct three-dimensional conformation could provide additional energy to pull the A-chain through the channel. In this connection it may be important that at neutral pH the *N*-terminal two-thirds of the A-chain has approximately the same number of positively and negatively charged groups, whereas the *C*-terminal third has an excess of negatively charged groups.

Metabolic energy is also required in other cases where proteins are transferred across membranes. Thus, the entry of colicin E2 and E3, which have intracellular sites of action, is energy-dependent (see Holland, 1976). Since in

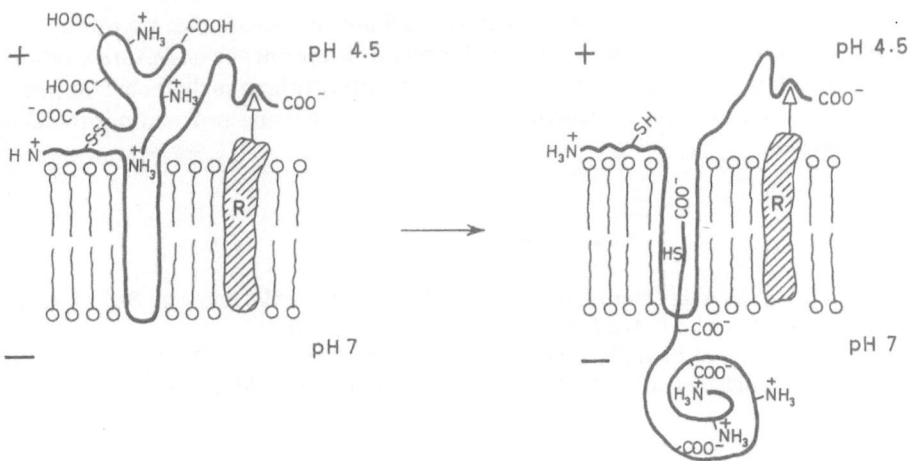

Fig. 8.15 Hypothetical scheme of the transfer of diphtheria toxin fragment A across membranes.

bacteria endocytosis does not occur, it is likely that the energy-dependent step is the transfer of the enzymatically active protein across the membrane. The entry of mitochondrial proteins from the cytosol into the mitochondria is dependent on high-energy phosphate bonds. Thus, if the mitochondrial matrix of intact yeast cells is depleted of ATP by suitable mutations or inhibitors, import of proteins into mitochondria stops and unprocessed protein precursors accumulate in the cytosol. Also, import of proteins into isolated mitochondria *in vitro* is inhibited in the presence of inhibitors which reduce the intramitochondrial ATP concentration (Neupert and Schatz, 1981). More recent data indicated that a membrane potential rather than ATP as such is required for entry (Schleyer *et al.*, 1982).

In addition to the membrane potential and the tendency of proteins to refold, it is also possible that linkage to certain ion-transport systems could provide energy for transfer of toxin A-moieties across membranes. The uptake of certain low-molecular-weight components such as sugars is linked to Na^+-ion transport, but this cannot be the driving force in toxin entry which occurs at a normal rate when no Na^+ is present in the medium (Sandvig and Olsnes, 1982a). However, the possibility exists that the entry of abrin and modeccin, and to a lesser extent, ricin, is somehow linked to Ca^{2+} influx. Although the influx of Ca^{2+} is very small compared to that of Na^+, it could still provide considerable amounts of energy, as there is a very steep concentration gradient of Ca^{2+} across the membrane.

Although our knowledge of the uptake mechanism of toxins is still fragmentary, it is clear that the mechanism is complex. In the course of evolution the toxins have acquired a structure which allows them to take advantage of cellular processes such as endocytosis and to be transported to defined vesicular compartments. Furthermore, the structure of the toxin A-moiety is such that it can exploit forces and transport mechanisms across the membrane which obviously must have been developed for other purposes. Future studies may elucidate whether physiologically important proteins such as certain protein hormones also enter the cytosol by these mechanisms and act on intracellular targets.

REFERENCES

Alving, C.R., Iglewski, B.H., Urban, K.A., Moss, J., Richards, R.L. and Sadoff, J.C. (1980), *Proc. Natl. Acad. Sci. U.S.A.*, **77**, 1986–1990.

Baenziger, J.U. and Fiete, D. (1979), *J. Biol. Chem.*, **254**, 9795–9799.

Basu, S.K., Goldstein, J.L., Andersson, R.G.W. and Brown, M.S. (1981), *Cell*, **24**, 493–502.

Boquet, P. (1979), *Eur. J. Biochem.*, **100**, 483–489.

Boquet, P. and Pappenheimer, A.M., Jr. (1976), *J. Biol. Chem.*, **251**, 5770–5778.

Boquet, P., Silverman, M.S., Pappenheimer, A.M. Jr. and Vernon, W.B. (1976), *Proc. Natl. Acad. Sci. U.S.A.*, **73**, 4449–4453.

Brehn, P. and Eckert, R. (1978), *Science*, **202**, 1203–1206.

Brodsky, W.A., Sadoff, J.C., Durham, J.H., Ehrenspeck, G., Schachner, M. and Iglewski, B.H. (1979), *Proc. Natl. Acad. Sci. U.S.A.*, **76**, 3562–3566.

Broström, M.A., Broström, C.O., Huang, S.-C. and Wolff, D.J. (1981), *Mol. Pharmacol.*, **20**, 59–67.

Chang, T.-M. and Neville, D.M., Jr. (1978), *J. Biol. Chem.*, **253**, 6866–6871.

Chung, D.W. and Collier, R.J. (1977a), *Infect. Immun.*, **16**, 832–841.

Chung, D.W., and Collier, R.J. (1977b), *Biochim. Biophys. Acta*, **483**, 248–257.

Craig, S.W. and Cuatrecasas, P. (1975), *Proc. Natl. Acad. Sci. U.S.A.*, **72**, 3844–3848.

Dobberstein, B., Blobel, G. and Chua, N.-H. (1977), *Proc. Natl. Acad. Sci. U.S.A.*, **74**, 1082–1085.

Donovan, J.J., Simon, M.I., Draper, R.K. and Montal, M. (1981), *Proc. Natl. Acad. Sci. U.S.A.*, **78**, 172–176.

Dorland, R.B., Middlebrook, J.L. and Leppla, S.H. (1979), *J. Biol. Chem.*, **254**, 11337–11342.

Dorland, R.B., Middlebrook, J.L. and Leppla, S.H. (1981), *Exp. Cell Res.*, **134**, 319–327.

Draper, R.K. and Simon, M.I. (1980), *J. Cell Biol.*, **87**, 849–854.

Duncan, J.L. and Groman, N.B. (1969), *J. Bacteriol.*, **98**, 963–969.

Eiklid, K. and Olsnes, S. (1980), *J. Receptor Res.* 1, 199–213.

Eiklid, K., Olsnes, S. and Píhl, A. (1980), *Exp. Cell Res.*, **126**, 321–326.

Fernández-Puentes, C. and Carrasco, L. (1980), *Cell*, **20**, 769–775.

Fishman, P.H. (1980), in *Secretory Diarrhea* (M. Field, J.S. Fordtran and S.G. Schultz, eds), American Physiological Society, Bethesda, pp. 85–106.

FitzGerald, D., Morris, R.E. and Saelinger, C.B. (1980), *Cell*, **21**, 867–873.

Gill, D.M. (1977), *Adv. Cyclic Nucleotide Res.*, **8**, 85–118.

Gill, D.M. (1978), in *Bacterial Toxins and Cell Membranes* (J. Jeljaszewicz and T. Wadstöm, eds), Academic Press, New York, pp. 291–332.

Gill, D.M., Hope, J.A., Meren, R. and Jacobs, D.S. (1981), in *Receptor-Mediated Binding and internalization of Toxins and Hormones* (J.L. Middlebrook and L.D. Kohn, eds), Academic Press, New York, pp. 113–121.

Gonatas, N.K., Kim, S.U., Stieber, A. and Avrameas, S. (1977), *J. Cell Biol.*, **73**, 1–13.

Gonatas, N.K., Stieber, A., Kim, S.U., Graham, D.I. and Avrameas, S. (1975), *Exp. Cell Res.*, **94**, 426–431.

Gonatas, J., Stieber, A., Olsnes, S. and Gonatas, N.K. (1980), *J. Cell Biol.*, **87**, 579–588.

Gordon, L.M., Sauerheber, R.D. and Esgate, J.A. (1978), *J. Supramol. Struct.*, **9**, 299–326.

Gottlieb, C. and Kornfeld, S. (1976), *J. Biol. Chem.*, **251**, 7761–7768.

Gupta, R.S. and Siminovitch, L. (1980), *Som. Cell Genet.*, **6**, 361–379.

Hagmann, J. and Fishman, P.H. (1981), *Biochem. Biophys. Res. Commun.*, **98**, 677–684.

Harper, C.G., Gonatas, J.O., Mizutani, T. and Gonatas, N.K. (1980), *Lab. Invest.*, **42**, 396–404.

Helenius, A., Kartenbeck, J., Simons, K. and Fries, E. (1980), *J. Cell Biol.* **84**, 404–420.

Holland, I.B. (1976), in *The Specificity and Action of Animal, Bacterial and Plant Toxins* (P. Cuatrecasas, ed.), Chapman and Hall, London, pp. 99–127.

Holmgren, J. (1981), *Nature (London)*, **292**, 413–417.

Houslay, M.D. and Elliott, K.R.F. (1981), *FEBS Lett.*, **128**, 289–292.

Houston, L.L. (1982), *J. Biol. Chem.*, **257**, 1532–1539.

Hsu, M.-C., Scheid, A. and Choppin, P.W. (1981), *J. Biol. Chem.*, **256**, 3557–3563.

Hyman, R., Lacorbiere, M., Stavarek, S. and Nicolson, G. (1974), *J. Natl. Cancer Inst.*, **52**, 963–969.

Ivins, B., Saelinger, C.B., Bonventre, P.F. and Woscinski, C. (1975), *Infect. Immun.*, **11**, 665–674.

Joseph, K.C., Stieber, A. and Gonatas, N.K. (1979), *J. Cell Biol.*, **81**, 543–554.

Kagan, B.L., Finkelstein, A. and Colombini, M. (1981), *Proc. Natl. Acad. Sci. U.S.A.*, **78**, 4950–4954.

Kayser, G., Lambotte, P., Falmagne, P., Capiau, C., Zanen, J. and Ruysschaert, J.-M. (1981), *Biochem. Biophys. Res. Commun.*, **99**, 358–363.

Kessel, M. and Klink, F. (1980), *Nature (London)*, **287**, 250–251.

Kim, K. and Groman, N.B. (1965), *J. Bacteriol.*, **90**, 1557–1562.

King, A.C., Hernaez-Davis, L. and Cuatrecasas, P. (1981), *Proc. Natl. Acad. Sci. U.S.A.*, **78**, 717–721.

Lai, C.-Y., Cancedda, F. and Duffy, L.K. (1981), *Biochem. Biophys. Res. Commun.*, **102**, 1021–1027.

Lambotte, P., Falmagne, P., Capiau, C., Zanen, J., Ruysschaert, J.-M. and Dirkx, J. (1980), *J. Cell Biol.*, **87**, 837–840.

Leppla, S.H., Dorland, R.B. and Middlebrook, J.L. (1980), *J. Biol. Chem.*, **255**, 2247–2250.

Leppla, S.H., Martin, O.C. and Muehl, L.A. (1978), *Biochem. Biophys. Res. Commun.*, **81**, 532–538.

Lory, S., Carroll, S.F., Bernard, P.D. and Collier, R.J. (1980a), *J. Biol. Chem.*, **255**, 12011–12015.

Lory, S., Carroll, S.F. and Collier, R.J. (1980b), *J. Biol. Chem.*, **255**, 12016–12019.

Lory, S. and Collier, R.J. (1980), *Proc. Natl. Acad. Sci. U.S.A.*, **77**, 267–271.

Maxfield, F.R. and Tycko, B. (1981), *J. Cell Biol.*, **91**, Abstr. 12026.

Meager, A., Ungkitchanukit, A. and Hughes, R.C. (1976), *Biochem. J.*, **154**, 113–124.

Mekada, E., Uchida, T. and Okada, Y. (1979), *Exp. Cell Res.*, **123**, 137–146.

Mekada, E., Uchida, T. and Okada, Y. (1981), *J. Biol. Chem.*, **256**, 1225–1228.

Michael, H.J., Bishayee, S. and Das, M. (1980), *FEBS Lett.*, **117**, 125–130.

Middlebrook, J.L. (1981), *J. Biol. Chem.*, **256**, 7898–7904.

Middlebrook, J.L. and Dorland, R.B. (1977), *Infect. Immun.*, **16**, 232–239.

Middlebrook, J.L., Dorland, R.B. and Leppla, S.H. (1978), *J. Biol. Chem.*, **253**, 7325–7330.

Middlebrook, J.L., Dorland, R.B. and Leppla, S.H. (1979), *Exp. Cell Res.*, **121**, 95–101.

Moehring, T.J. and Crispell, J.P. (1974), *Biochem. Biophys. Res. Commun.*, **60**, 1446–1452.

Moehring, J.M., Moehring, T.J. and Danley, D.E. (1980), *Proc. Natl. Acad. Sci. U.S.A.*, **77**, 1010–1014.

Moynihan, M.R. and Pappenheimer, A.M. Jr. (1981), *Infect. Immun.*, **32**, 575–582.

Neupert, W. and Schatz, G. (1981), *Trends. Biochem. Sci.*, **6**, 1–4.

Nicolson, G.L. (1974), *Nature (London)*, **251**, 628–630.

Nicolson, G.L. (1975), *Am. J. Clin. Pathol.*, **63**, 677–684.

Nicolson, G.L., Lacorbiere, M. and Ekhart, W. (1975a), *Biochemistry*, **14**, 172–179.

Nicolson, G.L., Lacorbiere, M. and Hunter, T.R. (1975b), *Cancer Res.*, **35**, 144–155.

Nicolson, G.L. and Poste, G. (1978), *J. Supramol. Struct.*, **8**, 235–245.

Nicolson, G.L., Robbins, J.C. and Hyman, R. (1976), *J. Supramol. Struct.*, **4**, 15–26.

Nicolson, G.L., Smith, J.R. and Hyman, R. (1978), *J. Cell Biol.*, **78**, 565–576.

Olsnes, S. (1978), in *Transport of Macromolecules in Cellular Systems* (S.C. Silverstein, ed.), Dahlem Konferenzen, Berlin, pp. 103–116.

Olsnes, S. (1981), *Nature (London)*, **290**, 84.

Olsnes, S. and Abraham, A.K. (1979), *Eur. J. Biochem.*, **93**, 447–452.

Olsnes, S. and Pihl, A. (1976), in *Receptors and Recognition, Series B. The Specificity and Action of Animal, Bacterial and Plant Toxins* (P. Cuatrecasas, ed.), Chapman and Hall, London, pp. 129–173.

Olsnes, S. and Pihl, A. (1982), in *The Molecular Action of Toxins and Viruses* (S. van Heyningen and P. Cohen, eds), Elsevier/North Holland, Amsterdam, pp. 57–105.

Olsnes, S. and Refsnes, K. (1978), *Eur. J. Biochem.*, **88**, 7–15.

Olsnes, S., Refsnes, K. and Pihl, A. (1974), *Nature (London)*, **249**, 627–631.

Olsnes, S., Reisbig, R. and Eiklid, K. (1981), *J. Biol. Chem.*, **256**, 8732–8738.

Olsnes, S. and Saltvedt, E. (1975), *J. Immunol.*, **114**, 1743–1748.

Olsnes, S., Sandvig, K., Eiklid, K. and Pihl, A. (1978), *J. Supramol. Struct.*, **9**, 15–25.

Olsnes, S., Sandvig, K., Refsnes, K. and Pihl, A. (1976), *J. Biol. Chem.*, **257**, 3985–3992.

Pappenheimer, A.M. Jr. (1977), *Annu. Rev. Biochem.*, **46**, 69–94.

Pappenheimer, A.M., Jr., Harper, A.A., Moynihan, M. and Bockes, J.P. (1982), *J. Infect. Dis.*, **145**, 94–103.

Pastan, I.H. and Willingham, M.C. (1981), *Science*, **214**, 504–509.

Pena, S.D.J., Mills, G. and Hughes, R.C. (1979), *Biochim. Biophys. Acta*, **550**, 100–109.

Poole, B. and Ohkuma, S. (1981), *J. Cell Biol.*, **90**, 665–669.

Proia, R.L., Eidels, L. and Hart, D.A. (1981), *J. Biol. Chem.*, **256**, 4991–4997.

Proia, R.L., Hart, D.A. and Eidels, L. (1979), *Infect. Immun.*, **26**, 942–948.

Proia, R.L., Wray, S.K., Hart, D.A. and Eidels, L. (1980), *J. Biol. Chem.*, **255**, 12025–12033.

Ray, B. and Wu, H.C. (1981a), *Mol. Cell Biol.*, **1**, 552–559.

Ray, B. and Wu, H.C. (1981b), *Mol. Cell Biol.*, **1**, 560–567.

Refsnes, K., Olsnes, S. and Pihl, A. (1974), *J. Biol. Chem.*, **249**, 3557–3562.

Reisbig, R., Olsnes, S. and Eiklid, K. (1981), *J. Biol. Chem.*, **256**, 8739–8744.

Robbins, J.C., Hyman, R., Stallings, V. and Nicolson, G.L. (1977), *J. Natl. Cancer Inst.*, **58**, 1027–1033.

Rosen, S.W. and Hughes, R.C. (1977), *Biochemistry*, **16**, 4908–4915.

Sanai, Y., Morihara, K., Tsuzuki, H., Homma, J. Y., and Kato, I. (1980), *FEBS Lett.*, **120**, 131–134.

Sando, G.N., Titus-Dillon, P., Hall, C.W. and Neufeld, E.F. (1979), *Expl. Cell. Res.*, **119**, 359–364.

Sandvig, K. (1978), *FEBS Lett.*, **89**, 233–236.

Sandvig, K. and Olsnes, S. (1979), *Exp. Cell Res.*, **121**, 15–25.

Sandvig, K. and Olsnes, S. (1980), *J. Cell Biol.*, **87**, 828–832.

Sandvig, K. and Olsnes, S. (1981a), *J. Biol. Chem.*, **256**, 9068–9076.
Sandvig, K. and Olsnes, S. (1981b), *Biochem. J.*, **194**, 821–827.
Sandvig, K. and Olsnes, S. (1982a), *J. Biol. Chem.*, **257**, 7495–7503.
Sandvig, K. and Olsnes, S. (1982b) *J. Biol. Chem.*, **257**, 7504–7513
Sandvig, K., Olsnes, S. and Pihl, A. (1976), *J. Biol. Chem.*, **251**, 3977–3984.
Sandvig, K., Olsnes, S. and Pihl, A. (1978a), *Eur. J. Biochem.*, **82**, 13–23.
Sandvig, K., Olsnes, S. and Pihl, A. (1978b), *Eur. J. Biochem.*, **88**, 307–313.
Sandvig, K., Olsnes, S. and Pihl, A. (1979), *Biochem. Biophys. Res. Commun.*, **90**, 648–655.
Schleyer, M., Schmidt, B. and Neupert, W. (1982), *Eur. J. Biochem.*, **125**, 109–116.
Seglen, P.O. (1975), *Biochem. Biophys. Res. Commun.*, **66**, 44–52.
Seglen, P.O. and Gordon, P.B. (1980), *Mol. Pharmacol.*, **18**, 468–475.
Standen, N.B. (1981), *Nature (London)*, **293**, 158–159.
Stanley, P., Narasimhan, S., Siminowitch, L. and Schlachter, H. (1975) *Proc. Natl. Acad. Sci. U.S.A.*, **72**, 3323–3327.
Tartakoff, A. and Vassalli, P. (1978), *J. Cell Biol.*, **79**, 694–707.
Tietze, C., Schlesinger, P. and Stahl, P. (1980), *Biochem. Biophys. Res. Commun.*, **93**, 1–8.
Tomasi, M. and Montecucco, C. (1981), *J. Biol. Chem.*, **256**, 11177–11181.
Tosteson, M.T. and Tosteson, D.C. (1978), *Nature (London)*, **275**, 142–144.
Uchida, N., Similowitz, H., Ledger, P.W. and Tanzer, M.L. (1980), *J. Biol. Chem.*, **255**, 8638–8644.
Van Ness, B.G., Howard, J.B. and Bodley, J.W. (1980), *J. Biol. Chem.*, **255**, 10710–10716.
Vasil. M.L., Kabat, D. and Iglewski, B.H. (1977), *Infect. Immun.*, **16**, 353–361.
Villafranca, J.E. and Robertus, J.D. (1981), *J. Biol. Chem.*, **256**, 554–556.
Walter, P. and Blobel. G. (1981), *J. Cell Biol.*, **91**, 557–561.
White, J., Matlin, K. and Helenius, A. (1981), *J. Cell Biol.*, **89**, 674–679.
Wisnieski, B.J. and Bramhall, J.S. (1981), *Nature (London)*, **289**, 319–321.
Yamaizumi, M., Mekada, E., Uchida, T. and Okada, Y. (1978), *Cell*, **15**, 245–250.
Yamaizumi, M., Uchida, T. and Okada, Y. (1979), *Virology*, **95**, 218–221.
Youle, R.J., Murray, G.J. and Neville, D.M., Jr. (1981), *Cell*, **23**, 551–559.
Youle, R.J. and Neville, D.M. Jr. (1979), *J. Biol. Chem.*, **254**, 11089–11096.

9 Maternal–Fetal Protein Transport

JOHN W. WOODS and THOMAS F. ROTH

Acknowledgements

We are indebted to Drs J. Daiss and J. Halpern for helpful discussions and comments on this manuscript. This work is supported in part by NIH grants HD 09549 and HD 11519 from the Institute for Child Health and Human Development and NSF grant PCM-8118717.

Receptor-Mediated Endocytosis
(*Receptors and Recognition*, Series B, Volume 15)
Edited by P. Cuatrecasas and T. F. Roth
Published in 1983 by Chapman and Hall, 11 New Fetter Lane, London EC4P 4EE
© 1983 Chapman and Hall

9.1 INTRODUCTION

The reproductive success of a species is dependent on the ability of the mother to provide her offspring with a variety of substances essential for the development, nutrition and immunological protection of the fetus and neonate. In this chapter we focus on the process by which essential maternal proteins are specifically transported from the mother to her offspring. Several distinct examples of this transport process are discussed.

Maternal–fetal protein transport results in the accumulation of selected maternal proteins in the oocyte, fetus and neonate. Depending on the species and stage of development, a variety of proteins are known to be transported from the mother to her offspring. Examples include: (1) vitellogenin and low-density lipoprotein which function as reservoirs of metabolically essential precursors; (2) transferrin and transcobalamin II which act as carriers of vitally important cofactors; and (3) immunoglobulins A and G which provide the neonate with passive immunity.

Species which bear live young often are able to carry out maternal–fetal protein transport as a one-step process. In these species this transport can take place across those cells of fetal origin which are in direct contact with maternal tissues. In many species the syncytiotrophoblastic cells of the placenta carry out this function. In other species the site of transport is at the endodermal cells of the placental yolk sac splanchnopleur.

It is important to realize, however, that maternal–fetal transport can also take place at maternal–fetal barriers that are not directly juxtaposed. For example, in some mammals, maternal–fetal transport can occur after birth as a two-step process. For instance, it is well known that in all mammals maternal proteins are secreted into the mother's milk by the cells of the mammary epithelium. After suckling, these proteins are sequestered by the cells of the small intestine of the neonate where most are degraded, but a select few are selectively transported and released into the circulation of the newborn. In animals which lay yolky eggs, maternal–fetal transport is also carried out as a two-step process. In this instance, the maternal proteins to be transported are first specifically moved from the maternal serum into the developing eggs where they are stored as components of the yolk. After embryonic development has begun, some of these maternal proteins are transported from the yolk, across the embryonic yolk sac epithelium and into the circulation of the developing embryo while others are degraded and serve to nourish the embryo.

All available evidence suggests that, regardless of the species, tissue or transported protein, such transport occurs by the general process of receptor-mediated endocytosis. As in other receptor-mediated endocytic events, this involves the initial binding of the transported protein to a specific

239

receptor located on the apical surface of a cell where storage or transport occurs. The specificity of the receptors determines which proteins will be transported. Either after or concomitant with receptor binding, the protein−receptor complexes become localized in coated pits, which are ubiquitous, morphological specializations of plasma membranes (see Chapter 2). These coated pits then pinch off, giving rise to endocytic coated vesicles containing the bound proteins. Except in the case of transport into the oocyte for storage, endocytic vesicles derived from the coated vesicles are translocated through the cytoplasm and ultimately fuse with the basolateral surface and release their contents into the extracellular space. The cells involved in this transport are organized into a polarized epithelium. The receptors are initially exposed to the maternal side of the epithelium and the endocytic vesicles release their contents on the fetal side.

9.2 PLACENTAL TRANSPORT

Maternal−fetal transport across the placenta provides the embryo with essential nutrients needed for growth *in utero*. Certain of these nutrients are transferred across the placenta on serum carrier proteins, most notable being the vitamin B_{12} carrier transcobalamin and iron-containing transferrin. These proteins are specifically recognized and endocytosed into the placenta (Fernandez-Costa and Metz, 1979; Gillian *et al.*, 1980). In addition, in some animals, most notably the human, maternal IgG is also transported across the placenta to provide the neonate with the panoply of IgG that is common to the mother (cf. Brambell, 1970).

9.2.1 Placental structure

Since the placenta is the interface between the fetus and mother, it is important to point out the principal morphological features that, on one hand, facilitate selective transport and, on the other hand, provide a barrier between the two circulations. In general, implantation of the early embryo in the uterine mucosa results in a localized breakdown of the uterine endometrium such that elements of the fetal and maternal circulation become juxtaposed, separated only by a single cellular layer. Depending upon the species, this layer of embryonic origin arises from either the chorionic or allantoic portions of the extraembryonic membranes. Blood vessels of embryonic origin populate one side of the layer while maternal vessels, derived from those that originally underlied the endometrium, are found on the other side (Ramsey, 1975).

The barrier between the two circulations consists of a single layer of trophoblast cells. Depending upon the animal, these trophoblasts can fuse to

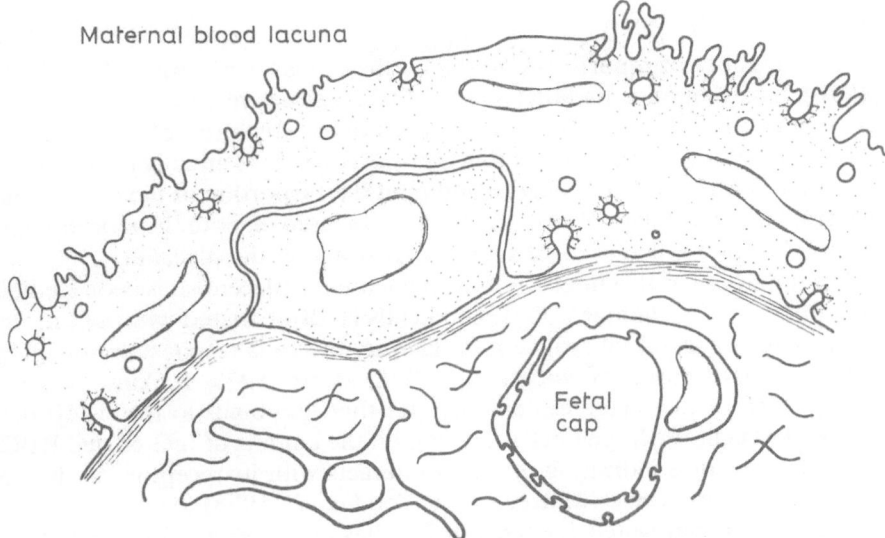
Maternal blood lacuna

Fetal cap

Fig. 9.1 Diagram of a portion of a chorionic villus of a hemochorial placenta where maternal–fetal protein transport takes place. Multinucleate syncytiotrophoblast cells border directly on the maternal blood lacuna. Receptors facing the lacuna mediate the endocytic uptake of maternal proteins via coated pits and vesicles into the syncytiotrophoblasts. These endocytic vesicles release their contents into the basal extracellular space from which the proteins are free to diffuse into the fetal capillaries.

form a syncytium. These cells are tightly apposed to one another thus preventing material leaking past the cells. This cellular membrane becomes progressively more convoluted and villous, providing a very large surface area for exchange and transport in a relatively small zone. Other cell types which are also found in this tissue include the cytotrophoblasts, occasional mesenchymal cells and the endothelial cells of the vascular elements (Fig. 9.1) (Ockleford and Whyte, 1977). These other cell types appear to play no direct role in IgG transport.

In the human, the work of Ockleford and co-workers has implicated the microvillous areas of the human syncytiotrophoblast as the cellular layer across which maternal–fetal protein transport takes place (Ockleford and Whyte, 1977; Ockleford and Clint, 1980). They observed that numerous coated vesicles occur at or near the cell surface where villi are located but only rarely where the surface is devoid of villi. In addition, a plethora of smooth surface vesicles populate the zone deeper beneath the microvilli. Large numbers of microtubules and microfilaments are also found in the zone. The route these vesicles take to transport proteins into and across the placenta is not known.

9.2.2 IgG transport

A large number of reports, prompted by the pioneering work of Brambell (1970), have demonstrated that maternal IgG is selectively transferred across the human placenta into the developing embryo. Much recent work, briefly described below, has shown that this transport process is initially mediated by specific receptors which recognize and bind the Fc portion of the circulating maternal IgG molecules. These studies have focused on both the morphological and biochemical aspects of IgG transport across the placenta.

The initial site of IgG interaction with the human placenta has been studied by a number of techniques. Red blood cells (RBCs) coated with IgG have been used by several investigators to identify the cells responsible for IgG transplacental transport (Wood *et al.*, 1978; Matre, 1977; Jenkinson *et al.*, 1976). Each of these studies demonstrates that IgG binds to the syncytiotrophoblast cells. Fc fragments of IgG inhibit the binding of IgG-coated RBCs to these cells, thus indicating that IgG interacts with its receptor via the Fc portion of the molecule (Matre, 1977; Wood *et al.*, 1978).

Fluorescein-conjugated IgG or anti-IgG has also been used to study the localization of the IgG Fc receptor. Fluorescein-conjugated IgG strongly stains the surface of certain placental syncytiotrophoblastic cells. In addition, intracellular staining is also seen. It is also noteworthy that within a single villus some regions of the syncytiotrophoblast are fluorescent while others are not, suggesting that not all cells transport IgG. A similar pattern of staining was seen when fluorescein-conjugated anti-IgG was used to localize endogenous IgG. In a similar experiment, anti-IgM and anti-IgA gave no staining, thus suggesting that little IgM or IgA is bound to this tissue (Wood *et al.*, 1978). When fluorescein-conjugated anti-IgG binding to first-trimester placentas is examined, little staining of the trophoblasts is seen (Johnson and Faulk, 1978). This observation is not unexpected, however, since significant placental transfer of maternal IgG does not occur until after the 22nd week of gestation (Gitlin and Biascussi, 1969).

When the localization of the IgG Fc receptor was studied using anti-(horseradish peroxidase) IgG–HRP as an immune complex, reaction product was found on the apical surface of the trophoblasts and endothelial cells (Matre and Haugen, 1978). At the higher resolution afforded by the electron microscope, Lin (1980) showed that HRP reaction product was predominantly localized to trophoblastic cells on microvilli and in endocytic vesicles. Some of these labeled vesicles appear to fuse with each other and with lysosomes. Others appear to fuse with the basolateral cell membrane, apparently discharging their contents which then enter the fetal circulation (Lin, 1980).

Of particular note are the recent findings that coated vesicles isolated from placenta contain transferrin and IgG (Ockleford and Clint, 1980; Booth and

Wilson, 1981; Pearse, 1982). These results, coupled with the morphological localization of the coated vesicles to the cellular layer which separates the two blood supplies (Ockleford and Whyte, 1977), suggest that coated vesicles mediate at least the initial events in maternal–fetal transport of these two proteins.

Only recently has it been possible to study IgG transport in *in vitro* incubations of the placenta where the system can be manipulated (Ockleford and Clint, 1980). In these experiments IgG uptake was specific, temperature-dependent and partially inhibited by colchicine but not by cytochalasin B. When homogenized and fractionated on sucrose gradients, the IgG co-migrated with a band containing coated vesicles.

Characterization of the biochemical properties of the placental IgG receptor has been the object of many recent studies. One approach to these studies has been to examine the binding of radiolabeled IgG to crude microsomal pellets of homogenized placental tissue. Specific IgG binding to these membranes is shown by comparing IgG binding in the presence and absence of homologous serum proteins (Gitlin and Gitlin, 1974). Using a similar crude membrane preparation, Balfour and Jones (1977) determined that binding was optimal between pH 6 and 6.5, that it was Fc-dependent and IgG-specific. In a subsequent publication Balfour and Jones (1978) demonstrated that the receptor was Pronase-sensitive but relatively insensitive to neuraminidase and collagenase. Using a similar membrane preparation McNabb *et al.* (1976) found that there is only one class of receptor when IgG binding results are plotted according to the Scatchard format. The apparent affinity constant is 4×10^6 M^{-1} for IgG1 = IgG3 > IgG4 > IgG2. Fc from IgG1 had the same affinity as IgG1. IgA and IgM did not bind. They also estimated that a near-term placenta has approximately 10^{15} receptors for IgG.

IgG binding to purified plasma membrane vesicles prepared from the human syncytiotrophoblast has also been studied. Such purified vesicles were shown to agglutinate IgG-coated red blood cells. The agglutination was Fc-specific since Fab, IgA and IgM did not inhibit agglutination, whereas IgG did (Van Der Mevlen *et al.*, 1980). It was further demonstrated that intact IgG1 and IgG3 and their Fc fragments inhibited agglutination more effectively than did IgG2 and IgG4. IgG from rabbit, mouse, guinea pig and dog were decreasingly less effective as inhibitors of agglutination, whereas IgG from sheep, goat and cow did not inhibit. In addition, a fragment corresponding to the C3 domain did not inhibit. These data suggest that both the C2 and C3 need to be intact for Fc recognition of the IgG receptor (Van Der Mevlen *et al.*, 1980). Brown and Johnson (1981) used a similar syncytiotrophoblast microvillus plasma membrane to further characterize the Fc receptor activity. They found that these membranes bound IgG optimally in 10 mM-Tris, pH 7.4, with little change in binding from pH 5.4 to 7.4. Binding was much enhanced if the preparations were first treated with 3 M-KCl to remove previously bound IgG. An apparent K_a of 4×10^7 M^{-1} was obtained at 37°C, with 1 mg of membrane protein yielding 1.5×10^{14} receptors.

9.2.3 Transcobalamin: vitamin B_{12}-carrier protein

Vitamin B_{12} (cobalamin) is transported from the maternal circulation to the fetal tissues (Ullberg *et al.*, 1967; Graber *et al.*, 1971; Friedman *et al.*, 1977; Fernandez-Costa and Metz, 1979). It is not found in the circulation as free cobalamin but rather as a complex with its binding proteins, transcobalamin I, II or III. However, only the transcobalamin II–cobalamin complex is recognized by surface receptors (Fernandez-Costa and Metz, 1979) and is taken into the cells by endocytosis. In turn, the endocytic vesicles fuse with the lysosomes, and cobalamin is released (Kolhouse and Allen, 1977; Youngdahl-Turner *et al.*, 1978). Little is known about the transport of cobalamin from the placenta into the fetus. However, the receptor responsible for the first step in transcobalamin II–cobalamin uptake in the placenta has been characterized (Friedman *et al.*, 1977) and, more recently, solubilized (Seligman and Allen, 1978; Nexo and Hollenberg, 1980).

Membrane preparations from the human placenta bind transcobalamin II–cobalamin complexes with twice the affinity that they bind apo-transcobalamin II. Free cobalamin does not bind to the receptor. Although there is a lack of agreement as to the precise apparent affinity constant, it appears to be approximately 2×10^9 M^{-1} (Seligman and Allen, 1978; Nexo and Hollenberg, 1980). Binding requires Ca^{2+} and is specific in that human transcobalamin I–B_{12}, bovine transcobalamin II–B_{12}, porcine intrinsic factor–B_{12} and human intrinsic factor–B_{12} do not bind (Seligman and Allen, 1978). Detergent-solubilized placental receptor binds to both wheatgerm agglutinin and phytohemagglutinin and is desorbed by the *N*-acetyl forms of the appropriate lectin-specific sugars. These data suggest that it is likely that the sugars on the receptor that normally recognize the lectins are *N*-acetylated (Nexo and Hollenberg, 1980).

9.2.4 Transferrin: iron-carrier protein

Iron transfer in near-term human placentae is achieved against a concentration gradient as shown by the fact that the iron concentration in fetal plasma exceeds that in maternal circulation (Fletcher and Suter, 1969). Iron is moved through the circulation bound principally to transferrin, an 80 000-mol.wt. protein (MacGillivray *et al.*, 1977). Receptors for the transferrin–iron complex exist on a variety of cell types including the syncytiotrophoblast cells of the human placenta (cf. Seligman *et al.*, 1979; Wada *et al.*, 1979; Loh *et al.*, 1980; Galbraith *et al.*, 1980). When localized using fluorescein-conjugated anti-transferrin, the transferrin receptors are found predominantly on the apical, maternally disposed surface of syncytiotrophoblasts (King, 1976; Faulk and Johnson, 1977; Johnson and Faulk, 1978). In organ culture this pattern persists for the first 2 days in culture and then diminishes rapidly thereafter, suggesting that the cells in culture lose their receptor (Gillian *et al.*, 1980).

In the syncytiotrophoblast, the concentration of receptor, as determined by radioimmunoassay, is about 35 μg/mg of membrane protein (Enns *et al.*, 1981), the highest of any reported cell and a quantity consistent with the fact that the fetus requires up to 300 mg of iron during gestation (Pribilla *et al.*, 1958). The receptor has an apparent affinity for transferrin of about 2×10^7 M^{-1} either on membrane vesicles or when solubilized with Triton X-100 (Wada *et al.*, 1979). The receptor does not bind apotransferrin (Sussman, 1982).

The receptor has been purified by immunoprecipitation with anti-receptor antibody. When analyzed on sodium dodecyl sulfate (SDS)/polyacrylamide gels, this receptor glycoprotein is found to have a mol.wt. of 94 000 (Seligman *et al.*, 1979; Wada *et al.*, 1979; Enns *et al.*, 1981; Enns and Sussman, 1981a,b). However, when analyzed by gel-exclusion chromatography, the transferrin-free receptor is calculated to have a mol.wt. of 213 000 which increases to 364 000 upon incubation with transferrin. This is consistent with the hypothesis that the active form of the receptor is a dimer capable of binding two transferrins (Enns and Sussman, 1981a).

Recently transferrin has been identified in coated vesicles isolated from human placenta (Booth and Wilson, 1981; Pearse, 1982). These results, and the work in other systems, give credence to the hypothesis that transferrin–receptor complexes are internalized by the coated-pit–coated-vesicle mechanism.

9.3 MAMMALIAN YOLK SAC

In rodents and rabbits both the chorioallantois and the yolk sac placentas are engaged in maternal fetal transport. These tissues differ in that the chorioallantois is juxtaposed to the maternal circulation and is the site of most maternal fetal exchange of small molecules, whereas the yolk sac is the site of IgG transport. This was initially shown in the rabbit by ligating only the vitelline vessels which serve the yolk sac but leaving the embryonic circulation through the chorioallantoic placenta uninterrupted (Fig. 9.2) (Brambell *et al.*, 1948; Brambell, 1958). Such ligature completely eliminated the transfer of maternal IgG into the fetal circulation. Similar observations have also been made in the guinea pig (Barnes, 1959).

An important aspect of the structure of the yolk sac is that it does not

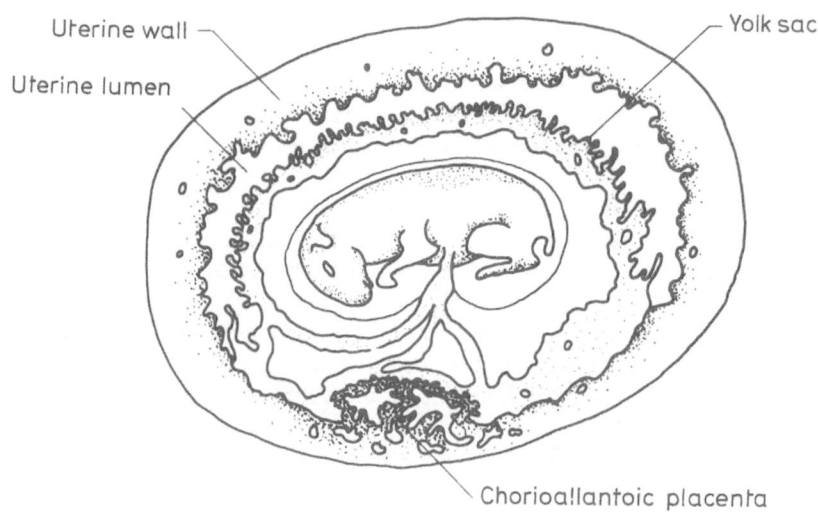

Fig. 9.2 A drawing of a mid-term fetal rabbit showing the organization of the extraembryonic tissues within the uterus. The fetal rabbit is nourished by both a chorioallantoic placenta where the syncytiotrophoblasts separate the maternal and fetal circulation as well as by a yolk sac which transports IgG from the uterine lumen into the fetal circulation. As yet, there is no information as to the cellular route by which maternal proteins are transported from the circulation into the uterine lumen.

directly contact maternal tissues or circulation. Rather, it begins as a closed sac which after undergoing selective degradation becomes exposed to the uterine lumenal contents (Fig. 9.2). Thus, it was inferred that the yolk sac must transport maternal IgG present in the uterine lumen into the embryonic circulation. This was demonstrated directly by injecting specific antibodies into the lumen and observing their appearance in the embryonic circulation (Brambell *et al.*, 1949). Non-IgG proteins which were injected as controls also appeared in the embryonic circulation, but at much reduced concentrations.

The mammalian yolk sac consists of a monolayer of columnar endothelial cells joined at their apical ends by tight junctions that prevent the diffusion of macromolecules between the cells (Fig. 9.3) (Morris, 1950). The apical surface of each cell is endowed with numerous microvilli, between which are many coated pits and canaliculi. The apical cytoplasm contains a large number of coated vesicles, vacuoles and canaliculi (Slade, 1970; Wild, 1976). The cells rest on a basement membrane beneath which is a loose mesenchymal connective tissue containing blood islands and fetal blood vessels.

Electron microscopic examination utilizing labeled ligands revealed that IgG is transported into the apical cytoplasm of the rabbit yolk sac endoderm

Uterine lumen

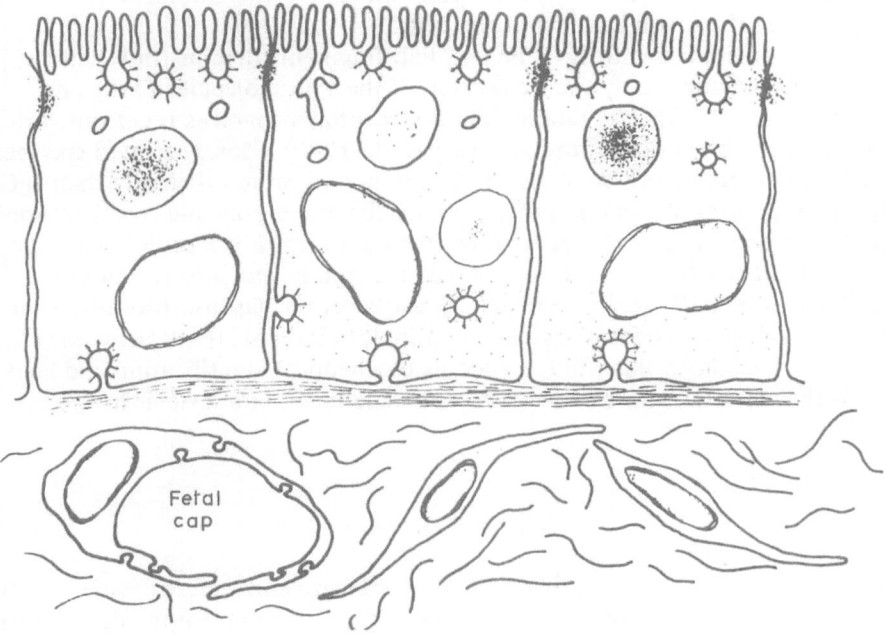

Fetal cap

Fig. 9.3 Diagram of a portion of a villous of the rabbit yolk sac. IgG present in the uterine lumen is bound to receptors located on the apical cell surface and endocytosed by coated pits and vesicles. The resulting endocytic vesicles ultimately release their contents into the basal extracellular space from which they diffuse into the fetal circulation.

via micropinocytic vesicles (Sonada *et al.*, 1973). Further studies indicate that the ligand first becomes associated with coated pits at the apical border of the cells (Slade, 1975; Moxen *et al.*, 1976) and is then transported within endocytic coated vesicles to the basal plasmalemma where it is released into the extracellular space (Moxen *et al.*, 1976). Similar observations have also been made in the guinea pig (King, 1977, 1982).

The biochemistry of yolk sac transport of maternal IgG has been extensively studied in the rabbit by Schlamowitz and co-workers (1975). Their early work demonstrated that ^{125}I-IgG was specifically sequestered by the yolk sac both *in vivo* and *in vitro* (Sonada and Schlamowitz, 1972a,b). More recently, they have shown that rabbit fluorescein-conjugated IgG binds to membrane vesicles prepared from yolk sac endodermal cells and to brush borders isolated from these cells (Schlamowitz, 1975; Schlamowitz *et al.*, 1975). Bovine fluorescein-conjugated IgG fails to bind to either. These results demonstrate both that the interaction between IgG and the membranes is selective and that the receptors mediating this interaction are exposed on the brush borders of the cells.

Inhibition studies with IgG, Fc and Fab fragments demonstrate that the receptor recognizes only the Fc portion of the IgG molecules (Tsay and Schlamowitz, 1978). The rabbit yolk sac receptor recognizes IgG from various species with various affinities (Tsay *et al.*, 1980). Closely related species, such as the guinea pig, mouse and rat, are bound more efficiently than IgG from more distantly related species, such as the horse, goat and cow. Enzyme pretreatments of the membrane show that the receptor is a protein and that associated carbohydrate plays no detectable role in receptor recognition (Hillman *et al.*, 1977). The rabbit yolk sac receptor has also recently been prepared in a soluble form using the non-ionic detergent NP-40 (Cobbs *et al.*, 1980). Presumably, this will facilitate its biochemical identification and lead to further insights into its role in receptor-mediated endocytosis in this system.

9.4 MAMMARY GLAND

The mammary glands serve to segregate, synthesize and secrete a wide variety of molecules which serve as an essential source of nutrients for the developing young. They also act as a selective filtration barrier that determines which maternal serum proteins will be transmitted between the maternal circulation and the offspring. As part of this function, the mammary gland synthesizes a particular protein called secretory component (SC) that mediates the transport of maternal IgA from the serum into the milk.

It is the early milk or colostrum of the mother that provides the newborn in many species with the principal or, in some cases, the only source of maternal immunoglobulins. For instance, in cows and pigs the newborn derives virtually all of its initial supply of antibodies from the colostrum during the initial few hours after birth. In contrast, rodents and carnivores obtain their humoral antibodies from both transplacental transport and from the milk during the first few days of life (Brambell, 1970).

Structurally, the lactating mammary gland consists of a highly vascularized constellation of separate glands joined by lactiferous ducts to the nipple. Each gland is a typical exocrine gland in which groups of cells form alveoli. These exocrine cells are structurally similar to other secretory cells in that they contain a basal zone of rough ER and a more apical region containing a well-developed Golgi apparatus with associated secretory granules (Sekhri *et al.*, 1967; Murad, 1970; Cowie and Tindal, 1971). At the basal and apical surfaces of these cells, coated pits and vesicles are located along the membrane and in the cortex (Fig. 9.4). What function these may serve in immuno-

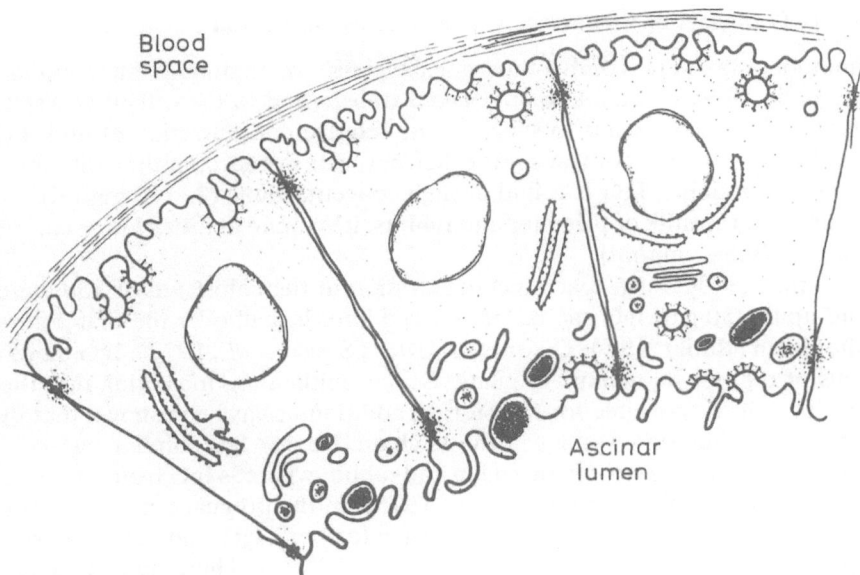

Fig. 9.4 This drawing illustrates the transport cells of the mammary acsinus. Maternal serum proteins in the blood space bind to receptors on the basal surface of the acsinar cells. These cells then transport the bound proteins through the cells releasing them into ascinar lumen. In addition, these cells also synthesize and secrete into the milk a variety of milk proteins including casein and lactalbumin.

globulin transport is not yet known; however, they have been implicated in the secretion of caseins into the milk (Kraehenbuhl *et al*., 1979; Franke *et al*., 1976). The secretory cells are joined to each other and to the duct cells by tight junctions to form the only intact barrier to small molecules that lies between the maternal blood and the milk (Pitelka *et al*., 1973). There appears to be only one class of exocrine cells, although this has not been confirmed rigorously by cytochemical criteria. It has been demonstrated in the rabbit that these cells both synthesize and secrete milk caseins as well as SC, a membrane protein that appears to mediate IgA binding and transport (Kuhn and Kraehenbuhl, 1982; Kraehenbuhl, 1981).

In different species the mammary glands differ not only in the proteins they transport, but also in the relative amounts they transport. The reader is referred to an excellent review that covers much of the earlier work on mammary transport (Kraehenbuhl *et al*., 1979). What is clear from this work is that the proteins secreted by the mammary include a number of maternal serum proteins, including albumin and immunoglobulins G and A. The other principal milk proteins are synthesized and secreted by the mammary and include casein, lactalbumin and lactoferrin.

9.4.1 Transport of IgG across the mammary epithelium

Immunoglobulin G is one of the principal classes of immunoglobulin found in the milk. However, the amount of IgG transmitted to the offspring by the milk varies from species to species. The relative concentration of milk IgGs can be a crude index of transport efficiency. For instance, in the milk of pigs, cows and rodents, IgG is found at high concentrations (2–150 mg/ml), whereas in the milk of primates and rabbits, it is found at much lower concentrations (0.5–2 mg/ml).

In the cow IgG can reach a concentration in the colostrum of 150 mg/ml and up to 250 g of IgG can be transferred into the calf over the first 2 days after birth (Butler, 1974; Dixon *et al.*, 1961; Sasaki *et al.*, 1977). IgG binds to cells of the bovine mammary gland ascinar epithelium, indicating that these are the cells responsible for transfer. In addition, it has been shown that the specificity of binding resides in the H chain. Bovine IgG binding can be inhibited by IgG from human, sheep and rabbit, whereas IgG from turtle and chicken do not inhibit (Kemler *et al.*, 1975). In the pregnant cow, the mammary cells have independent binding sites for both IgG1 and IgG2 with apparent affinity constants of 5×10^8–10×10^8 M^{-1}. There appear to be 9000 sites per cell for IgG1 and 3000 for IgG2. About 2 weeks before parturition, about 5000 new sites for IgG1 appear with a K_a of about 4.5×10^9 M^{-1} (Sasaki *et al.*, 1977).

In the mouse, approximately 30% of the maternal IgG passes into the milk each day. However, not all mouse IgGs were equally well transported, as shown by the observation that some myeloma IgGs were poorly transported (Gitlin *et al.*, 1976; Halsey *et al.*, 1980). The rat mammary also transports IgG (Jordan and Morgan, 1967). IgG antibodies have been detected in the human colostrum but are less prominent than IgA (Ahlstedt *et al.*, 1975; Sewell *et al.*, 1979). No detailed analysis of the transport rates and kinetic parameters of IgG transport across the human mammary is available.

Human colostrum also contain IgM and IgD (Ahlstedt *et al.*, 1975; Sewell *et al.*, 1979). Both are found in relatively low concentration with IgD ranging from 1 to 50 μg/ml in the colostrum. Their presence in the colostrum of the human raises the possibility of both being transmitted at low levels into the serum of the neonate. At present no studies detail the cellular transport of these immunoglobulins across the mammary tissue in the human.

9.4.2 Transport of IgA across the mammary epithelium

The principal humoral mediator of immunity in mucosal surfaces of all higher animals is IgA (Halsey *et al.*, 1980). IgA is found in tears, saliva and secretions in the lungs and gut. It is typically an 11 S dimer composed of two IgA molecules joined at their Fc regions by a J chain. IgA in mucosal secretions also contains another polypeptide, secretory component (SC), which is absent from IgA in the serum. Suckling neonates receive large quantities of maternal IgA from the milk and especially the colostrum where it is present in concentrations as high as 40 mg/ml (Kraehenbuhl *et al.*, 1979). Most of this orally transferred IgA is not adsorbed and retained in the serum, but rather provides a passive secretory immunity in the gut.

The selective transport of IgA from the maternal serum to the milk is mediated by SC. That serum IgA binds to SC present on the basolateral surface of the mammary epithelium is suggested by the observation that Fab fragments of anti-SC antibodies block this interaction (Kuhn and Kraehenbuhl, 1979). Goat anti-(rabbit SC) conjugated to Sepharose removes from detergent extracts of rabbit mammary epithelium four polypeptides, a doublet M_r = 120 000 and 116 000 and a second doublet of M_r = 96 000 and 91 000, all considerably larger than SC found in the milk but sharing collectively the size heterogeneity of secreted SC (58 000–83 000 daltons; Kuhn and Kraehenbuhl, 1981). Further, immunoprecipitation with anti-SC serum of the *in vivo* translation products of mRNA from rabbit mammary epithelial cells yields four similar polypeptides that can be core-glycosylated and integrated into dog pancreas microsomes *in vitro* (Mostov *et al.*, 1980). Taken together, these observations suggest that SC is synthesized by mammary epithelial cells as a set of four related, transmembrane glycoproteins that are expressed on the basolateral membrane. Upon association with serum dimeric IgA, the IgA–membrane–SC complex is internalized, possibly in coated vesicles, and transported to the lumenal surface of the cell where SC is cleaved to yield a soluble IgA–SC complex in the milk. In this view, SC is an unusual cell surface receptor in that it functions once and is cleaved in the process. Thus, IgA transport and SC synthesis must be tightly linked. Similar mechanisms may operate in the intestinal epithelium and the liver where large quantities of serum IgA are transported into the gut lumen and bile respectively.

9.5 TRANSPORT ACROSS THE INTESTINE

The small intestine serves to transport maternal immunoglobulins from the maternal milk and/or colostrum into the neonatal circulation in a variety of species (Brambell, 1970). In rodents this tissue continues the transport process of acquiring passive immunity that was previously begun by the placenta during embryonic development. In other species such as the cow, pig, horse and wallaby, the intestine provides the principal route by which these animals obtain maternal IgG (Brambell, 1970). Transport is demonstrated by the appearance of maternal IgG in the serum of the neonate after suckling.

In the neonatal mouse and rat, IgG transport across the intestine is restricted to the duodenum and proximal region of the jejunum (Rodewald, 1970, 1976a, 1980; MacKenzie, 1972; Morris and Morris, 1974, 1976; Morris, 1975; Waldman and Jones, 1976; Jones, 1976). In the neonate, the absorptive epithelial cells of this region which transport IgG are morphologically similar to adult cells, but with several important differences. In neonatal cells there are numerous invaginations and tubules at the base of the brush border. The apical cytoplasm of these cells contains many irregularly shaped tubular vesicles and 100–200 nm-diameter coated vesicles (Fig. 9.5). Both of these vesicle types appear to have a clathrin coating on their cytoplasmic surfaces (Clark, 1959; Rodewald, 1973).

Distal to the section of the small intestine which carries out IgG transport,

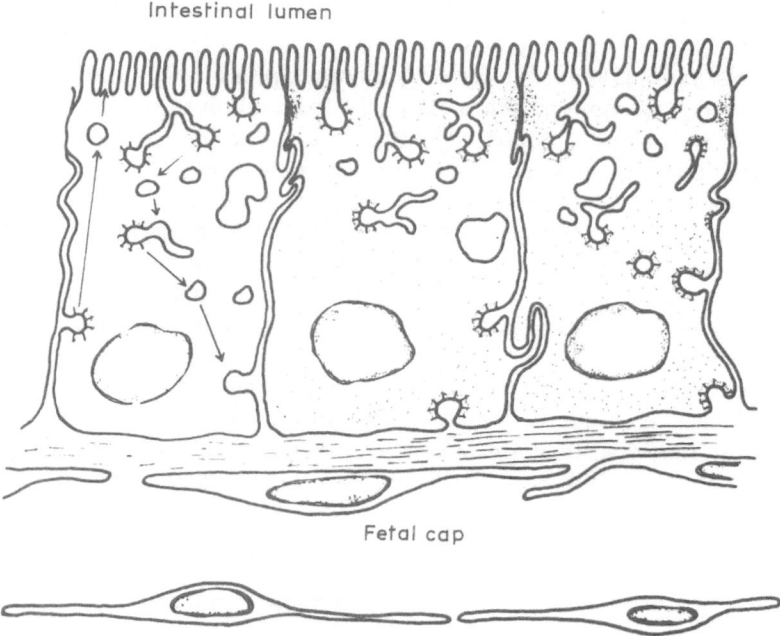

Intestinal lumen

Fetal cap

Fig. 9.5 Diagrammatic representation of a portion of a villus from the proximal portion of a neonatal rat intestine. Receptors located on the apical membranes bind the maternal IgG found in the intestinal lumen and mediate its endocytic uptake into the intestinal epithelium cells. Following initial internalization, the endocytic vesicles apparently fuse with intracellular membrane structures where sorting of receptor-bound and non-specifically entrapped contents takes place. After sorting, receptor-bound IgG is transferred by another membrane vesicle to the basolateral plasma membrane where the extracellular pH of 7.5 mediates the release of bound IgG from the receptors. This IgG is subsequently transported throughout the embryo by the fetal circulation.

the cells lack the specialized coated regions and deep tubular vesicles at the base of the microvillae associated with IgG transport. In addition, these distal cells are phagocytically very active, degrading engulfed material in lysosomes with no detectable transport of intact proteins into the neonatal circulation (Grancy, 1968; Hugan, 1971; Rodewald, 1973).

The morphological route used to transport IgG across the neonatal rat intestine has been followed using ferritin-conjugated IgG and HRP-conjugated IgG. In the proximal regions of the small intestine, direct binding studies with IgG or Fc fragments illustrated that IgG first binds to the microvilli and at the small tubular invaginations between microvilli. Binding is more concentrated on cells in the apical half of the villus with less binding toward the villus crypts. Within 15 minutes the IgG becomes localized in the tubular vesicles and vacuoles deeper in the apical cytoplasm. In addition, IgG was noted in vesicles close by the Golgi complex and near the lateral cell margins. Thirty minutes after administration, the conjugate is noted in many more vesicles in the cytoplasm and between cells near coated pits at the lateral surfaces. Both Fc fragments and IgG showed this pattern of localization; however, the Fab fragments did not bind. With time, the Fab became localized only in a few apical vesicles which bore a resemblance to lysosomes. From these elegant studies, Rodewald proposed that the route of IgG transport involved: (1) binding to surface receptors, (2) internalization via specialized tubular vesicles which were frequently coated on their cytoplasmic face, and (3) sorting of IgG from heterologous proteins in the region of the Golgi where coated vesicles ultimately bud from these vesicles to deliver IgG to the abluminal surface (Rodewald, 1970, 1973, 1976a,b, 1980).

Release of IgG from the receptor at the abluminal surface appears to be regulated by pH. IgG binds to both luminal and abluminal surface receptors at pH 6.0 but not at pH 7.4 (Rodewald, 1976a; Wild and Richardson, 1979). Since the gut is at pH 6.0 and the blood at pH 7.4, this pH-dependence provides a ready mechanism for loading and unloading at opposite surfaces of the gut epithelial cells (Rodewald, 1976a).

Recycling of membrane and presumably IgG receptors from the abluminal to luminal surface was monitored by injecting HRP into the circulation. The HRP was taken into the cells via coated pits on the basolateral margins which, in turn, gave rise to tubular vesicles. No HRP was seen in the Golgi region but with time it was found in large apical vacuoles (Rodewald, 1980). Morphological observations on the transport of immunoglobulins across the intestine in other species are rudimentary in that tracer studies are lacking. To date the data on the rat gut are the most complete and compelling.

Direct biochemical evidence for the existence of receptors for IgG in the neonatal gut exists for the mouse (Guyer *et al.*, 1976) and the rat. Brambell *et al.* (1958) postulated that receptors exist for IgG on the epithelial cells of the gut to account for the highly selective protein transport observed in rats and mice. Subsequently, unlabeled rat IgG was shown to compete with iodinated rat IgG for transfer into the serum of the neonate which supported the presence of receptors (Jones and Waldman, 1972; Waldman and Jones, 1973). Further studies by Borthistle *et al.* (1977) showed specific binding of IgG to cells of the proximal small intestine. However, the most definitive evidence for IgG receptors in the rodent gut is that of Wallace and Rees (1980). In this study brush border membranes were prepared and binding parameters established, which demonstrate that only IgG and the Fc fragment of IgG bind specifically. Both equilibrium and kinetic methods show that there are receptors with a high-affinity site (K_a 1.3×10^8 M^{-1}) and a low-affinity site (K_a 5×10^6 M^{-1}), a pH optimum of 6.0 and an accelerated off rate in the presence of excess IgG. Biochemical evidence for IgG receptors in the neonatal gut of the piglet (Leary and Lecce, 1979) and cow (Husband *et al.*, 1978) has also been presented.

Immunoglobulin other than IgG appears to be transported across intestinal epithelia in some animals. For instance, IgM and IgA are markedly elevated in the serum of the calf immediately after suckling. However, the IgA remains elevated for only 4 days. Its loss from the serum of the neonate could be due to secretion of IgA across the various mucous membranes. Although IgG is not transmitted across the gut in the human (Husband *et al.*, 1978), IgA can be absorbed from the colostrum if fed during the first 24 hours after birth; thereafter, transport does not occur. In these experiments infants were fed colostrum rich in anti-(poliovirus IgA) antibody (Ogra *et al.*, 1977). Earlier work by Mohr (1973) had demonstrated transfer of specific immunity against tuberculosis after breast feeding with tuberculin-positive mothers. Thus, evidence is beginning to appear which suggests that different immunoglobulins can be transported across the gut at early times after birth. Many of the data are rudimentary, since in most species the full complement of immunoglobulins is yet to be used to probe for specific transport.

9.6 TRANSPORT ACROSS THE OOCYTE PLASMA MEMBRANE

Oocytes in most animals selectively store from the maternal circulation a variety of proteins that will, after fertilization, provide the developing embryo with the amino acids, sugars and lipids needed for embryonic growth. Since this transport occurs before fertilization, the oocyte plasma membrane is the initial site of selection between the maternal circulation and the embryo. Vitellogenin, a female-specific phospholipoglycoprotein, is the principal yolk protein. It is sequestered into the oocyte after being synthesized in the maternal liver or equivalent organ. However, it is not the only protein transported from the circulation into the egg. In some species such as birds, maternal immunoglobulin G is also transported into the oocyte and provides the hatchling with the panoply of circulating IgGs that are common to the mother. Other serum proteins such as transferrin, transcobalamin, low-density lipoprotein and very-low-density lipoproteins are transported into the avian oocyte.

The process of maternal–fetal transport across the oocyte plasma membrane has been most intensively studied in birds, amphibians and insects. It may also occur in egg-laying mammals but is insignificant in most mammals where the placenta, mammary and neonatal gut provide alternative routes.

9.6.1 Insect oocytes

The process by which the developing insect oocyte accumulates yolk was first described by Telfer (1961, 1965). He observed the distribution within the ovary of the moth of fluorescein-conjugated antibodies to various blood proteins. These studies demonstrated that the oocytes take up proteins selectively from the blood. Antibodies to blood proteins were first observed in the intercellular spaces between the follicle cells, then in association with the brush border at the oocyte surface, and finally within the oocyte in the yolk spheres. Biochemical analysis of the yolk spheres confirmed that the blood proteins accumulated within these structures. Telfer (1961) proposed, '. . . that blood proteins reach the surface of the oocyte by an intracellular route, that they combine with some component of the brush border, and that they are transformed into yolk spheres by a process akin to pinocytosis.'

The fine structure of the process by which blood proteins are transported into the developing mosquito oocyte was described in detail by Roth and Porter (1962, 1964). During the rapid phase of oocyte maturation the oocytes rapidly sequester vitellogenin, a 500 000-mol.wt. blood protein (Atlas *et al.*, 1978). This protein percolates between the follicle cells which overlay the oocyte thus diffusing from the hemolymph to the surface of the oocyte. Roth and Porter reported that the number of coated pits on the oocyte surface increased 15-fold during the deposition of protein yolk in the oocyte. Such coated pits are 140 nm pit-like depressions on the oocyte surface formed by invagination of the surface membrane. Their cytoplasmic surface is characterized by a 20 nm bristle-like coat (Fig. 9.6). A layer of protein which is apparently adherent on the concave extracellular side is interpreted to be the vitellogenin selectively absorbed from the extraoocyte space. The coated pits appear to pinch off the cell membrane to become coated vesicles carrying the absorbed protein into the oocyte. After losing their coats, the vesicles appear to fuse to form small paracrystalline yolk droplets, which subsequently coalesce to form the large proteid yolk bodies of the mature oocyte. Similar observations on the involvement of coated pits and coated vesicles in vitellogenesis have been reported in other insect species, including the moth (Stay, 1965), cockroach (Anderson, 1969) and the fruit fly (Georgi and Jacob,

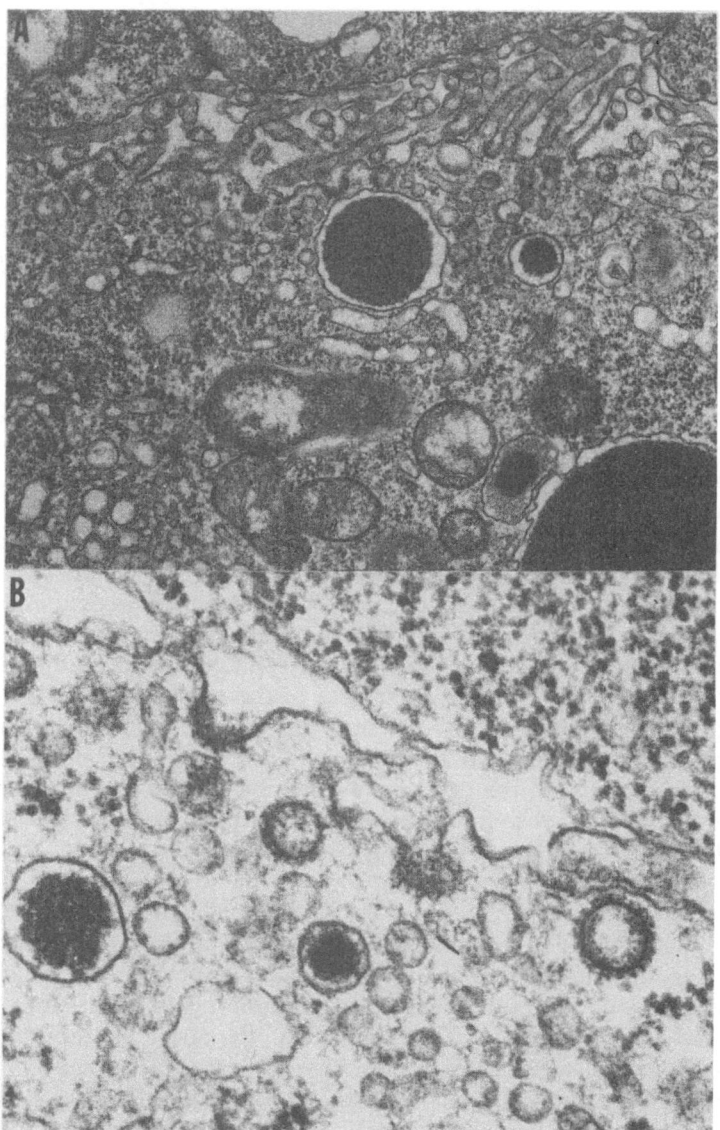

Fig. 9.6 Electron micrograph of the surface of the developing mosquito oocyte.
(A) The plasma membrane is very irregular owing to the many microvilli and
numerous coated pits located along the base of the microvilli. Coated pits give rise to
the many coated vesicles (CVs) present in the cortical cytoplasm adjacent to the
plasma membrane. Note the dense contents of the CVs and resultant smooth surfaced
vesicles. Sequential fusion of these smooth vesicles results in increasingly larger
intermediate sizes of presumptive yolk granules deeper in the cytoplasm. (B) Detailed
view of the plasma membrane showing coated pits with adherent dense extracellular
material, coated vesicles with membrane-bound content and uncoated smooth vesicles
in which the dense material has been released from the membrane and is condensed in
the vesicle lumen.

1977). This literature has been reviewed by Telfer (1965) and Engelmann (1979).

Numerous lines of biochemical evidence demonstrate that vitellogenin is selectively transported into the developing insect ovary. A number of studies have shown that of the proteins present in the female hemolymph, only vitellogenin is sequestered and concentrated in the developing oocyte (Engelmann, 1979). The binding and uptake of vitellogenin has been studied directly using radiolabelled molecules. Typical of this type of study is the report by Kunkel and Pan (1976). They isolated vitellogenin from two insect species, *Hyalophora cecropia* and *Blattella germanica*, and compared their rates of uptake. They found that *Blattella* oocytes take up their own vitellogenin at a rapid rate and that of *Hyalophora* at a lower rate which is comparable to that of other non-vitellogenin proteins. Similarly, *Hyalophora* oocytes take up their own vitellogenin rapidly and *Blattella* vitellogenin only at a negligible rate. Related experiments, which led to similar results, were carried out with a series of 18 different species of cockroach (Bell, 1972). When ovaries were transplanted between the various animals, it was found that host yolk was deposited principally in those ovaries incubated in members of the same subfamily. The results of these two experiments strongly suggest that the oocytes of each species must have receptors on their surfaces that recognize and bind principally the homologous vitellogenin.

In the mosquito the uptake of vitellogenin can be prevented by a wide variety of inhibitors of glycolysis. In addition, there appears to be a sensitive sulfhydryl which, if reacted with iodoacetamide, prevents uptake at concentrations that have little effect on glycolysis (Roth and Jackson, 1972).

9.6.2 Avian oocytes

The avian oocyte performs one of the most dramatic examples of receptor-mediated endocytosis. Within a period of 7 days, the oocyte of the chicken increases in diameter from 8 to 37 mm and from 0.08 g to 14 g in weight. It is not possible to explain such a dramatic growth rate for a simple cell on the basis of *in situ* synthesis alone. Rather, this rapid growth results from the selective uptake of yolk precursors from the circulation (Schjeide *et al.*, 1963, 1970; Cutting and Roth, 1973; Roth *et al.*, 1976; Perry *et al.*, 1978). Vitellogenin, the precursor of the major yolk proteins phosvitin and lipovitellin, is synthesized in the liver, carried by the circulation and accumulated in the developing oocyte (Chagaff, 1942; Greengard *et al.*, 1965; Heald and McLachlan, 1965; Deeley *et al.*, 1975; Christmann *et al.*, 1977). As might be expected, vitellogenin is not found in the circulation of roosters or non-laying hens (Bergink *et al.*, 1974).

The ovarian follicle of the chicken facilitates the transport of maternal serum proteins to the growing oocyte and provides mechanical support for this exceptionally large cell (Perry *et al.*, 1978; Bellairs, 1965). The follicles, suspended from the dorsal body wall, each contain an oocyte surrounded by many layers of cells. The outermost layer of connective tissue is a highly vascularized extension of the follicle stalk that overlays the thecal layer. The thecum is comprised of sheets of fibroblasts separated by wide intercellular spaces. These layers are also traversed by numerous blood islands and nerves. Interior to the thecal layers, a single layer of cuboidal granulosa cells (follicle cells) envelops the maturing oocyte. This layer of cells is covered by an extracellular base lamella and is separated from the oocyte by a coarsely fibrous perivitelline layer. Intercellular spaces between the cells of all of these layers permit macromolecules to diffuse from the circulation to the surface of the developing oocyte. During the stages of maximum growth, the cells of the granulosa layer become somewhat separated from each other thus facilitating the rapid diffusion of macromolecules from the circulation to the oolemma (Perry *et al.*, 1978). Thus, the oocyte plasma membrane provides the only membrane barrier between the mother and her potential offspring.

An examination of the fine structure of the oocyte reveals that the oocyte plasma membrane has many deep enfoldings and tortuous crypts that greatly increase the surface area (Perry *et al.*, 1978a,b; Roth *et al.*, 1976). The highly convoluted oolemma is also characterized by numerous coated pits and coated vesicles (Fig. 9.7). Grazing sections through these surfaces show that the coat material is organized into the polyhedral network characteristic of clathrin-coated structures (Heuser, 1980). In addition to coated pits and vesicles, a large proportion of the total membrane area also appears to be coated. As much as 60% of the total membrane area can be coated at those stages of maximal growth (Zajac and Roth, unpublished observation). The cortical cytoplasm contains a large number of yolk granules of varying sizes, as well as a few mitochondria and Golgi bodies (Perry *et al.*, 1978). Unfortunately, even in the best preserved specimens the cytoplasmic contents are often badly extracted, thus making it difficult to visualize cytoplasmic organelles except those closest to the oolemma.

In addition to the morphological evidence for endocytic uptake, there are a variety of biochemical experiments which demonstrate the existence of a number of different receptors on the oocyte membrane. Undoubtedly, these receptors mediate the specific uptake of molecules by receptor-mediated endocytosis. On the basis of the selective uptake of intravenously injected radiolabeled molecules by the oocyte, it has been suggested that the oolemma has receptors for vitellogenin (Cutting and Roth, 1973), IgG (Paterson *et al.*, 1962; Cutting and Roth, 1973), very-low-density lipoprotein (VLDL) (Holdsworth *et al.*, 1974) and biotin-binding protein (White *et al.*, 1976). In addition, it has been postulated that there must also be receptors for retinol-binding protein since the concentration in the yolk is elevated relative to the serum concentration (Heller, 1976). Further evidence for the existence of

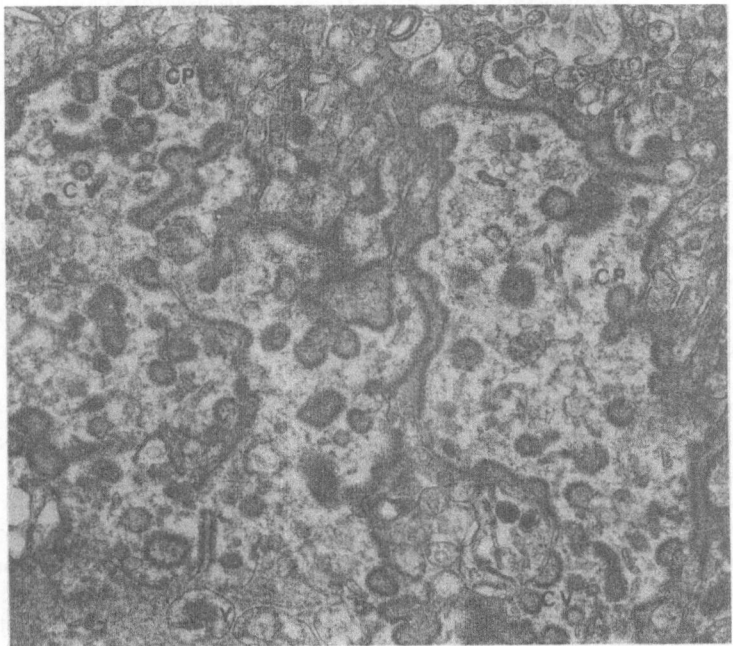

Fig. 9.7 Electron micrograph of the plasma membrane of a developing chicken oocyte. The oocyte surface, shown here in a somewhat oblique section, has many deep crypts and infoldings, as well as a large number of microvilli. Virtually the entire plasma membrane surface is coated on its cytoplasmic surface and the many coated pits (CP) appear to be budding off to form coated vesicles (CV). Large numbers of coated vesicles, smooth membrane vesicles and nascent yolk granules (Y) populate the cytoplasm.

certain of these receptors has been obtained from experiments that directly assayed the binding of radiolabeled proteins to isolated fragments of oocyte membranes. From such studies specific receptors for vitellogenin (Yusko and Roth, 1976; Woods and Roth, 1977, 1978; Yusko *et al.*, 1980), IgG (Roth *et al.*, 1976; Jackson *et al.*, 1983) and LDL and VLDL (Krumins and Roth, 1981) have been demonstrated. In general these receptors appear to bind their respective ligands with affinities that are commensurate with the ligands' concentration in the circulation. For example, the vitellogenin receptor binds its ligand with an apparent $K_d = 2$ μM (Woods and Roth, 1977). Thus, since the serum concentration of vitellogenin is 2 μM, at any particular moment 50% of the receptors should be occupied.

The vitellogenin receptor from the chicken oocyte has been extensively studied. Competition binding studies have shown that this receptor recognizes determinants on the phosvitin portion of the circulating vitellogenin molecule (Woods and Roth, 1981). After binding and endocytic uptake, vitellogenin is specifically processed to yield phosvitin and lipovitellin which are stored in the yolk granules until needed during embryogenesis. The vitellogenin receptor can be solubilized with non-ionic detergents and still retain its ability to specifically bind phosvitin (Woods and Roth, 1979). The solubilized receptor has been tentatively identified as a 116K-mol.wt. protein found in soluble extracts of isolated oocyte membranes (Woods and Roth, 1980). Enzymatic treatments of isolated oocyte membranes with pronase reduce the binding of phosvitin by 90% indicating that the receptor is a protein. Treatment of isolated membranes with fucosidase reduces phosvitin binding by 50% suggesting that this protein contains at least one fucose residue (Yusko *et al.*, 1981).

A number of investigators have applied a combined morphological–biochemical approach to study the process of receptor-mediated endocytic uptake across the oolemma. Coated vesicles have been isolated from developing chicken oocytes and their protein compositions studied (Woods *et al.*, 1978; Pearse, 1978). It has been suggested that at least one component found in these isolated coated vesicles may represent the protein component of lipovitellin, one of the processing products of vitellogenin (Woods *et al.*, 1978). However, this interpretation has been questioned (Pearse, 1978). In a separate type of experiment, the receptors for IgG were localized using IgG–ferritin conjugates (Roth *et al.*, 1976). In these experiments, ferritin-conjugated IgG binds to about 20% of the coated pits, whereas unconjugated ferritin did not appear to bind to any regions of the membrane. These results suggest that the IgG receptor is predominantly localized in the coated pits. In related experiments endogenous 27 nm particles, identified morphologically as VLDL, are observed to associate in large numbers in coated pits and coated vesicles (Perry and Gilbert, 1979). The clear implication from these observations is that VLDL binds first to the surface of the oolemma, accumulates in coated pits, and is internalized via coated vesicles.

9.6.3 Xenopus oocytes

Yolk formation in the amphibian *Xenopus laevis* has been extensively studied by Wallace and co-workers. As in most oocytes, the principal components of the mature amphibian oocyte are the yolk platelets formed during the later stages of oocyte maturation concomitant with an increase in oocyte volume (Grant, 1953; Kemp, 1953). This observation led Wallace and co-workers to investigate where the yolk proteins were made and how they were accumulated by the developing oocyte. Initially, they determined that the yolk platelets consisted of two proteins, phosvitin and lipovitellin (Wallace, 1963a,b), derived from the serum phospholipoglycoprotein, vitellogenin

(Bergink and Wallace, 1974). Subsequently, they established that this serum protein, vitellogenin, was found only in females with developing oocytes (Wallace and Dumont, 1968; Wallace, 1972; Wallace and Bergink, 1974) and not in females in which the oocytes are quiescent nor in males. However, vitellogenin synthesis could be induced in either males or non-vitellogenic females by estrogen (Wallace, 1972; Wallace and Bergink, 1974).

Serum vitellogenin was shown to be sequestered by developing oocytes (Wallace and Dumont, 1968; Wallace and Bergink, 1974). Females injected with sodium [^{32}P]phosphate incorporated it into newly synthesized vitellogenin which was, in turn, selectively transported into the developing oocytes. After uptake, vitellogenin was selectively processed to yield phosvitin and lipovitellin.

A concomitant study of the structure of the amphibian oocyte provided a morphological framework from which a further understanding of the transport process was obtained. In *Xenopus* the ovary is a hollow sac-like organ. Oogonia take up residence in the thick thecal tissue which forms the bulk of the organ. This highly vascularized layer consists primarily of fibroblasts and is lined on the inner lumen by the inner epithelium. Exterior to the thecal tissue is an outer epithelium. As the oogonia develop and enlarge, they protrude into the inner or ovarian lumen while still surrounded by thecal cells. Each oocyte is enveloped by a closely apposed acellular vitelline envelope of loose fibers and a single layer of flat, stellate follicle cells. Although the adjacent follicle cells appear to be attached to each other by desmosomal junctions, there are also large channels between neighboring cells which provide a route for molecules such as vitellogenin to diffuse to the oocyte surface (Fig. 9.8).

The surface of the oocyte plasma membrane changes as the oocyte matures (Dumont, 1972). Prior to the appearance of yolk platelets, the surface is fairly smooth with few microvilli and no apparent coated pits and vesicles. Trypan Blue is excluded by oocytes at this stage indicating that there is little or no endocytic activity. However, as the first yolk platelets appear in the peripheral cytoplasm, the oocyte microvilli increase in number and micropinocytic pits and vesicles become more prevalent. During the stages of maximum yolk accumulation, the surface has numerous microvilli and an abundance of coated pits. At this stage Trypan Blue is readily sequestered. In contrast, in the fully mature oocyte, no additional yolk is accumulated, few microvilli are seen, no coated pits are noted, and once again Trypan Blue is no longer sequestered. In addition, a thick extraoocyte layer occludes the surface (Dumont, 1978). It is also observed that during the rapid oocyte growth the endocytic vesicles in the cortex appear to fuse to form primordial yolk platelets which, in turn, fuse to form mature yolk platelets (Wallace and Dumont, 1972).

Concurrently, a number of biochemical studies to determine how oocytes

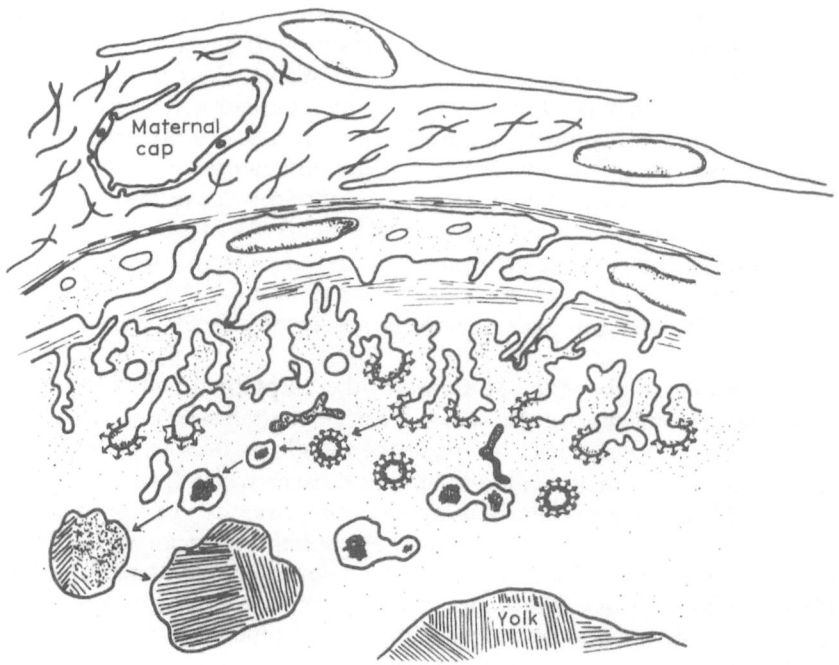

Fig. 9.8 Diagram of a portion of a *Xenopus laevis* oocyte. Maternal serum proteins, including vitellogenin, percolate between the monolayer of follicle cells which overlay the oocyte surface. Receptors localized in the oocyte plasma membrane specifically bind certain maternal serum proteins and mediate their uptake by coated pits and coated vesicles. The resultant endocytic vesicles transfer their contents to primordial yolk bodies, which ultimately form mature yolk granules.

accumulate yolk were carried out. These experiments were simplified by the ability to culture oocytes *in vitro*. These oocytes retained the ability to internalize proteins at a constant rate for 24 h (Wallace *et al.*, 1973). Recent advances have extended the viable time in culture to 28 days (Wallace and Misulovin, 1978).

Uptake experiments using ^{125}I-labelled ligands demonstrated that vitellogenin is selectively transported into isolated oocytes with a K_m of 1.5×10^{-6} M (Wallace and Jared, 1976). Bovine serum albumin (BSA), at concentrations up to 10 mg/ml, did not appear to inhibit uptake. BSA alone is incorporated with a much lower affinity into isolated oocytes. However, in the presence of 2 mg of vitellogenin/ml, BSA incorporation is promoted by about 40%. Thus, it appears that BSA is most likely incorporated by adventitious engulfment during the receptor-mediated micropinocytic uptake of vitellogenin (Wallace and Jared, 1976). These results demonstrate that vitellogenin uptake must be mediated by a specific receptor which selects particular extra-cellular proteins for endocytic uptake.

Further work has shown that once endocytosed, vitellogenin is processed and stored in the yolk platelets. The fate of sequestered vitellogenin was compared to microinjected vitellogenin. Only sequestered vitellogenin was converted to phosvitin and lipovitellin and stored. In contrast, the microinjected vitellogenin is degraded (Dehn and Wallace, 1973). Thus, the processing to phosvitin and lipovitellin appears to take place within an endocytic membrane compartment since microinjected vitellogenin is simply degraded. Further insight into the mechanism of vitellogenin incorporation into yolk platelets is obtained (Opresko *et al.*, 1980) when oocytes are first incubated for 15 min with labeled protein and then cultured for various periods of time. The conversion of labeled vitellogenin into phosvitin and lipovitellin is assayed by first homogenizing and fractionating the oocytes on a sucrose gradient and then analyzing the resultant fractions by SDS/polyacrylamide gel electrophoresis. The fractions are also characterized by electron microscopy. Within 15 minutes, the vitellogenin is incorporated into a membrane compartment, most probably derived from the endocytic vesicles. After 1 hour, the vitellogenin is incorporated into an intermediate structure the authors term a 'transitional yolk body' (TYB) where it is processed to yield phosvitin and lipovitellin. Within 2 h the phosvitin and lipovitellin are found in yolk platelets where they remain intact until embryonic development begins.

If ^{125}I-BSA is incubated with oocytes for 2 h in the absence of extracellular vitellogenin, labeled BSA sediments in the same fraction as the yolk platelets but not with the TYB. After the same 2 hour incubation period in the presence of extracellular vitellogenin, labeled BSA appeared in both the TYB and the yolk platelets. Co-internalization experiments using both ^{125}I-BSA and ^{32}P-vitellogenin showed that after 1.5 h both ligands were located in the same relative amounts in both the TYB and the yolk platelets. From these results the authors hypothesize that subsequent to endocytosis, sorting takes place. If the receptors in the endocytic vesicles are occupied by vitellogenin, the vesicle is directed to the TYB compartment where vitellogenin is processed before fusion with the forming yolk body. However, if the vitellogenin receptors are unoccupied, the endocytic vesicles fuse directly with the forming yolk platelets.

9.7 TRANSPORT BY THE AVIAN YOLK SAC

The avian yolk sac is an embryonic tissue which, during development, comes to envelop the yolk. Its primary function is to transport nutritive materials from the yolk into the circulation of the developing embryo. In addition, it also serves to specifically transport maternal IgG from the yolk into the embryonic circulation. In doing so, it carries out the final step in the transport of the maternal IgG which began when maternal IgG was sequestered from the maternal circulation into the developing oocyte.

The early experimental evidence for this process came from a number of studies which introduced specific antibodies into the yolk and subsequently assayed the level of circulating antibody in the embryo. Such investigations have shown that only IgG is transported intact from the yolk into the embryonic circulation (Buxton, 1952; Brierley and Hemmings, 1956; Kramer and Cho, 1970; Heller, 1975). These studies also suggested that this transport process is selective since although the injection of either immune pigeon and chicken serum into the chicken yolk would give rise to circulating antibody in the embryo, the levels of activity of the pigeon antibodies were consistently lower. This selectivity is also demonstrated by the observation that the injection of immune rabbit, cow or horse serum into the yolk did not result in any

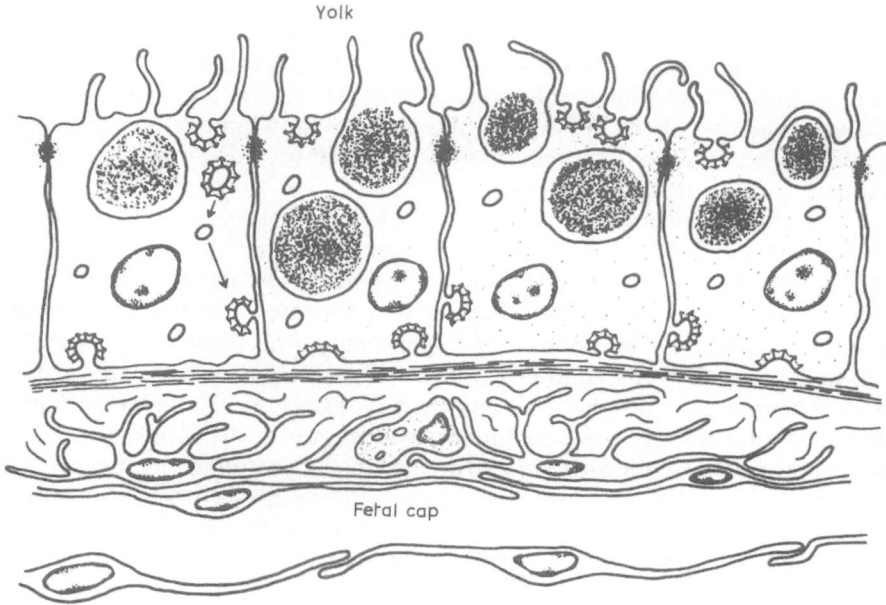

Fig. 9.9 Diagram of a portion of chick yolk sac. Receptors on the apical plasma membrane bind IgG found in the yolk. Coated pits–coated vesicles mediate IgG uptake into the epithelial cells of the yolk sac. The resulting vesicles ultimately release their contents into the basolateral extracellular space from which the IgG is free to diffuse into the fetal circulation. The epithelial cells also sequester large amounts of yolk by non-specific phagocytosis. This material is broken down and released into the circulation.

detectable antibody activity in the embryonic circulation (Brierley and Hemmings, 1956). Evidence for Ig isotype selectivity was obtained from subsequent experiments where a laying hen was hyperimmunized with multiple injections of *E. coli* (Heller, 1975). Although the hen produced both anti-(*E. coli* IgG) and -(IgM), only the anti-(*E. coli* IgG) was found in the yolks of the subsequently laid eggs. When these eggs were allowed to develop, anti-(*E. coli* IgG) was present in the circulation of the resultant embryos.

Morphological studies of the chicken yolk sac reveal that this is an extremely rugose tissue which provides a large surface area for transport from the yolk into the embryonic circulation. Fine structure analysis indicates that this tissue is organized into a monolayer of columnar epithelial cells which are separated by a basement lamella from the fibroblastic pericapillary tissue and a venous capillary network (Willier, 1968; Lambson, 1970). The absorptive epithelial cells are joined at their apical surfaces by tight junctions, desmosomes are prominent along the lateral plasmalemma (Fig. 9.9). Thus, the only route from the yolk to the embryonic circulation is across these cells. The apical surfaces of these cells exhibit sparse microvilli (Willier, 1968). Coated membrane invaginations are also observed at the base of the microvilli (Lambson, 1970). Numerous endocytic and phagocytic vesicles are also apparent in the apical cytoplasm.

The biochemical mechanism of IgG transport across the chicken yolk sac is just beginning to be probed. Linden and Roth (1978) have demonstrated the presence of receptors specific for chicken IgG on isolated chick yolk sac. Their results showed that IgG binds to these receptors with an apparent $K_a = 5 \times 10^5$ M^{-1} and a distinct optimum at pH 6.0. Heterologous proteins compete less efficiently than yolk IgG for chicken ^{125}I-IgG binding.

9.8 DISCUSSION

In this chapter we have attempted to provide the reader with an introduction to selected portions of the literature on maternal–fetal protein transport. This has permitted us to form a number of conclusions about certain aspects of this process, as well as point out a number of areas where questions remain.

This literature provides overwhelming evidence for the role of receptors in determining which proteins are transported. In some tissues, most notably the human placenta, rabbit yolk sac and the oocyte plasma membrane, the receptors have been studied by direct binding experiments using radiolabeled ligands. The net result of these experiments has been to characterize the affinities, specificities and number of receptors in these tissues. In other systems, the binding data are less complete. However, it is very clear that regardless of species, tissue or ligand, receptors mediate the selection of particular proteins for transport.

The evidence which indicates that coated pits and coated vesicles mediate the endocytosis of receptor-bound ligands also appears to be very convincing. The elegant studies of IgG transport across the neonatal intestine by Rodewald, using electron-dense conjugates of IgG, are particularly compelling. Similar, though less complete, studies have been reported for IgG transport across the rabbit yolk sac and IgG transport across the oocyte plasma membrane of the chicken. Direct evidence that coated vesicles mediate the internalization of maternal proteins is provided by the recent observation that transferrin and IgG are found in preparations of coated vesicles isolated from human placenta. Indirect evidence has long been present which implicates coated pits and coated vesicles in such protein transport since they are commonly found in every tissue which mediates maternal–fetal transport. Very often their initial appearance, relative numbers and disappearance correlates with periods of intense uptake. It is likely that additional studies will provide further direct evidence for the role of coated vesicles in these tissues.

The question of the precise intracellular route used in maternal–fetal protein transport has thus far been largely unstudied. Only in two systems, the neonatal gut and *Xenopus* oocyte, is there any direct evidence as to the intracellular route. Rodewald has suggested that IgG is initially internalized by coated tubular vesicles. The resulting vesicles fuse with a set of intracellular vesicles in the region of the Golgi where sorting of IgG from heterologous proteins takes place. Vesicles containing IgG bud off from these intracellular organelles and ultimately fuse with the basolateral membrane and release their contents. Wallace and associates have suggested that in the *Xenopus* oocyte vitellogenin is initially internalized in coated vesicles. These vesicles fuse with one another or with intracellular vesicles to give rise to transitional yolk bodies, where processing of vitellogenin to phosvitin and lipovitelline takes place. The transitional yolk bodies eventually give rise to mature yolk platelets which are stored until embryonic development begins. Evidence for the intracellular routes traversed by other proteins transported from mother to young awaits further study.

An important aspect of maternal–fetal transport yet to be defined is the ultimate fate of the receptors involved in the initial selection process. In a number of cases of receptor-mediated endocytosis, it has been suggested that following endocytosis receptors are unloaded and recycled back to the cell surface. In other cases, the receptors appear to be degraded following internalization. Thus far, there are limited data on the fate of only two of the receptors involved in maternal–fetal protein transport. It has been suggested that in the neonatal gut the receptors for IgG are recycled from the abluminal surface where ligand release takes place back to the cortical region of the cell. However, at present, the only evidence for this recycling is the indirect observation of uptake of a void marker from the basolateral surface followed by its appearance in the cortical cytoplasmic vesicles. In another system, the fate of secretory component, which acts as a receptor for IgA in the rabbit mammary gland, has been well characterized. The fate of this receptor is unlike that of any other receptor studied thus far. Secretory component appears to irreversibly bind to its ligand. Ligand release is affected by the proteolytic cleavage of secretory component which results in the release of IgA into the milk as a complex with a proteolytic fragment of secretory component. This apparently unique form of receptor processing precludes the possibility of receptor recycling.

As in most fields of cell biology, much has been determined. The problems are often clear and well defined. Unfortunately, numerous questions remain to be answered. Chief among them are the intracellular routes used in maternal–fetal transport, receptor and ligand sorting, receptor recycling, and membrane-bound 'signals' which determine vesicle fusion sites.

REFERENCES

Ahlstedt, S., Carlsson, B., Hanson, L.A. and Goldblum, R.G. (1975), *Scand. J. Immunol.*, **4**, 535.

Anderson, E. (1969), *J. Microsc.*, **8**, 721–738.

Atlas, S.J., Roth, T.F. and Falcone, A.J. (1978), *Insect Biochem.*, **8**, 111–115.

Balfour, A.H. and Jones, E.A. (1977), *Clin. Sci. Mol. Med.*, **52**, 383–394.

Balfour, A.H. and Jones, E.A. (1978), *Int. Arch. Allergy Appl. Immunol.*, **56**, 435–442.

Barnes, J.M. (1959), *J. Pathol. Bacteriol.*, **77**, 371–380.

Bell, W.J. (1972), *J. Exp. Zool.*, **181**, 41–48.

Bellairs, R. (1965), *J. Embryol. Exp. Morphol.*, **17**, 267–281.

Bergink, E.W. and Wallace, R.A. (1974), *J. Biol. Chem.*, **249**, 2897–2903.

Bergink, E.W., Wallace, R.A., VandeBerg, J., Bos, E., Gruber, M. and Ab, G. (1974), *Am. Zool.*, **14**, 1177–1193.

Booth, A.G. and Wilson, M.J. (1981), *Biochem. J.*, **196**, 355–362.

Borthistle, B.K., Kubo, R.T., Brown, W.R. and Grey, H.M. (1977), *J. Immunol.*, **119**, 471–476.

Bourne, J. and Curtis, J. (1973), *Immunol.*, **24**, 157–162.

Brambell, F.W.R. (1958), *Biol. Rev.*, **33**, 488–531.

Brambell, F.W.R. (1970), *The Transmission of Passive Immunity from Mother to Young*. North-Holland Publishing Co., Amsterdam.

Brambell, F.W.R., Halliday, R. and Morris, I.G. (1958), *Proc. R. Soc. London, Ser. B*, **149**, 1–11.

Brambell, F.W.R., Hemmings, W.A., Henderson, M., Perry, H.J. and Rowlands, W.T. (1949), *Proc. R. Soc. London, Ser. B*, **136**, 131–144.

Brambell, F.W.R., Hemmings, W.A. and Rowlands, W.T. (1948), *Proc. R. Soc. London, Ser. B*, **135**, 390–403.

Brierley, J. and Hemmings, W.A. (1956), *J. Embryol. Exp. Morphol.*, **4**, 34–41.

Brown, P.J. and Johnson, P.M. (1981), *Immunol.*, **42**, 313–319.

Butler, J.E. (1974), in *Lactation: A Comprehensive Treatise* (B.L. Larson and V.R. Smith, eds), Academic Press, New York, Vol. III, p. 217.

Buxton, A. (1952), *J. Gen. Microbiol.*, **7**, 268–282.

Chargaff, E. (1942), *J. Biol. Chem.*, **142**, 505–511.

Christmann, J.L., Grayson, M.J. and Huang, R.C. (1977), *Biochemistry*, **16**, 3250–3256.

Clark, S.L. (1959), *J. Biophys. Biochem. Cytol.*, **5**, 41–50.

Cobbs, C.S., Shaw, A.R., Hillman, K. and Schlamowitz, M. (1980), *J. Immunol.*, **124**, 1648–1655.

Cowie, A.T. and Tindal, J.S. (1971), *The Physiology of Lactation*, Edward Arnold, Ltd., London, pp. 24–33.

Cutting, J.A. and Roth, T.F. (1973), *Biochim. Biophys. Acta*, **298**, 951–955.

Deeley, R.G., Mullinix, K.P., Weteham, W., Kronenberg, H.M., Meyers, M., Eldridge, J.P. and Goldberger, R.F. (1975), *J. Biol. Chem.*, **250**, 9060–9066.

Dehn, P.F. and Wallace, R.A. (1973), *J. Cell Biol.*, **58**, 721–724.

Dixon, F.J., Weigle, W.O. and Vazquez, J.J. (1961), *Lab. Invest.*, **10**, 216–237.

Dumont, J.N. (1972), *J. Morphol.*, **136**, 153–180.

Dumont, J.N. (1978), *J. Exp. Zool.*, **204**, 193–218.

Engelmann, F. (1979), *Adv. Insect Physiol.*, **14**, 49–108.

Enns, C.A., Shindelman, J.E., Tonik, S.E. and Sussman, H.H. (1981), *Proc. Natl. Acad. Sci. U.S.A.*, **78**, 4222–4225.

Enns, C.A. and Sussman, H.H. (1981a), *J. Biol. Chem.*, **256**, 9820–9823.

Enns, C.A. and Sussman, H.H. (1981b), *J. Biol. Chem.*, **256**, 12620–12623.

Faulk, W.P. and Johnson, P.M. (1977), *Clin. Exp. Immunol.*, **27**, 365–375.

Fernandez-Costa, F. and Metz, J. (1979), *Br. J. Hematol.*, **43**, 625–630.

Fletcher, J. and Suter, P.E.N. (1969), *Clin. Sci.*, **36**, 209–220.

Franke, W.W., Luder, M.R., Kartenbeck, J., Zerban, H. and Keenan, T.W. (1976), *J. Cell Biol.*, **69**, 173–195.

Friedman, P.A., Shia, M.A. and Wallace, J.K. (1977), *J. Clin. Invest.*, **59**, 51–58.

Galbraith, G.M.P., Galbraith, R.M., Temple, A. and Faulk, W.P. (1980), *Blood*, **55**, 240–242.

Georgi, F. and Jacob, J. (1977), *J. Embryol. Exp. Morphol.*, **38**, 125–138.

Gillian, M.P., Galbraith, G.M.P., Galbraith, R.M. and Faulk, W.P. (1980), *Placenta*, **1**, 33–46.

Gitlin, B. and Biasucci, A. (1969), *J. Clin. Invest.*, **48**, 1433–1438.

Gitlin, J.D. and Gitlin, D. (1974), *J. Clin. Invest.*, **54**, 1156–1166.

Gitlin, J.D., Gitlin, J.I. and Gitlin, D. (1976), *Am. J. Physiol.*, **230**, 1594–1602.

Graber, S.E., Schettel, U., Hodkinson, B. and McIntyre, P.A. (1971), *J. Clin. Invest.*, **50**, 1000–1004.

Graney, D.O. (1968), *Am. J. Anat.*, **123**, 227–254.

Grant, P. (1953), *J. Exp. Zool.*, **124**, 513–543.

Greengard, O., Sentenac, A. and Acs, G. (1965), *J. Biol. Chem.*, **240**, 1687–1691.

Guyer, R.L., Koshland, M.E. and Knopf, P.M. (1976), *J. Immunol.*, **117**, 587–593.

Halsey, J.F., Johnson, B.H. and Cebra, J.J. (1980), *J. Exp. Med.*, **151**, 767–772.

Heald, P.J. and McLachlan, P.M. (1975), *Biochem. J.*, **94**, 32–39.

Heller, E.D. (1975), *Res. Vet. Sci.*, **18**, 117–120.

Heller, J. (1976), *Dev. Biol.*, **51**, 1–9.

Heuser, J. (1980), *J. Cell Biol.*, **88**, 560–583.
Hillman, K., Schlamowitz, M. and Shaw, A.R. (1977), *J. Immunol.*, **118**, 782–788.
Holdsworth, G., Mitchell, R.H. and Finean, J.B. (1974), *FEBS Lett.*, **39**, 275–277.
Hugon, J.S. (1971), *Histochemie*, **26**, 19–27.
Husband, A.J., Brandon, M.R. and Lascelles, A.K. (1978), in *Antigen Adsorption by the Gut* (W.A. Hemmings, ed), University Park Press, Baltimore.
Jackson, J. and Roth, T.F. (1983), submitted.
Jenkinson, E.J., Billington, W.D. and Elson, J. (1976), *Clin. Exp. Immunol.*, **23**, 456–461.
Johnson, P.M. and Faulk, W.P. (1978), *Immunol.*, **34**, 1027–1035.
Jones, E.A. and Waldman, T.A. (1972), *J. Clin. Invest.*, **51**, 2916–2927.
Jones, R.E. (1976), in *Maternal–Fetal Transmission of Immunoglobulins* (W.A. Hemmings, ed.), Cambridge University Press, pp. 325–339.
Jordan, S. and Morgan, E. (1967), *Q. J. Exp. Physiol. Cogn. Med. Sci.*, **52**, 422.
Kemler, R., Mossmann, H., Strohmaier, U., Kickhofen, B. and Hammer, D.K. (1975), *Eur. J. Immunol.*, **5**, 603–608.
Kemp, N.E. (1953), *J. Morphol.*, **92**, 487–506.
King, B.F. (1976), *Anat. Rec.*, **186**, 151–160.
King, B.F. (1977), *Am. J. Anat.*, **148**, 447–456.
King, B.F. (1982), *J. Ultrastruct. Res.*, **79**, 273–284.
Kolhouse, J.F. and Allen, R.H. (1977), *Proc. Natl. Acad. Sci. U.S.A.*, **74**, 921–925.
Kraehenbuhl, J.P. (1981), Private communication.
Kraehenbuhl, J.P., Bron, C. and Sordat, B. (1979), *Curr. Top. Pathol.*, **66**, 105–157.
Kramer, T.T. and Cho, H.C. (1970), *Immunol.*, **19**, 157–167.
Krumins, S.A. and Roth, T.F. (1981), *Biochem. J.*, **196**, 481–488.
Kuhn, L.C. and Kraehenbuhl, J.P. (1979a), *J. Biol. Chem.*, **254**, 11066–11071.
Kuhn, L.C. and Kraehenbuhl, J.P. (1979b), *J. Biol. Chem.*, **254**, 11072–11081.
Kuhn, L.C. and Kraehenbuhl, J.P. (1981), *J. Biol. Chem.*, **256**, 12490–12495.
Kuhn, L.C. and Kraehenbuhl, J.P. (1982), *Trans. Int. Biochem. Soc.*, (in press).
Kunkel, J.G. and Pan, M.L. (1976), *J. Insect Physiol.*, **22**, 809–818.
Lambson, R.O. (1970), *Am. J. Anat.*, **129**, 1–20.
Leary, H.L. and Lecce, J.G. (1979), *J. Nutr.*, **109**, 458–466.
Lin, C.T. (1980), *J. Histol. Cytol. Chem.*, **28**, 339–346.
Linden, C.D. and Roth, T.F. (1978), *J. Cell Sci.*, **33**, 317–328.
Loh, T.T., Higuchi, D.A., van Bockxmeer, F.M., Smith, C.H. and Brown, E.B. (1980), *J. Clin. Invest.*, **65**, 1182–1191.
MacGillivray, R., Mendez, E. and Brew, K. (1977), in *Proteins of Iron Metabolism* (E.B. Brown, P. Aisen, J. Fielding and R.R. Crichton, eds), Grune and Stratton, New York, pp. 133–151.
MacKenzie, D.D.S. (1972), *Am. J. Physiol.*, **223** 1286–1295.
Matre, R. (1977), *Scand. J. Immunol.*, **6**, 953–958.
Matre, R. and Haugen, A. (1978), *Scand. J. Immunol.*, **8**, 187–193.
McNabb, T.C., Koh, T.Y., Dorrington, K.J. and Painter, R.H. (1976), *J. Immunol.*, **117**, 880–888.
Mohr, J.A. (1973), *J. Pediatr.*, **82**, 1062–1068.
Morris, B. (1975), *J. Physiol. (London)*, **245**, 249–259.
Morris, B. and Morris, R. (1974), *J. Physiol. (London)*, **241**, 761–770.

Morris, B. and Morris, R. (1976), *J. Physiol. (London)*, **255**, 619–634.

Morris, I.G. (1950), *Q. J. Microsc. Sci.*, **91**, 237–249.

Mostov, K.E., Kraehenbuhl, J.P. and Blobel, G. (1980), *Proc. Natl. Acad. Sci. U.S.A.*, **77**, 7257–7261.

Moxen, L.A., Wild, A.E. and Slade, B.S. (1976), *Cell. Tissue Res.*, **171**, 175–193.

Murad, T.M. (1970), *Anat. Rec.*, **167**, 17–25.

Nexo, E. and Hollenberg, M.D. (1980), *Biochim. Biophys. Acta*, **628**, 190–200.

Ockleford, C.D. and Clint, J.M. (1980), *Placenta*, **1**, 91–111.

Ockleford, C.D. and Whyte, A. (1977), *J. Cell Sci.*, **25**, 293–312.

Ogra, S.S., Weintraub, D. and Ogra, P.L. (1977), *J. Immunol.*, **119**, 245–248.

Opresko, L., Wiley, H.S. and Wallace, R.A. (1980), *Cell*, **22**, 47–57.

Paterson, R., Younger, S., Weigle, W.O. and Devon, F.J. (1962), *J. Gen. Physiol.*, **45**, 501–513.

Pearse, B.M.F. (1978), *J. Mol. Biol.*, **126**, 803–812.

Pearse, B.M.F. (1982), *Proc. Natl. Acad. Sci. U.S.A.*, **79**, 451–455.

Perry, M.M. and Gilbert, A.B. (1979), *J. Cell Sci.*, **39**, 257–272.

Perry, M.M., Gilbert, A.B. and Evans, A.J. (1978a), *J. Anat.*, **127**, 379–392.

Perry, M.M., Gilbert, A.B. and Evans, A.J. (1978b), *J. Anat.*, **125**, 481–497.

Pitelka, D.R., Hamamoto, S.T., Duafala, J.G. and Nemanic, M.K. (1973), *J. Cell Biol.*, **56**, 797–818.

Pribilla, W., Bothwell, T.H. and Finch, C.A. (1958), in *Iron in Clinical Medicine* (R.O. Wallerstein and S. Mattier, eds), University of California Press, Berkeley, pp. 58–64.

Ramsey, E.M. (1975), *The Placenta of Laboratory Animals and Man*, Holt Rinehart and Winston.

Rodewald, R. (1970), *J. Cell Biol.*, **45**, 635–640.

Rodewald, R. (1973), *J. Cell. Biol.*, **58**, 198–211.

Rodewald, R. (1976a), *J. Cell Biol.*, **71**, 666–670.

Rodewald, R. (1976b), in *Maternofoetal Transmission of Immunoglobulins* (W.A. Hemmings, ed.), Cambridge University Press, pp. 137–149.

Rodewald, R. (1980), *J. Cell Biol.*, **85**, 18–32.

Roth, T.F., Cutting, J.A. and Atlas, S. (1976), *J. Supramol. Struct.*, **4**, 527–548.

Roth, T.F. and Jackson, R. (1972), *J. Cell Biol.*, **55**, 221a.

Roth, T.F. and Porter, K.R. (1962), in *Electron Microscopy*, (S.S. Breese, ed.), Academic Press, New York, pp. LL-4.

Roth, T.F. and Porter, K.R. (1964), *J. Cell Biol.*, **20**, 313–332.

Sasaki, M., Larson, A.L. and Nelson, D.R. (1977), *Biochim. Biophys. Acta*, **497**, 160–170.

Schjeide, O.A., Galey, F., Grellert, E.A., Lin, I., DeVillis, J. and Mead, J.F. (1970), *Biol. Reprod. Suppl.*, **2**, 14–43.

Schjeide, O.A., Wilkens, M., McCandless, R.G., Munn, R., Peterson, M. and Carlsen, E. (1963), *Am. Zool.*, **3**, 167–188.

Schlamowitz, M. (1975), in *Maternofoetal Transmission of Immunoglobulins* (W.A. Hemmings, ed.), Cambridge University Press, pp. 179–197.

Schlamowitz, M., Hillman, K., Lichliger, B. and Ahearn, M.J. (1975), *J. Immunol.*, **115**, 296–302.

Sekhri, K.K., Pitelka, D.R. and Deome, K.B. (1967), *J. Natl. Cancer Inst.*, **39**, 459–469.

Seligman, P.A. and Allen, R.H. (1978), *J. Biol. Chem.*, **253**, 1766–1772.
Seligman, P.A., Schleicher, R.B. and Allen, R.H. (1979), *J. Biol. Chem.*, **254**, 9943–9946.
Sewell, H.F., Matthews, J.B., Flack, V. and Jefferis, R. (1979), *Clin. Exp. Immunol.*, **36**, 183–188.
Slade, B.S. (1970), *J. Anat.*, **107**, 531–545.
Slade, B.S. (1975), *IRCS Med. Sci.*, **3**, 235.
Sonada, S. and Schlamowitz, M. (1972a), *J. Immunol.*, **108**, 807–818.
Sonada, S. and Schlamowitz, M. (1972b), *J. Immunol.*, **108**, 1345–1352.
Sonada, S., Shigematsu, T. and Schlamowitz, M. (1973), *J. Immunol.*, **110**, 1682–1692.
Sussman, H.H. (1982), Private communication.
Stay, B. (1965), *J. Cell Biol.*, **26**, 49–62.
Telfer, W.H. (1961), *J. Biophys. Biochem. Cytol.*, **9**, 747–759.
Telfer, W. H. (1965), *Annu. Rev. Entomol.*, **10**, 161–184.
Tsay, D.D., Ogden, D. and Schlamowitz, M. (1980), *J. Immunol.*, **124**, 1562–1565.
Tsay, D.D. and Schlamowitz, M. (1978), *J. Immunol.*, **121**, 520–525.
Ullberg, S., Kristofferson, H., Flodh, H. and Hanngren, A. (1967), *Arch. Inst. Pharmacodyn. Ther.*, **167**, 431–449.
Van Der Meulen, J.A., McNabb, T.C., Haeffner-Cavillon, N., Klein, M. and Dorrington, K.J. (1980), *J. Immunol.*, **124**, 500–507.
Wada, H.G., Hass, P.E. and Sussman, H.H. (1979), *J. Biol. Chem.*, **254**, 12629–12635.
Waldman, T.A. and Jones, E.A. (1973), *Ciba Found. Symp.*, **9**, 5–23.
Waldman, T.A. and Jones, E.A. (1976), in *Maternofoetal Transmission of Immunoglobulins* (W.A. Hemmings, ed.), Cambridge University Press, pp. 123–136.
Wallace, K.H. and Rees, A.R. (1980), *Biochem. J.*, **188**, 9–16.
Wallace, R.A. (1963a), *Biochim. Biophys. Acta*, **74**, 495–504.
Wallace, R.A. (1963b), *Biochim. Biophys. Acta*, **74**, 505–518.
Wallace, R.A. (1972), *The Role of Protein Uptake in Vertebrate Oocyte Growth and Yolk Formation in Oogenesis* (J. Biggers and A.V. Schutz, eds), University Park Press, pp. 339–359.
Wallace, R.A. and Bergink, E.W. (1974), *Am. Zool.*, **14**, 1159–1175.
Wallace, R.A. and Dumont, J.N. (1968), *J. Cell Physiol.*, **72** (Suppl), 73–89.
Wallace, R.A. and Jared, D.W. (1976), *J. Cell Biol.*, **69**, 345–351.
Wallace, R.A., Jared, D.W., Dumont, J.N. and Sega, M.W. (1973), *J. Exp. Zool.*, **184**, 321–334.
Wallace, R.A. and Misulovin, Z. (1978), *Proc. Natl. Acad. Sci. U.S.A.*, **75**, 5534–5538.
White, H.B., Dennison, B.A., Della Fera, M.A., Whitney, C.J., McGuire, J.C., Meslar, H.W. and Lammelwitz, P.H. (1976), *Biochem. J.*, **157**, 395–400.
Wild, A.E. (1976), in *Maternofoetal Transmission of Immunoglobulins* (W.A. Hemmings, ed.), Cambridge University Press, pp. 155–167.
Wild, A.E. and Richardson, L.J. (1979), *Experientia*, **35**, 838–840.
Willier, B.H. (1968), *Wilhelm Roux Arch.*, **161**, 89–117.
Wood, G., Reynard, J., Krishnan, E. and Racela, L. (1978), *Cell. Immunol.*, **35**, 191–204.

Woods, J.W. and Roth, T.F. (1979), *J. Supramol. Struct.*, **12**, 491–504.
Woods, J.W. and Roth, T.F. (1980), *J. Supramol. Struct.*, **14**, 473–481.
Woods, J.W. and Roth, T.F. (1981), *J. Cell Biol.*, **91**, 219a.
Woods, J.W., Woodward, M.P. and Roth, T.F. (1978), *J. Cell. Sci.*, **30**, 87–97.
Youngdahl-Turner, P., Rosenberg, L.E. and Allen, R.H. (1978), *J. Clin. Invest.*, **61**, 133–141.
Yusko, S.C. and Roth, T.F. (1976), *J. Supramol. Struct.*, **4**, 89–97.
Yusko, S., Roth, T.F. and Smith, T. (1981), *Biochem. J.*, **200**, 43–50.

10 The Formyl Peptide Chemotactic Receptor: Cellular Processing of Peptide and Receptor

JAMES E. NIEDEL

Receptor-Mediated Endocytosis
(*Receptors and Recognition*, Series B, Volume 15)
Edited by P. Cuatrecasas and T. F. Roth
Published in 1983 by Chapman and Hall, 11 New Fetter Lane, London EC4P 4EE
© 1983 Chapman and Hall

10.1 INTRODUCTION

An integrated series of complex cellular functions of polymorphonuclear leukocytes, monocytes and macrophages comprise the inflammatory response. After recognition of an inflammatory stimulus, these phagocytic cells become adherent, migrate in a directed manner through the blood vessel wall and/or the tissue space, ingest opsonized particulate material, generate toxic oxygen free radicals and release lysosomal enzymes. These processes can ultimately result in the killing of invading micro-organisms, clearance of necrotic debris or injury to normal tissue. Because of the importance of inflammation in health and disease, the functional components of the response have been the subjects of intensive investigation. However, the lack of a pure, chemically defined inflammatory stimulus had hampered development of an understanding of these processes at a molecular level. This deficiency was recently corrected by the development of a series of synthetic N-formylated oligopeptides which are potent stimuli for phagocytic cells.

The development of these peptides by Schiffmann followed from his characterization of a partially purified bacterial product which was chemotactic for phagocytes (Schiffmann *et al.*, 1975a,b). The bacterial factor was a small, oxidation-sensitive peptide, with a blocked N-terminus and a free C-terminus. Schiffmann reasoned that because prokaryotes initiate protein synthesis with formylmethionine, whereas eukaryotes initiate with unformylated methionine, formylmethionyl peptides may be recognized by leukocytes as a signal of the presence of bacteria. In rapid succession, these peptides were shown to induce phagocyte adherence, chemokinesis, chemotaxis, lysosomal enzyme release, production of oxygen free radicals, cellular deactivation and to bind to a specific cell surface receptor. In fact, these peptides stimulated all the components of the inflammatory response except phagocytosis and thus provided an experimentally accessible model system for the study of inflammation.

Several recent manuscripts have reviewed our increased understanding of one or more of the components of the inflammatory response based on studies with formyl peptides (Niedel and Cuatrecasas, 1980; Zigmond, 1978; Schiffmann and Gallin, 1979; Snyderman and Goetzl, 1981; Smolen and Weissmann, 1981; Becker *et al.*, 1981). The reader is referred to these manuscripts for a comprehensive treatment of the subject, particularly with regard to possible transduction mechanisms. Based on the title of this volume, this review will be limited to a discussion of three areas: (1) receptor structure, (2) peptide endocytosis and processing, and (3) receptor modulation. Most of these data are new and were not presented in the aforementioned reviews.

10.2 RECEPTOR STRUCTURE

10.2.1 Model for the binding site

Following Schiffmann's initial report of the chemotactic potency of several formylmethionyl peptides, Richard Freer and his colleagues at the Medical College of Virginia and Elmer Becker and his co-workers at the University of Connecticut began a long and productive collaboration to determine the structural features of the peptides essential for receptor binding and biological activity. The data in their initial paper first suggested that a specific cell surface formyl peptide receptor existed (Showell *et al.*, 1976; Becker, 1976). Freer synthesized 24 di-, tri- and tetra-peptides. These were compared for potency and efficacy as induced of chemotaxis and lysosomal enzyme release. The following conclusions were drawn. Among the peptides, greater than a 10^7-fold difference in potency was seen. A blocked *N*-terminus was required; *N*-formylation was preferred. Methionine in the first position and hydrophobic amino acids in positions 2 and 3 gave the highest activity. The peptide *N*-formylmethionylleucylphenylalanine (fMet-Leu-Phe) was the most potent and has become the standard against which new analogs are measured.

A subsequent manuscript investigated in detail the structural requirements for positions 1 and 3 in the peptides (Freer *et al.*, 1980). All of the 30 analogs were tested for potency as inducers of lysosomal enzyme release and a smaller number were also tested for chemotactic potency and inhibition of binding to the formyl peptide receptor. Again the formyl group was essential. Met-Leu-Phe, acetyl-Met-Leu-Phe, desamino-Met-Leu-Phe and 2-ethylhexanoyl-Leu-Phe were all approximately 5000-fold less active than formyl-Met-Leu-Phe. Substitution of norleucine, ethionine, α-aminoheptanoic acid or isoleucine for methionine resulted in at least a 5-fold decrease in activity. A four-carbon side chain gave the greatest activity when linear aliphatic substitutions were introduced at position 1.

Substituents coupled at the *N*-terminus via a urethan linkage had been shown previously to produce antagonist peptides (Aswanikumar *et al.*, 1978). The *t*-butyloxycarbonyl group produced the most potent antagonists when coupled to the tetrapeptide Phe-Leu-Phe-Leu.

In their most recent paper, Freer *et al.* (1982) have investigated the structural requirements for positions 2 and 3 with 20 new peptides. Leucine was preferred at position 2 and was 4- to 10-fold more potent than the closely related analogs, isoleucine, norvaline, valine or methionine. Branched-chain aliphatics were readily accepted by the receptor but branching at the γ carbon was preferred.

A much more interesting finding was that esterification of the preferred phenylalanine at position 3 greatly increased potency in the enzyme release assay. A 10- to 100-fold increase in activity was seen when the benzyl ester was compared with the unsubstituted tripeptide. However, only a 2- to 5-fold increase in receptor affinity was noted. These derivatives were the first to deviate from the strict correlation between binding affinity and biological potency. The reason for this deviation is unclear.

From these data and others (Niedel *et al.*, 1979a), Freer *et al.* proposed a model of the formyl peptide binding site as follows. The model is based on a β-pleated sheet peptide conformation with a rigid backbone (Becker *et al.*, 1979). Five areas of interaction are postulated: (1) the proton of the formyl group participates in hydrogen-bonding with an acceptor on the receptor, (2) the methionine side chain sits in a hydrophobic pocket with an electron-deficient area to interact with the electron-rich sulfur atom, (3) and (4) both leucine and phenylalanine also interact with hydrophobic pockets and (5) the phenylalanine carbonyl is also involved in hydrogen-bonding.

10.2.2 Affinity-labeled receptor

The structure–function studies just presented allow a hypothetical model of the receptor binding site to be proposed. However, they do not address the question of the overall structure of the receptor. Toward this end, we have used the technique of covalent affinity labeling to define some of the physical characteristics of the receptor (Niedel *et al.*, 1980b).

The chemotactic hexapeptide N-formyl-Nle-Leu-Phe-Nle-^{125}I-Tyr-Lys (^{125}I-labeled fNLPNTL) tolerated covalent modifications on the ε amino group of lysine without significant loss of binding affinity or specificity (Niedel *et al.*, 1979b). We prepared and characterized several hexapeptide derivatives including the N^{ε}-bromoacetyl-, N^{ε}-4-azido-2-nitrophenyl- and N^{ε}-benzoyl-maleimide. These derivatives, or the native hexapeptide following cross-linking with a bifunctional imidate or succinimide ester (Niedel, 1981), specifically radiolabeled protein of the human neutrophil membrane which migrated as a broad band on sodium dodecyl sulfate–polyacrylamide gel electrophoresis (SDS–PAGE) with an apparent molecular weight of approximately 55 000 to 70 000 (Fig. 10.1). Covalent cross-linking of the peptide to this protein met the same criteria of affinity, specificity and saturability established for the formyl peptide receptor with standard radioligand binding assays (Fig. 10.2). Additionally, a protein with a similar molecular weight was labeled on human neutrophils, monocytes and differentiated HL60 cells, all of which bear the formyl peptide receptor. In contrast, no specific labeling was seen with several receptor-negative cells including erythrocytes, lymphocytes and undifferentiated HL60. A similar protein was affinity-labeled in digitonin extracts of neutrophil membranes and led to the development of a soluble-receptor assay (Niedel, 1981).

Recently, Bart Dolmatch, a third-year medical student in my laboratory, prepared and characterized a high-affinity and high-efficiency photo-activatable analog N^{ε}-6-(4′-azido-2′-nitrophenylamino)hexanoate. Using this

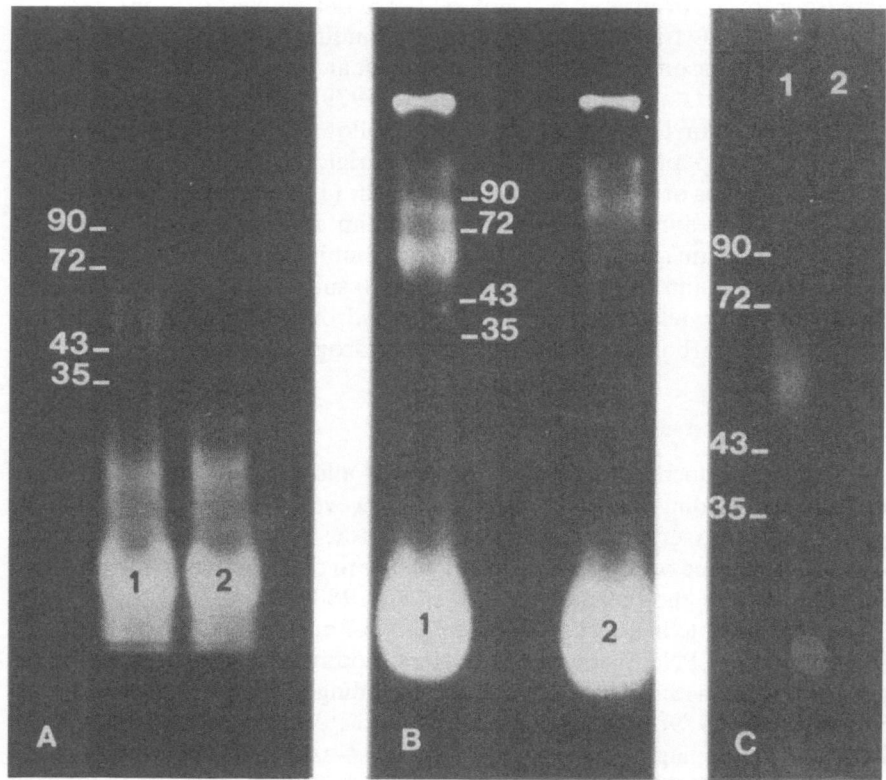

Fig. 10.1 Radioautographs demonstrating covalent affinity radiolabeling of the 55 000 to 70 000-dalton formyl peptide receptor of human neutrophil membranes. (A) ^{125}I-labeled fNLPNTL cross-linked with dimethyl suberimidate. (B) ^{125}I-labeled fNLPNTL-N^ε-bromoacetyl. (C) ^{125}I-labeled fNLPNTL-N^ε-arylazide. Lane 1 was in the absence and lane 2 was in the presence of 500 nM-unlabeled fNLPNTL. Molecular-weight standards of 90 000, 72 000, 43 000 and 35 000 are marked. Reproduced with permission from Niedel *et al.* (1980b).

ligand he has been able to label the receptor on living cells at 4°C without damage to the cells. Papain and pronase treatment defined the protease domains of the receptor binding site. Initially a 35 000-dalton fragment is produced, followed by fragments of 28 000, 23 000 and 18 000. All of these fragments remain embedded in the membrane and at least the 35 000- and 28 000-molecular-weight fragments retain binding activity without a change in affinity. This latter observation explains why protease treatment does not decrease receptor binding activity (Schiffman *et al.*, 1980). The 35 000-molecular-weight fragment not only retains binding activity, but is able to

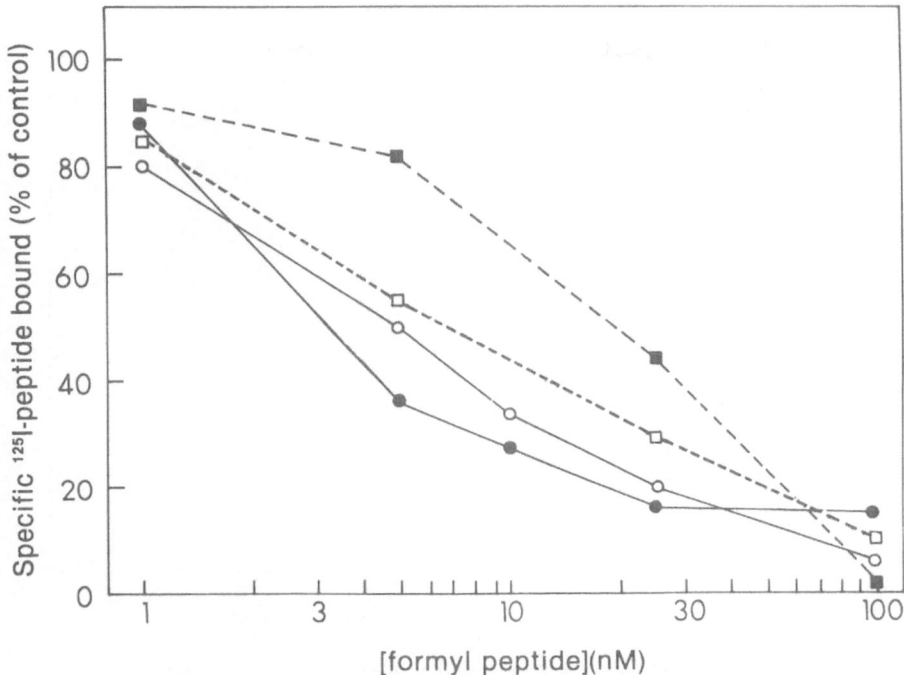

Fig. 10.2 ^{125}I-labeled fNLPNTL bound to the membrane receptor in the presence of the indicated concentrations of unlabeled fNLPNTL correlated with ^{125}I-labeled fNLPNTL cross-linked to the 55 000- to 70 000-dalton receptor band in the presence of the same concentrations of unlabeled fNLPNTL. Specific binding of ^{125}I-labeled fNLPNTL (○——○); specific cross-linking of ^{125}I-labeled fNLPNTL with suberimidate (●——●). Specific binding of ^{125}I-labeled fNLPNTL-N^ε-bromoacetyl (□——□); specific cross-linking of ^{125}I-labeled fNLPNTL-N^ε-bromoacetyl (■——■). Reproduced with permission from Niedel *et al.* (1980b).

transduce the signal for peptide-induced lysosomal enzyme release and receptor-mediated peptide uptake. Neuraminidase treatment did not alter receptor mobility on SDS–PAGE, while chymotrypsin and trypsin caused smearing of the labeled receptor without production of discrete fragments.

10.2.3 Detergent-soluble receptor

The formyl peptide receptor has recently been solubilized in an active form by workers in two independent laboratories (Niedel, 1981; Goetzl *et al.*, 1981). Using a strategy based on covalent affinity labeling, we were able to demonstrate an active, soluble receptor in digitonin extracts of human neutrophil membranes. Approximately 70% of the receptor initially present in the mem-

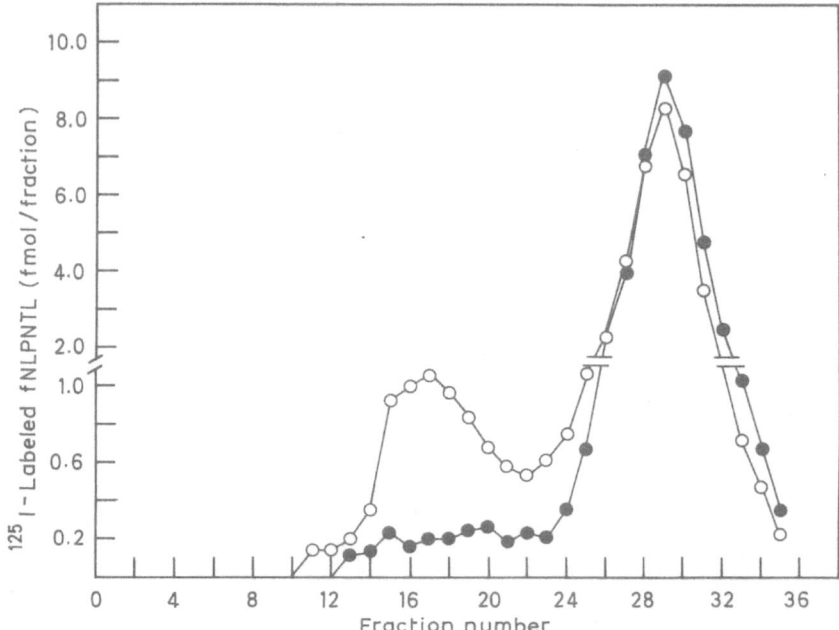

Fig. 10.3 Separation of receptor-bound and free ^{125}I-labeled fNLPNTL by gel
filtration. The 0.3% digitonin extract of neutrophil membranes was incubated with
4 nM-^{125}I-fNLPNTL in the absence (O——O) or presence (●——●) of
200 nM-unlabeled fNLPNTL and fractionated at 4°C on a 1.6 ml AcA 44 column.
Reproduced with permission from Niedel (1981).

branes was recovered in the digitonin supernatant. An assay based on separation of receptor-bound from free peptide by gel filtration was used to determine
the binding characteristics of the soluble receptor (Fig. 10.3). Binding was
rapid, saturable, dependent on *N*-formylation of the peptide and of high
affinity (K_d = 2.2 nM for *N*-formyl-Nle-Leu-Phe-Nle-^{125}I-Tyr-Lys). The soluble receptor, when compared with the membrane-bound receptor, displayed
about a 3-fold lower affinity for the peptide. This may result from an alteration of the lipid environment of the receptor or may be due to a falsely high
value for the free peptide concentration because of partitioning of the hydrophobic peptide into the detergent micelles. Binding was maximal at pH 6.5
and was sulfhydryl-dependent. A protein in the digitonin extract, with the
same apparent molecular weight as the membrane-bound receptor, was specifically affinity-labeled with a bifunctional succinimide ester (Fig. 10.4). The
properties of the soluble receptor are therefore similar to the well-
documented properties of the receptor on whole cells and isolated membranes (Niedel *et al.*, 1979a; Schiffmann *et al.*, 1980).

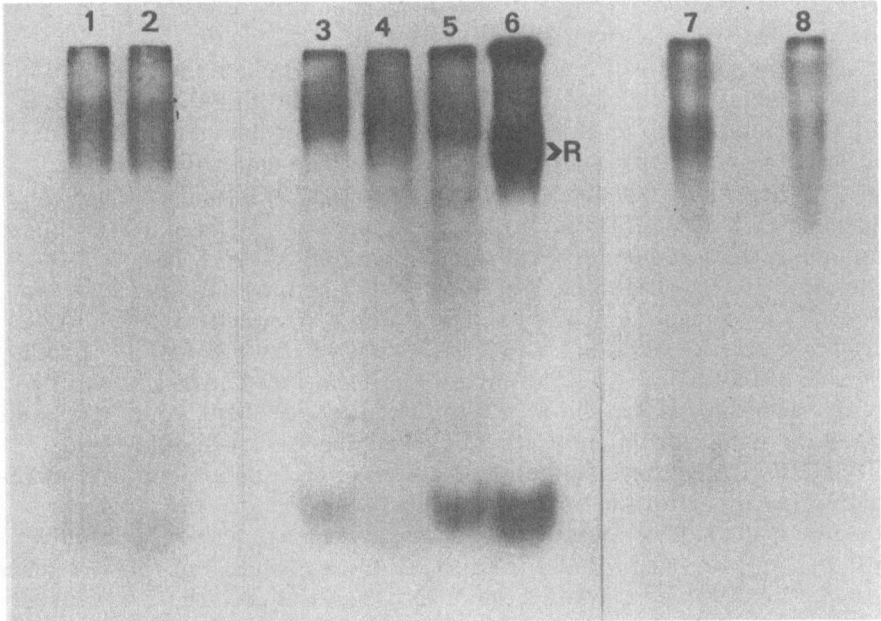

Fig. 10.4 Affinity labeling of the detergent-solubilized formyl peptide receptor. ^{125}I-labeled fNLPNTL was cross-linked with ethyleneglycolbis(succinimidyl succinate) to the various detergent extracts in the absence or presence of 200 nM-unlabeled fNLPNTL. Lane 1, 0.5% deoxycholate, total labeling. Lane 2, 0.5% deoxycholate, labeling in the presence of 200 nM-unlabeled fNLPNTL (non-saturable). Lane 3, 0.2% Triton X-100, total labeling. Lane 4, 0.5% Triton X-100, total labeling. Lane 5, 0.5% Triton X-100, non-saturable labeling. Lane 6, (standard) ethyleneglycolbis(succinimidyl succinate) affinity labeling of the membrane-bound receptor (>R). Lane 7, 0.2% digitonin, total labeling. Lane 8, 0.2% digitonin, non-saturable labeling. Reproduced with permission from Niedel (1981).

Our attempts at purification of the soluble receptors have been repeatedly unsuccessful. Even the best digitonin preparations are quite impure and we have experienced considerable variation between different batches, especially with regard to the yield of soluble receptor and the solubility of the digitonin itself. We have prepared several different formyl-peptide–agarose columns, but none would specifically adsorb the soluble receptor. Gel filtration chromatography was of little value because the recovery of receptor was low (<20%) and resolution was poor owing to digitonin crystallization. The receptor was adsorbed to wheatgerm-agglutinin–agarose and specifically eluted with N-acetylglucosamine, but this effected little purification. In our hands, the yield of soluble receptor with other detergents (cholate, deoxycholate, CHAPS, NP-40, Triton X-100, Brig 35 and octyl glucoside) has been too low to warrant further work.

Goetzl and his co-workers have been more successful in their purification of the receptor (Goetzl *et al.*, 1981). They applied the NP-40 extract of neutrophil membranes to an fMet-Leu-Phe–Sepharose column and eluted the adsorbed proteins with soluble fMet-Leu-Phe. The major protein bands of 94 000 (MP-1), 68 000 (MP-2) and 40 000 (MP-3) molecular weight were further purified by gel filtration chromatography on P-150 in a Tris buffer containing 0.5% sodium dodecyl sulfate. By equilibrium dialysis, MP-2 demonstrated high-affinity ($K_a = 9 \times 10^8$ M^{-1}) and low-affinity ($K_a = 3 \times 10^7$ M^{-1}) sites for [^{3}H]fMet-Leu-Phe with a valence of 0.2–0.3. MP-3 displayed only low-affinity sites ($K_a = 2 \times 10^7$ M^{-1} and 2×10^6 M^{-1}) with a valence of 0.3–0.5. MP-1 did not bind [^{3}H]fMet-Leu-Phe as assessed by equilibrium dialysis, but did rebind to fMet-Leu-Phe–Sepharose. MP-2 and MP-3 did not adsorb significantly to C5a-Sepharose nor did they bind [^{3}H]HETE (5-hydroxyeicosatetraenoic acid) or [^{3}H]leukotriene B, demonstrating specificity for the formyl peptides.

Amino acid analysis indicated an above average content of hydrophobic residues for MP-1 and MP-2 and an above average content of acidic residues for MP-2 and MP-3. Carbohydrate analysis was not performed, but for all three proteins, the molecular weight calculated from the amino acid analysis agreed exactly with the molecular weight estimated by SDS–PAGE and gel filtration. Also, unlike most glycoproteins, all three proteins migrated as tight bands on SDS–PAGE and stained well with Coomassie Blue. The gels were not stained for carbohydrate. This suggests that MP-1, MP-2 and MP-3 may not be glycoproteins or may contain a relatively small proportion of carbohydrate.

10.2.4 Anti-receptor antibody

MP-2, purified by affinity chromatography and gel filtration, was used to raise antisera in goats (Goetzl *et al.*, 1982). IgG prepared from immune animals, but not pre-immune animals, bound to neutrophils as assessed by a ^{125}I-protein A assay. ^{125}I-Fab was specifically bound to approximately 65 000 sites per neutrophil and was inhibited by MP-2 in solution. Unlabeled fMet-Leu-Phe did not significantly inhibit ^{125}I-Fab binding, nor did unlabeled Fab significantly inhibit [^{3}H]fMet-Leu-Phe binding, indicating that the Fab was not directed against the receptor binding site. Neither IgG nor Fab elicited neutrophil chemotaxis or chemokinesis. However, IgG, but not Fab, evoked release of significant quantities of β-glucuronidase and lysozyme. Pre-incubation of neutrophils with Fab inhibited the subsequent chemotactic response to fMet-Leu-Phe, but did not alter chemotaxis in response to C5a or leukotriene B$_4$. Prior incubation of neutrophils with fMet-Leu-Phe caused a dose-dependent down regulation of [^{3}H]fMet-Leu-Phe binding while ^{125}I-Fab binding was significantly increased by the same procedures. The authors speculated that the discordance between [^{3}H]fMet-Leu-Phe binding and ^{125}I-Fab binding following receptor down regulation may indicate that binding site internalization leaves MP-2 'framework' antigenic determinants exposed on the membrane or that non-binding receptor precursors, which share common antigenic sites (perhaps MP-1), are expressed on the membrane.

Because the antisera are not monospecific, it is also possible that unrelated antigens contaminating the MP-2 preparation elicited an antibody response and that these antigens may have been altered during formyl-peptide-induced down regulation.

Marasco *et al.* (1982) and Marasco and Becker (1982) used a different approach to produce anti-receptor antibodies. They coupled fMet-Leu-Phe to several carrier proteins and used these covalent conjugates as immunogens to elicit antibodies in rats and rabbits. Detailed studies of antibody specificity were performed with a high-titer rabbit immunoglobulin fraction. Using selected peptides from the series synthesized by Freer to characterize the receptor binding site (see Section 10.2.1), they demonstrated that the anti-(fMet-Leu-Phe) antibody and the formyl peptide receptor recognized identical structural properties of the peptides. This raised the possibility that the anti-(fMet-Leu-Phe) antibody and the peptide receptor shared common structural features. With this in mind, anti-idiotypic antibodies were raised in several species. After demonstrating that the monospecific antibody interacted with the anti-(fMet-Leu-Phe)-binding site and was therefore anti-idiotypic, they assessed its interaction with neutrophils. A direct interaction with the formyl peptide receptor was suggested by (1) direct binding of anti-idiotypic antibody to neutrophils, (2) a 25–35% inhibition of [^{3}H]fMet-Leu-Phe receptor binding by anti-idiotype F(ab')2, and (3) loss of anti-idiotype and anti-receptor activity after passage over an anti-(fMet-Leu-Phe) column. This interesting series of experiments suggests that at least a subpopulation of anti-(fMet-Leu-Phe) antibody-binding sites must be analogous to the fMet-Leu-Phe site on the receptor, such that an antibody to the former also recognizes the latter. The biological activity of the antibody was not reported.

10.3 PEPTIDE ENDOCYTOSIS AND PROCESSING

Receptor-mediated binding, aggregation and internalization appear to be universal mechanisms by which eukaryotic cells process polypeptide ligands. This has been demonstrated by electron microscopic, fluorescent microscopic and biochemical techniques with many different cells and ligands.

We prepared and purified *N*-formyl-Nle-Leu-Phe-Nle-Tyr-Lys-*N*$^\varepsilon$-rhodamine and showed that this fluorescent conjugate bound specifi-

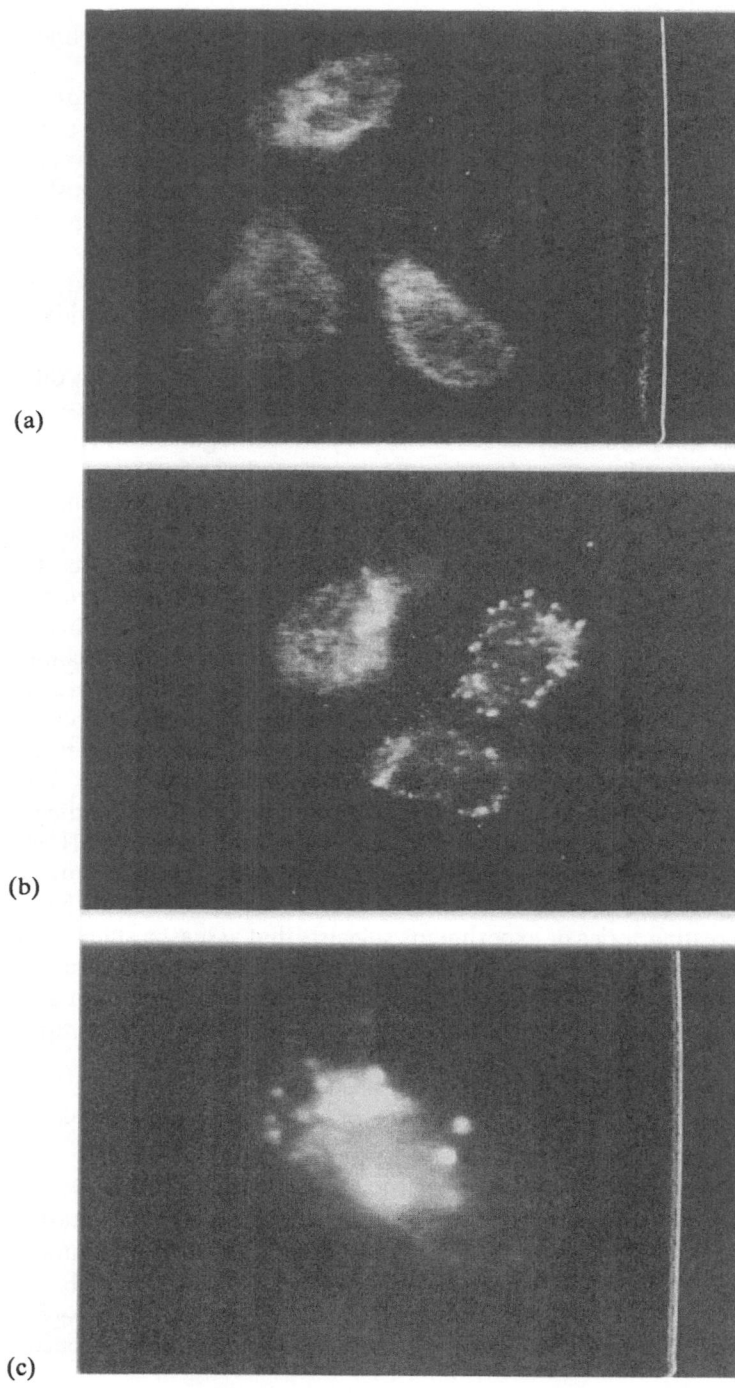

(a)

(b)

(c)

cally to the human neutrophil receptor (IC_{50} = 3 nM) and was biologically active (EC_{50} = 1 nM for chemotaxis). These values indicated that rhodamine conjugation caused about a 2-fold decrease in potency when compared to the underivated hexapeptide (Niedel *et al.*, 1979b). This fluorescent peptide and video intensification microscopy were used to study cellular processing by human neutrophils, monocytes (Weinberg *et al.*, 1981) and differentiated promyelocytic leukemia cells (Niedel *et al.*, 1980a). With each cell type, a similar sequence of events was observed.

Using peptide processing by human monocytes as an example (Fig. 10.5), relatively homogeneous surface fluorescence was observed when the cells were exposed to low concentrations (2.5 nM) of the fluorescent peptide for 1 minute at 37°C. The bound peptide rapidly aggregated to form multiple (50–200) brightly fluorescent patches on the membrane within 2 to 3 minutes. Between 3 and 10 minutes, these aggregates were internalized within the cell. The fluorescent endocytic vesicles displayed the saltatory motion characteristic of cytoplasmic organelles and were seen to stream with other cytoplasmic particles. In neutrophils and monocytes, these small endocytic vesicles then fused to form larger vesicles which were evident by phase-contrast microscopy without the aid of image intensification. Once inside the differentiated promyelocytic leukemia cells, the small endocytic vesicles did not coalesce but remained as multiple small vesicles. The leukemia cells do not differentiate fully to neutrophils, but remain as myelocytes and metamyelocytes. The ability to fuse endocytic vesicles may develop late in the differentiation pathway and be absent at the myelocyte and metamyelocyte stage. The mechanism or importance of vesicle coalescence is unknown. All of these fluorescent events were abolished in the presence of 100 nM non-fluorescent peptide indicating that they were receptor-mediated.

Once internalized, the fluorescent vesicles remain stable for at least 30 minutes, usually for several hours. They then begin to break down into smaller vesicles, but are still visible even after 12 hours. However, observing the fluorescent events for extended periods of time was of little value, because by 30 minutes, 30% of the cell-associated fluorescent peptide had been degraded and fragments began to appear in the medium.

Fig. 10.5 Internalization of rhodamine-labeled fNLPNTL by human blood monocytes. Frames a, b, and c show photographs of monocytes that were incubated with 2.5 nM-rhodamine-labeled formyl peptide and fixed after 1, 3 and 15 min respectively. At 1 min the peptide was distributed diffusely over the entire membrane; at 3 min, the fluorescent peptide appeared as aggregates in the plane of the membrane; and at 15 min it appeared in internalized vesicles. The magnifications on the TV monitor in (a) and (b) were 2080 and 5200 in (c). Reproduced with permission from Weinberg *et al.* (1981).

Although some differences in intensity were observed, all viable peripheral blood neutrophils and all adherent peripheral mononuclear cells were shown to bind and internalize the fluorescent peptide, demonstrating that these cells are homogeneous with regard to expression of the formyl peptide receptor. Using the same peptide labeled with fluorescein and the Ortho Cyto-fluorgraph, we have recently confirmed that >92% of peripheral granulocytes are receptor-positive.

In unpublished experiments, we have also used the rhodamine-labeled peptide to study receptor distribution and processing by neutrophils responding to a chemotactic gradient in the Zigmond chamber. The fluorescent peptide was seen to be initially homogeneously distributed over the cell membrane. Although the cells were exposed to a gradient of the peptide, we could never discern the small differences in fluorescent intensity predicted by the differential receptor occupancy theory. As the cells became oriented and began to move up the gradient, the fluorescent peptide was swept back along the cell to eventually form multiple aggregates just ahead of the uropod. These aggregates were continuously internalized into endocytic vesicles which migrated within the cytosol towards the front of the cell to a place just behind the multi-lobed nucleus. Fluorescent peptide accumulated in this region for the 5 to 10 minute interval when the cells were exposed to the gradient.

Biochemical evidence in support of peptide endocytosis has come from several laboratories. We used N-formyl-Nle-Leu-Phe-Nle-^{125}I-Tyr-Lys to characterize receptor-mediated peptide processing by human neutrophils (Niedel *et al.*, 1979a). Dissociation of cell-bound peptide could only be demonstrated following binding at low temperature (4°C) for extended time periods or following short duration binding incubations at 37°C (< 5 min). When binding was allowed to reach a plateau value at 24°C or 37°C, dissociation of native ^{125}I-labeled fNLPNTL did not occur at any temperature. However, slow release of proteolytic fragments was evident at 37°C. By 2 hours, release had reached a plateau with about 50% of the ligand remaining cell-associated. Twelve percent of the soluble radioactivity was unaltered peptide, while the majority was monoiodotyrosine. Of the radioactivity which remained cell-associated, 35% was unaltered peptide and the remainder was a single unidentified proteolytic fragment. Production of the proteolytic fragment was presumably due to an internal or plasma membrane protease because the supernatant from a 2-hour cell incubation did not contain a protease which would produce the fragment.

Receptor-mediated peptide uptake was maximal at pH 6.75 and enhanced by divalent cations, especially Ca^{2+}. An $EC_{50} = 1.3$ nM was estimated for the process of binding and uptake by human neutrophils at 24°C.

Using both fluorescent-labeled and ^{125}I-labeled fNLPNTL, we have been unable to find non-toxic inhibitors of receptor-mediated endocytosis. Cytochalasin B, colchicine, azide, cyanide, 2-deoxyglucose, cation chelators, ammonium chloride, methylamine, dansylcadaverine, chloroquine, sulfhydryl reagents and dithiothreitol have all been pushed to significant toxicity without inhibition of peptide uptake by human neutrophils. Low temperature was the only reliable inhibitor.

Vitkauskas *et al.* (1980) also demonstrated peptide endocytosis using [³H]fNle-Leu-Phe and rabbit neutrophils. They observed rapid dissociation of the radioligand following binding to cells at 4°C or membranes at 4°C or 37°C. However, following binding to cells at 37°C, the radiolabel was slowly released at 37° as [³H]Phe, indicating cell-dependent proteolysis had occurred. Subcellular distribution experiments following binding at 4°C demonstrated that most of the specifically bound peptide was found in the microsomal supernatant (i.e. it had dissociated during the fractionation procedure). The same fractionation procedure following binding at 37°C for 10 minutes demonstrated 27% of the radioligand in the lysosomal pellet and 30% in the microsomal pellet. Because the radiolabel did not dissociate from the microsomal pellet, this fraction may have contained the inside-out endocytic vesicles seen with the fluorescent peptide.

Sullivan and Zigmond (1980) made similar observations regarding endocytosis with rabbit neutrophils and [³H]fNle-Leu-Phe. In addition, they called attention to peptide-induced non-specific pinocytosis. Using radiolabeled sucrose, even at 4°C, pinocytosis increased from 1.7 to 6.6 nl/min/10⁷ cells for 5 minutes after cell exposure to the peptide. However, fluid-phase pinocytosis appears to make a significant contribution to total uptake only with relatively low-affinity peptide ligands or with high concentrations of ligand in solution. At 100 nM-[³H]fNle-Leu-Phe, fluid-phase pinocytosis accounted for approximately 30% of total uptake by rabbit cells. At a concentration of 10 nM, which saturates approximately 85% of available receptor sites, fluid-phase pinocytosis accounts for less than 4% of the total uptake of either ¹²⁵I-labeled or rhodamine-labeled fNLPNTL by human neutrophils. Fletcher *et al.* (1982) were unable to demonstrate stimulated uptake of radiolabeled sucrose by human neutrophils after exposure to fMet-Leu-Phe.

A recent study has carefully analyzed the kinetics of peptide uptake and processing (Zigmond *et al.*, 1982). [³H]fNle-Leu-Phe uptake by rabbit neutrophils was measured as a function of peptide concentration at 37°C over a 45-minute interval when the surface receptor number was at steady-state. A linear relationship was evident between calculated receptor occupancy and uptake. This first-order relationship suggested that receptor aggregation was not rate-limiting for the uptake process.

In order to calculate a rate constant for uptake from the above experiments, an estimate of simultaneous release was necessary. Release was shown to be a pseudo-first-order process and the observed rate constant was independent of the concentration of peptide present during either uptake or release. Forty percent of the internalized peptide was rapidly released, while the remaining 60% appeared to be stored intracellularly for periods in excess of 4 hours. The 40% was released as digested fragments, while the intracellular material remained substantially intact. A surprising finding was that peptides taken up by non-saturable fluid-phase pinocytosis were handled in an identical fashion to those taken up by the receptor-mediated process. A rate constant for peptide release was estimated to be between 0.013 and 0.022 min^{-1}, and a constant of between 0.12 and 0.18 min^{-1} was derived for receptor-mediated uptake. This latter rate constant appeared to be independent of peptide concentration.

10.4 RECEPTOR MODULATION

10.4.1 Down regulation

Independently, Vitkauskas *et al.* (1980), Sullivan and Zigmond (1980) and Nelson *et al.* (1980) noted a loss of surface receptors concomitant with peptide endocytosis. This phenomenon, termed receptor down regulation, was shown to be dependent on the concentration of unlabeled fMet-Leu-Phe or fNle-Leu-Phe used to treat the cells. With rabbit neutrophils, a measurable decrease in binding was seen following treatment with 10^{-9} M-fNle-Leu-Phe and was complete with peptide concentrations about 10^{-7} M. Approximately 5-fold lower concentrations of fMet-Leu-Phe produced the same degree of down regulation. However, even following treatment with very high peptide concentrations, 15–20% of the surface receptors remained available for binding. With both peptides, down regulation was rapid, being essentially complete by 15 minutes. Down regulation was temperature-dependent; it was pronounced at 37°C, somewhat decreased at 23°C and nearly absent at 4°C. Following down regulation, rabbit neutrophils were shown to recover their surface receptors during incubation in a balanced salt solution at 37°C. Recovery was complete by 20 minutes and independent of protein synthesis. Cbz-Phe-Met, an antagonist of the formyl peptides, did not induce down regulation, but did inhibit down regulation induced by fNle-Leu-Phe. Sullivan and Zigmond showed that the pretreatment receptors and the receptors remaining on the surface following down regulation were indistinguishable in terms of affinity for [^{3}H]fNle-Leu-Phe.

Peptide-induced receptor down regulation did not appear to fully explain the loss of biological response to a subsequent exposure of the peptide. Vitkaukas *et al.* (1980) noted a 20% greater decrease in stimulated lysosomal enzyme release than in surface receptor number. Nelson *et al.* (1980) also noted a discrepancy between the percentage decrease in receptor number and the decrease in subsequent chemotactic response of human neutrophils. In these experiments, however, loss of receptor binding exceeded loss of the biological response. Much higher fMet-Leu-Phe concentrations were necessary to induce down regulation in human cells. At 10^{-7} M, a 15% increase in surface receptors was noted. 10^{-4} M-peptide down-regulated approximately 80% of the surface receptors. No receptor recovery was noted over an 18-hour period.

Different methodologies, especially with regard to the buffer temperature used for cell washing after down regulation, may account for some of the differences between human and rabbit neutrophils. Receptor recovery in human monocytes for instance was dependent on both the concentration of peptide and length of time used to induce down regulation (Weinberg *et al.*, 1981). Receptor down regulation in human monocytes was also concentration-dependent and essentially identical to uptake of the same concentrations of ^{125}I-labeled peptide, suggesting that one receptor was down-regulated for each receptor-mediated uptake event (Fig. 10.6). As with the rabbit neutrophils, 5 to 20% of the original surface receptor number remained after prolonged incubations with peptide concentrations more than 100-fold above the K_d. These remaining receptors were indistinguishable from the original receptors (Fig. 10.7).

Receptor recovery was measured after down regulation with peptide con-

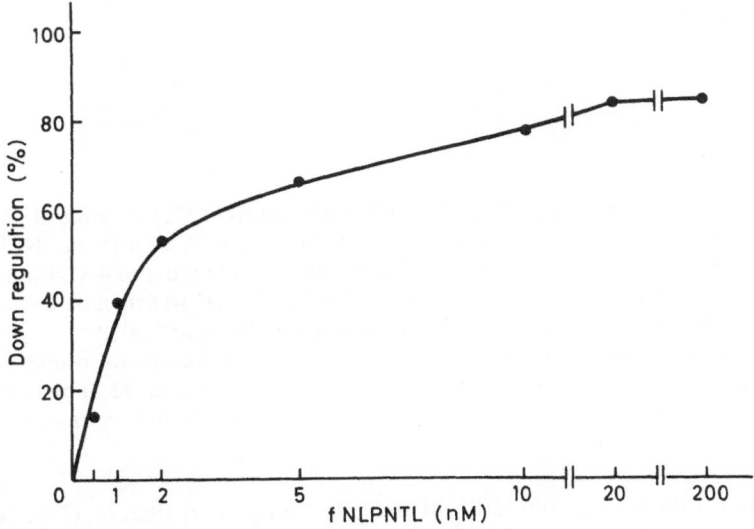

Fig. 10.6 Down regulation of ^{125}I-labeled fNLPNTL receptors on monocytes as a function of peptide concentration used to pretreat monocytes. The cells were treated 30 minutes at 37°C with the designated peptide concentrations and washed at 4°C. Peptide binding was then determined after a 2-h incubation with 2 nM-^{125}I-labeled fNLPNTL at 4°C. Reproduced with permission from Weinberg *et al.* (1981).

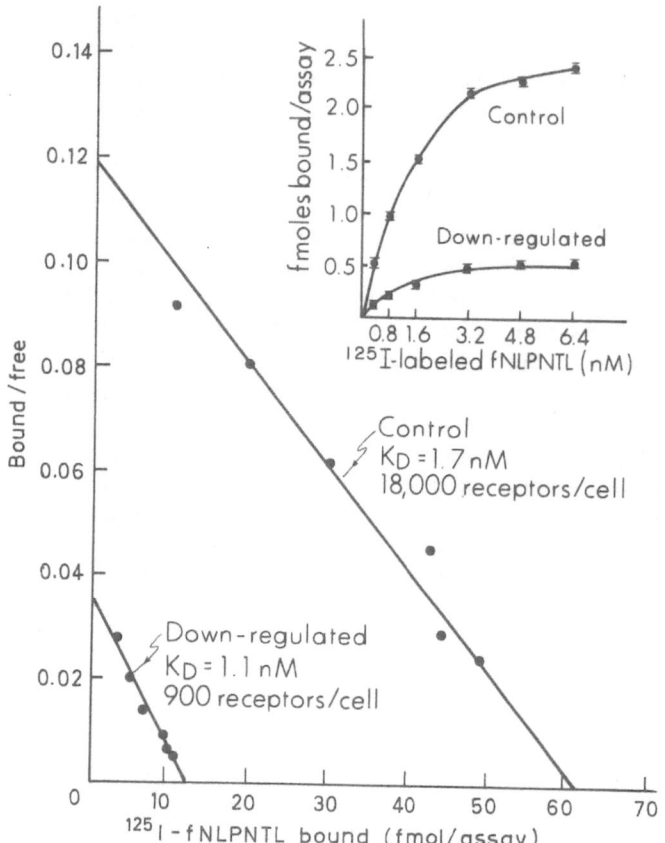

Fig. 10.7 Scatchard analysis of binding of ^{125}I-labeled fNLPNTL to control and down-regulated cells. Cells were treated for 30 minutes at 37°C with 10 nM-fNLPNTL (down-regulated) or Gey's-BSA (control) and washed three times at 4°C. Binding at the indicated concentrations of ^{125}I-labeled fNLPNTL (inset) was then determined at 4°C for 2 h with 2×10^6 cells/300 μl (control) or 8×10^6 cells/300 μl (down-regulated) and the mean values plotted ± 1 S.E.M. These data were analyzed by the method of Scatchard and the line that best fit the points was calculated by the method of least squares. Reproduced with permission from Weinberg *et al.* (1981).

centrations that were 2-fold, 20-fold and 200-fold greater than K_d (Fig. 10.8). Recovery after treatment with the lowest concentrations was rapid and frequently exceeded the pretreatment receptor number. At high concentrations, recovery did not occur over the 2-hour incubation period, but required culture of the cells for periods of from 8 to 18 hours. The rapid recovery was insensitive to inhibition with non-toxic concentrations of cycloheximide sug-

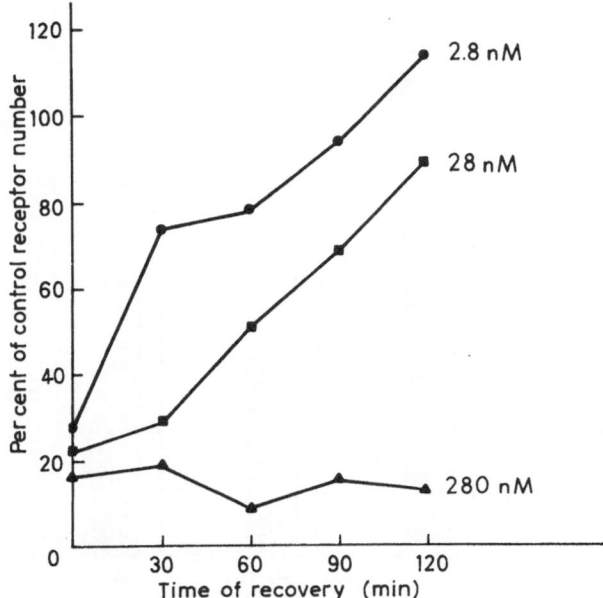

Fig. 10.8 Time of monocyte receptor recovery after down regulation with various concentrations of fNLPNTL. The cells were treated with 0, 2.8, 28 or 280 nM-peptide for 30 minutes at 37°C, washed at 4°C, and incubated at 37°C. At the designated times, cells were assayed for binding after 2-h incubation at 4°C with 2 nM-^{125}I-labeled fNLPNTL. Reproduced with permission from Weinberg *et al.* (1981).

gesting that new protein synthesis was not required. Recovery was also dependent on the length of pretreatment time used to induced down regulation. After 15 minutes of exposure to 20 nM-unlabeled fNle-Leu-Phe-Nle-Tyr-Lys, complete recovery occurred within 60 minutes. But when a 60 minute pretreatment was followed by a 60 minute recovery incubation, less than 50% of the receptors were replenished (Fig. 10.9).

The kinetics of rabbit neutrophil receptor down regulation and recovery have been studied in detail (Zigmond *et al.*, 1982). As with human monocytes, the extent of recovery from down regulation was dependent on the peptide concentration used to induce receptor loss. With concentrations near the K_d, receptor recovery often exceeded 100%, while with high peptide concentrations, recovery was incomplete. The rate of receptor recovery appeared to be a first-order function of the number of receptors missing. The number of receptors missing was defined as the difference between the receptor number present immediately after down regulation and the receptor number following an 80-minute recovery incubation. The observed rate constant for recovery was highest when low peptide concentrations were used to

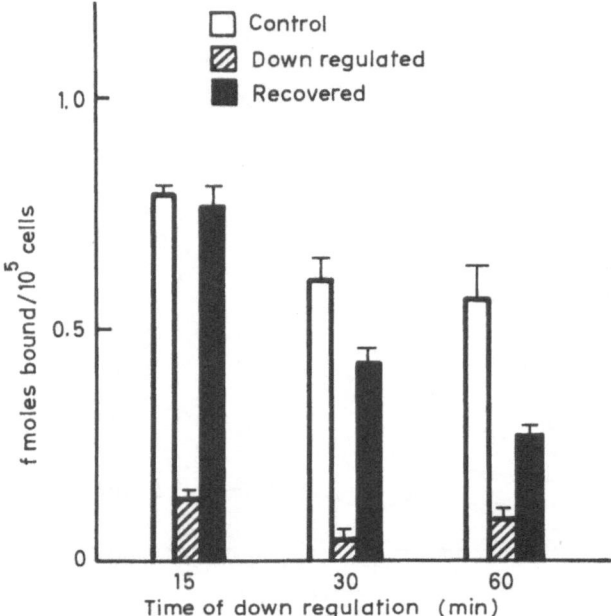

Fig. 10.9 Monocyte peptide receptor recovery as a function of time of down regulation. Monocytes were treated at 37°C with 0 (control) or 20 nM-(down-regulated) fNLPNTL for 15, 30 or 60 minutes, washed at 4°C, and then incubated in the absence of peptide for 60 minutes at 37°C. Binding of 2 nM-^{125}I-labeled fNLPNTL at 4°C for 2 h was determined immediately after the pretreatment and after the 60-min recovery incubation (recovered). Error bars represent 1 s.e.m. Reproduced with permission from Weinberg *et al.* (1981).

induce receptor loss and the rate constant decreased when higher peptide concentrations were used. At low peptide concentrations, rabbit neutrophils could be cycled repeatedly through down regulation and recovery, although recovery was less efficient with each cycle. A rate constant of 0.34 min^{-1} was estimated for receptor down regulation.

Zigmond (1981) has modeled the consequences of peptide-induced receptor down regulation for neutrophil orientation. Cell orientation along a concentration gradient of a formyl peptide is the necessary first step in the chemotactic response. It is believed that differential receptor occupancy across the cell's dimensions results in orientation and therefore that the orientation response may be sensitive to changes in surface receptor number. The observed orientation response appeared to be proportional to the difference in occupied receptors and the net effect of receptor down regulation appeared to be a decreased sensitivity to gradients at high peptide concentrations.

Although receptor down regulation has been extensively studied, the mechanism of down regulation is unknown. Internalization of the peptide–receptor complex as a unit is an attractive hypothesis which we have recently been able to directly test.

Human neutrophil receptors were photoaffinity-labeled at 4°C. If the cells were kept at 4°C, all of the affinity-labeled receptor remained accessible to proteolysis by papain and was therefore located on the cell surface (see Section 10.2.2). However, if these cells were warmed to 37°C, the labeled receptors became progressively inaccessible to proteolysis by papain and were therefore no longer exposed on the cell surface. By 10 minutes, approximately 50% of the labeled receptors were no longer accessible and by 30 minutes, only 13% were cleaved by papain (Dolmatch and Niedel, unpublished).

Surface receptor number can be decreased by other treatments. Lane *et al.* (1981) have shown that phagocytosis of latex or opsonized zymosan by human neutrophils caused a dose-dependent decrease in [^{3}H]fMet-Leu-Phe binding. This effect appeared to be due to both released proteases and toxic oxygen species. At low concentrations of opsonized zymosan, receptor affinity was decreased without a change in receptor number, but at higher zymosan concentrations, receptor number was also markedly decreased.

In a subsequent manuscript these authors demonstrated that cysteine and dithiothreitol protected against and reversed phagocytosis-induced receptor modulation (Lane and Lamkin, 1982). They concluded that (1) intact thiol groups are required for formyl peptide receptor binding, (2) the thiol groups may not be externally exposed, (3) thiol oxidation–reduction induces reversible receptor modulation and (4) phagocytosis-induced receptor modulation is due in part to thiol oxidation. Thiol oxidation did not appear to be involved in peptide-induced receptor down regulation. Thiol oxidation may be partially responsible for the decreased receptor affinity observed during storage of granulocyte concentrates (Lane and Windle, 1981).

Spilberg and Mehta (1981) have shown that cytochalasin B treatment will decrease available surface receptors on human neutrophils. Koo and Snyderman (1980) have reported similar observations.

10.4.2 Up regulation

Gallin and his co-workers were the first to call attention to the phenomenon of receptor up regulation (Gallin *et al.*, 1978). They noted that treatment of human neutrophils with low concentrations of the calcium ionophore A23187 or the tumor promoter phorbol myristate acetate (PMA) increased receptor binding by approximately 20%. The maximal increase in binding correlated with the release of 10 to 15% of the total cellular lysozyme, suggesting that the specific granule membrane may serve as an intracellular store of preformed receptors.

In a subsequent manuscript, Fletcher and Gallin (1980) investigated the relationship between degranulation and receptor up regulation in greater detail. Human neutrophils were treated with A23187 (50 nM), PMA (0.5 ng/ml) or *E. coli* endotoxin-activated serum plus cytochalasin B for 30 minutes at 37°C. These conditions resulted in release of approximately 17% of the specific granule enzyme lysozyme and no release of the primary granule enzyme β-glucuronidase or the cytosolic marker, lactate dehydrogenase. Scatchard analysis of [^{3}H] fMet-Leu-Phe binding to control and degranulated cells at 4°C, demonstrated approximately a 2-fold increase in surface receptor number. The control cells displayed approximately 10 000 receptors per cell, while the receptor number had increased to approximately 20 000 per cell following limited degranulation. These newly expressed receptors showed the same specificity for a series of unlabeled peptides as did the receptors on control cells. The receptors on degranulated cells exhibited a slight, but reproducible decrease in affinity from $K_a = 0.151$ nM^{-1} to 0.123 nM^{-1}. In a series of 13 experiments, the K_a for the control cells varied from approximately 0.04 nM^{-1} up to 0.28 nM^{-1} and receptor number per cell varied from 3380 to 15 600.

In their most recent manuscript, Fletcher *et al.* (1982) studied the functional consequences of pretreatment with degranulating stimuli. fMet-Leu-Phe-induced superoxide and H_2O_2 production were studied in control and A23187- or PMA-pretreated cells. As both A23187 and PMA stimulated superoxide and H_2O_2 production directly, low concentrations were used during the pretreatment. Pretreated cells produced 2- to 3-fold more superoxide in response to maximal stimulation with fMet-Leu-Phe without a shift in the dose response curve. These data indicated that the increased receptors elicited by degranulating stimuli were functionally coupled to the enzymatic mechanism for superoxide production and that cells with increased receptors were capable of increased maximal superoxide production when compared to control cells.

The increased binding of [^{3}H] fMet-Leu-Phe at 4°C elicited by degranulating stimuli was separated into rapidly dissociable and non-dissociable components. Control and degranulated cells demonstrated both components. Whereas both components were of approximately equal magnitude and plateaued by 5 minutes in control cells, in stimulated cells the non-dissociable component continued to increase linearly throughout the 20 minute incubation period. Unlike rabbit neutrophils, the human cells did not demonstrate a burst of pinocytic activity and the non-dissociable component could therefore not be explained by fluid-phase, non-receptor-mediated uptake. Whether the non-dissociable component represents peptide internalization or increased affinity of surface receptors is unknown. Above 18 to 20% lysozyme release, increasing inhibition of binding was observed.

The ionophore A23187 required extracellular Ca^{2+} in order to stimulate lysosomal enzyme release, and receptor enhancement was not observed in the presence of Ca^{2+} chelators. PMA is active as a secretagogue in the absence of Ca^{2+}. PMA-mediated receptor enhancement was unaffected by Ca^{2+} chelation. Receptor enhancement was not dependent on *de novo* protein synthesis. The specific granule constituents, lactoferrin and lysozyme, did not enhance binding.

These observations may have relevance *in vivo*. Tsung *et al.* (1980) have shown that elicited rabbit peritoneal neutrophils displayed approximately 7-fold more surface receptors than did rabbit peripheral blood neutrophils. The peritoneal cells also displayed distinctly greater chemotactic responsiveness to fMet-Leu-Phe, whereas no difference in chemotactic response to C5a was noted. This supports the hypothesis that the peritoneal cells had been induced to accumulate in the peritoneal cavity by locally produced C5a and that partial degranulation in response to C5a had caused enhancement of formyl peptide receptors. *In vitro* treatment of human neutrophils with C5a has recently been shown to up regulate the number of fMet-Leu-Phe receptors without a change in receptor affinity (Donabedian and Gallin, 1981).

Liao and Freer (1980) have studied the effect of aliphatic alcohols on the formyl peptide receptor. Normal butanol at 2.5% doubled the number of surface receptors on rabbit neutrophils. Whether the aliphatic alcohols increase surface receptors by inducing degranulation is unknown. Not surprisingly 2.5% butan-1-ol produced a loss of membrane integrity (Fletcher *et al.*, 1982).

The effect of enzymatic treatment of whole cells has also been studied. Schiffmann *et al.* (1980) showed that although several proteases were without effect, phospholipases A_2 and C increased binding of [^{3}H]fMet-Leu-Phe. Nelson *et al.* (1982) investigated the mechanism of phospholipase C mediated increased binding in detail. Phospholipase C-induced lysozyme release was paralleled by increased formyl peptide surface receptors. The final common pathway of receptor up regulation therefore appears to be specific granule release. Although direct evidence to support the specific granule as the reservoir of intracellular formyl peptide receptors is lacking, a large body of indirect evidence makes this an attractive hypothesis.

10.4.3 Pharmacological regulation

Two observations regarding the interaction of anti-inflammatory drugs with the formyl peptide receptor may be of clinical importance. Corticosteroids are reported to be efficacious in the treatment of adult respiratory distress syndrome and endotoxic shock. These clinical disorders are postulated to be due in part to neutrophil activation and aggregation and sequestration in the lung. Skubitz *et al.* (1981) have recently shown that corticosteroids cause a dose-dependent decrease in the association rate between [^{3}H]fMet-Leu-Phe and the human neutrophil receptor. Receptor number and dissociation kinetics were unaffected. The potencies of a series of steroids to decrease the association rate correlated with their clinical efficacy and their potencies as inhibitors of *in vitro* neutrophil aggregation (Skubitz *et al.*, 1981).

Two non-steroidal anti-inflammatory drugs also interact with the formyl peptide receptor (Dahinden and Fehr, 1980). Phenylbutazone and sulfinpyrazone selectively inhibited peptide-induced acute neutropenia *in vivo* and cellular hyperadhesiveness, lysosomal enzyme release, hexose monophosphate shunt activity and superoxide production *in vitro*. Phenylbutazone has also been shown to protect human neutrophils from peptide-induced chemotactic deactivation and receptor down regulation (Nelson *et al.*, 1981). Phenylbutazone did not inhibit binding of ^{125}I-labeled C5a nor did it alter C5a-induced chemotaxis.

Inhibitors of transmethylation reactions also modulate formyl-peptide-receptor-dependent responses. Pike and Snyderman (1982) have presented data which suggest that an ongoing methylation reaction is required for maintenance of the macrophage receptor in a high-affinity state. Inhibition of methylation resulted in a 4.5-fold decrease in receptor affinity and uncoupled the receptor from chemotaxis, superoxide production and arachidonate release.

10.5 CONCLUDING REMARKS

Our understanding of phagocyte activation has grown exponentially as a result of the development of the formyl oligopeptides. In this manuscript, I have reviewed progress through February, 1982 in three active research areas: receptor structure, peptide processing and receptor modulation. But many questions remain unanswered.

Work on receptor structure is just beginning, but already some discrepancies are evident. Affinity-labeling studies indicate that the receptor is heterogeneous with regard to size, displaying a range of apparent molecular weights between 55 000 and 70 000. This size heterogeneity, and the fact that the receptor binds to wheatgerm agglutinin–agarose, is evidence in support of its glycoprotein nature. In contrast, the 68 000 mol.wt. protein partially purified by affinity chromatography, although of similar molecular weight, does not display heterogeneity by SDS–PAGE analysis, and amino acid composition studies suggest that carbohydrate may not be present on the molecule.

Two explanations for the discrepancy are possible. All affinity-labeling studies are plagued by the possibility that the labeled protein is adjacent to the actual receptor. This is unlikely for two reasons. The formyl hexapeptide is a very small ligand, the first four amino acids of which are required for receptor binding. The reactive moiety on the smallest affinity label, N^ϵ-bromoacetyl, would therefore be no more than three amino acids removed from the actual binding site. Additionally, although nearest-neighbor labeling is a theoretical possibility, in practice, the insulin receptor, epidermal growth factor receptor and acetylcholine receptor were all correctly identified by affinity labeling.

Alternatively, the 68 000 mol.wt. protein eluted from the fMet-Leu-Phe–Sepharose column, although the major protein identified by Coomassie Blue staining, may not be responsible for the high-affinity peptide binding. Other, poorly staining proteins were evident by SDS–PAGE, some of which had apparent molecular weights of approximately 55 000 to 70 000 mol.wt. It is possible, therefore, that these 55 000- to 70 000-mol.wt. proteins eluted from the affinity column are identical to those identified by affinity labeling. If so, they may be responsible for high-affinity peptide binding. Clearly, further work is needed to unequivocably identify the receptor.

Peptide endocytosis and processing following receptor binding are well documented, but very little is known about the mechanisms or biological importance of these events. In many other systems, polypeptide ligands have been shown to be internalized via clathrin-coated plasma membrane pits. Although coated pits have been identified on neutrophils and macrophages, synthesis of an electron-dense ligand for the formyl peptide receptor has not been reported. Studies of peptide processing and intracellular sorting await the development of a ligand suitable for identification with electron microscopy.

The closely related processes of receptor down regulation, recovery and receptor up regulation have also been studied extensively at a phenomenological level. Biochemically, little is known about these processes. Circumstantial evidence points to receptor internalization as the mechanism for down regulation, but definitive studies using anti-receptor antibody for electron microscopic localizations or endocytic vesicle purification have not been performed.

The demonstration of receptor up regulation is most compatible with a pool of latent or cryptic receptors which can be expressed on the plasma membrane. The specific granule membrane may provide this cryptic pool, but again the evidence is indirect.

Although many other phagocyte stimuli have been recognized, and several have been purified and characterized, the formyl peptides still provide the most experimentally accessible model for study. As we continue to exploit this unique system, we will obtain answers to specific questions regarding phagocytic cell function and insight into general questions of receptor-mediated responses.

REFERENCES

Aswanikumar, S., Schiffmann, E., Corcoran, B.A., Pert, C.B., Morell, J.L. and Gross, E. (1978), *Biochem. Biophys. Res. Commun.*, **80**, 464–471.

Becker, E.L. (1976), *Am. J. Pathol.*, **85**, 385–394.

Becker, E.L., Bleich, H.E., Day, A.R., Freer, R.J., Glasel, J.A. and Visintainer, J. (1979), *Biochemistry*, **18**, 4656–4668.

Becker, E.L., Naccache, P.H., Showell, H.J. and Walenga, R.W. (1981), *Lymphokines*, **4**, 297–334.

Dahinden, C. and Fehr, J. (1980), *J. Clin. Invest.*, **66**, 884–891.

Dolmatch, B. and Niedel, J.E. (1983), *J. Biol. Chem.* (in press, June).

Donabedian, H. and Gallin, J.I. (1981), *J. Immunol.*, **127**, 839–844.

Fletcher, M.P. and Gallin, J.I. (1980), *J. Immunol.*, **124**, 1585–1588.

Fletcher, M.P., Seligmann, B.E. and Gallin, J.I. (1982), *J. Immunol.*, **128**, 941–948.

Freer, R.J., Day, A.R., Muthukumaraswamy, N., Pinion, D., Wu, A., Showell, H.J. and Becker, E.L. (1982), *Biochemistry*, **21**, 257–263.

Freer, R.J., Day, A.R., Radding, J.A., Schiffmann, E., Aswanikumar, S., Showell, H.J. and Becker, E.L. (1980), *Biochemistry*, **19**, 2404–2410.

Gallin, J.I., Wright, D.G. and Schiffmann, E. (1978), *J. Clin. Invest.*, **62**, 1364–1370.

Goetzl, E.J., Foster, D.W. and Goldman, D.W. (1981), *Biochemistry*, **20**, 5717–5722.

Goetzl, E.J., Foster, D.W. and Goldman, D.W. (1982), *Immunology*, **45**, 249–256.

Koo, C. and Snyderman, R. (1980), *Clin. Res.*, **28**, 373a.

Lane, T.A., Lamkin, G.E. and Windle, B.E. (1981), *Blood*, **58**, 228–236.

Lane, T.A. and Lamkin, G.E. (1982), *Blood*, **59**, 1337–1343.

Lane, T.A. and Windle, B.E. (1981), *Transfusion*, **21**, 450–456.

Liao, C.S. and Freer, R.J. (1980), *Biochem. Biophys. Res. Commun.*, **93**, 566–571.

Marasco, W.A. and Becker, E.L. (1982), *J. Immunol.*, **128**, 963–968.

Marasco, W.A., Showell, H.J., Freer, R.J. and Becker, E.L. (1982), *J. Immunol.*, **128**, 956–962.

Nelson, R.D., Fiegel, V.D. and Chenoweth, D.E. (1982), *Am. J. Pathol.*, **107**, 202–211.

Nelson, R.D., Fiegel, V.D., Herron, M.J. and Simmons, R.L. (1980), *J. RES.* **28**, 285–293.

Nelson, R.D., Gracyk, J.M., Fiegel, V.D., Herron, M.J. and Chenoweth, D.E. (1981), *Blood*, **58**, 752–758.

Niedel, J.E. (1981), *J. Biol. Chem.*, **256**, 9295–9299.

Niedel, J.E. and Cuatrecasas, P. (1980), *Curr. Top. Cell. Regul.*, **17**, 137–170.

Niedel, J.E., Davis, J. and Cuatrecasas, P. (1980b), *J. Biol. Chem.*, **255**, 7063–7066.

Niedel, J.E., Kahane, I. and Cuatrecasas, P. (1979b), *Science*, **205**, 1412–1414.

Niedel, J.E., Kahane, I., Lachman, L. and Cuatrecasas, P. (1980a), *Proc. Natl. Acad. Sci. U.S.A.*, **77**, 1000–1004.

Niedel, J.E., Wilkinson, S. and Cuatrecasas, P. (1979a), *J. Biol. Chem.*, **254**, 10700–10706.

Pike, M.C. and Snyderman, R. (1982), *Cell*, **28**, 107–114.

Schiffmann, E., Aswanikumar, S., Venkatasubramanian, K., Corcoran, B.A., Pert, C.B., Brown, J., Gross, E., Day, A.R., Freer, R.J., Showell, H.J. and Becker, E.L. (1980), *FEBS Lett.*, **117**, 1–7.

Schiffmann, E., Corcoran, B.A. and Wahl, S.M. (1975b), *Proc. Natl. Acad. Sci. U.S.A.*, **72**, 1059–1062.

Schiffmann, E. and Gallin, J. (1979), *Curr. Top. Cell. Regul.*, **15**, 203–261.

Schiffmann, E., Showell, H.J., Corcoran, B.A., Ward, P.A., Smith, E. and Becker, E.L. (1975a), *J. Immunol.*, **114**, 1831–1837.

Showell, H.J., Freer, R. J., Zigmond, S.H., Schiffmann, E., Aswanikumar, S., Corcoran, B.A. and Becker, E.L. (1976), *J. Exp. Med.*, **143**, 1154–1169.
Skubitz, K.M., Craddock, P.R., Hammerschmidt, D.E. and August, J.T. (1981), *J. Clin. Invest.*, **68**, 13–20.
Smolen, J.E. and Weissmann, G. (1981), in *Lysosomes and Lysosomal Storage Diseases* (J.W. Callahan and J.A. Lowden, eds), Raven Press, New York, pp. 31–62.
Snyderman, R. and Goetzl, E.J. (1981), *Science*, **213**, 830–837.
Sullivan, S.J. and Zigmond, S. (1980), *J. Cell Biol.*, **85**, 703–711.
Tsung, P.K., Showell, H.J. and Becker, E.L. (1980), *Inflammation*, **4**, 271–277.
Vitkauskas, G., Showell, H.J. and Becker, E.L. (1980), *Mol. Immunol.*, **17**, 171–180.
Weinberg, J.B., Muscato, J.J. and Niedel, J.E. (1981), *J. Clin. Invest.*, **68**, 621–630.
Zigmond, S.H. (1978), *J. Cell Biol.*, **77**, 269–287.
Zigmond, S.H. (1981), *J. Cell Biol.*, **88**, 644–647.
Zigmond, S.H., Sullivan, S.J. and Lauffenburger, D.A. (1982), *J. Cell Biol.*, **92**, 34–43.

Index